Advances in Acoustics

The book series Advances in Acoustics, endorsed by the Acoustic Society of China, publishes both monographs and edited volumes. It covers all research fields of acoustics, including physical acoustics, underwater acoustics, solid-state acoustics, environmental acoustics, detection acoustics, power ultrasound, ultrasonic electronics, audio engineering, biomedical ultrasonic engineering, speech and hearing acoustics, and musical acoustics. By bridging theories, experiments, and practical applications, this series serves as an invaluable resource for researchers, engineers, and students engaged in the diverse aspects of acoustics.

Xiaozhou Liu · Hairong Zheng

Acoustic Radiation Force

Principles and Application in Medical Ultrasound

Science Press
Beijing

Springer

Xiaozhou Liu
Nanjing University
Nanjing, China

Hairong Zheng
Nanjing University
Nanjing, China

ISSN 3091-3373 ISSN 3091-3381 (electronic)
Advances in Acoustics
ISBN 978-981-95-3151-6 ISBN 978-981-95-3152-3 (eBook)
https://doi.org/10.1007/978-981-95-3152-3

Jointly published with Science Press
The print edition is not for sale in China (Mainland). Customers from China (Mainland) please order the print book from: Science Press.
ISBN of the Co-Publisher's edition: 978-7-03-074629-0

Translation from the Chinese Simplified language edition: "Acoustic Radiation Force: Principles and Application in Medical Ultrasound" by Xiaozhou Liu and Hairong Zheng, © Science Press 2023. Published by Science Press. All Rights Reserved.

The original submitted manuscript has been translated into English. The translation was done using artificial intelligence. A subsequent revision was performed by the author(s) to further refine the work and to ensure that the translation is appropriate concerning content and scientific correctness. It may, however, read stylistically different from a conventional translation.

This Springer imprint is published by the registered company Springer Nature Singapore Pte Ltd.
The registered company address is: 152 Beach Road, #21-01/04 Gateway East, Singapore 189721, Singapore

Preface

With the rapid development of modern technology, the application of ultrasound in scientific research and industrial fields, such as medical ultrasound examination and non-destructive testing, is becoming increasingly mature. The use of acoustic radiation force can achieve cell screening and drug delivery in the biomedical field, as well as non-contact manipulation of microparticles. This is currently one of the hotspots in medical ultrasound research. The study of acoustic radiation force on particles has significant practical implications for the design of new acoustic tweezers.

Sound waves carry momentum and energy. When they encounter an object in the sound field, the object will produce effects such as reflection, absorption, and scattering on the sound waves, leading to an exchange of momentum and energy. This causes changes in the energy density and momentum of the propagating wave, causing the object to move under the action of a force. This force is the "acoustic radiation force", which can be used to manipulate objects in the sound field. This technology is known as acoustic manipulation, and the corresponding equipment is called acoustic tweezers.

Acoustic radiation force manipulation of particles has non-contact, non-destructive, unmarked, and multifunctional characteristics. It does not require special conductivity or transparency of the medium. The corresponding device is simple, easy to integrate, and miniaturize. Therefore, this technology has great application value and prospects in many fields such as medicine, biophysics, medical ultrasound, life sciences, and materials science.

In recent years, with the development of microfluidic technology, particle capture systems using standing waves, surface acoustic waves, and single (double) beam have been able to transport and separate nanoparticles, biological particles, and droplets. Compared to light waves, sound waves are mechanical waves. Acoustic manipulation technology does not require special conductivity or transparency of the medium, and sound waves have strong penetration and low loss in liquid media. Acoustic manipulation devices are small, simple in structure, low in manufacturing cost, and multifunctional. Therefore, acoustic manipulation technology has great application value in many fields such as biomedicine, biophysics, and medical ultrasound.

Acoustic manipulation technology, which manipulates particles based on acoustic radiation force, has the advantages of non-contact, good biocompatibility, no need for chemical biological labeling of microparticles, and easy integration of devices. Acoustic radiation force has a wide range of applications in precision manufacturing, precision medical diagnosis, evaluation of the viscoelastic properties of biological tissues and liquids, and other ultrasound-related fields. The acoustic radiation force acts as a virtual finger for remote detection of internal anatomical structures and obtaining diagnostic information. Another actively explored field is the manipulation of biological cells and particles.

Since the 1950s, the application of ultrasound in medicine has surged, and the proportion of medical ultrasound areas related to acoustic radiation force has greatly increased. Most of the studies on radiation force published before 1990 were related to various biological effects, and subsequent developments have opened up new and extensive biomedical application fields, such as vibration acoustic imaging, shear wave elasticity imaging (SWEI), acoustic radiation force impulse (ARFI) imaging, magnetic resonance elasticity imaging using acoustic radiation force, ultrasonic rheological variation and evaluation of the viscoelastic properties of biofluids and tissues, assisting targeted drug and gene delivery and promoting ultrasound molecular imaging of radiation force, as well as many microfluidic applications.

This book mainly introduces the principles of acoustic radiation force and its applications in medical ultrasound from the following aspects:

Chapter 1 introduces the study of the acoustic radiation force on microspheres based on the acoustic ray method. It mainly introduces the method of geometric acoustics approximation. Through this method, the characteristics of the acoustic radiation force on microspheres in the Gaussian sound field are studied. It also considers the effect of sound attenuation in particles on the acoustic radiation force, analyzes the conditions for the generation of acoustic potential traps, and discusses the effects of the initial position of particles and particle size on the magnitude of the acoustic radiation force. The results show that when particles are at certain specific positions, they can be affected by the acoustic trap force. Considering attenuation does not affect the capture properties of acoustic tweezers, it only affects their capture ability. Under the theory of geometric acoustics, a model of the radiation force on a double-layer sphere is proposed, and the effects of the thickness of the shell of the double-layer sphere and the acoustic properties of each layer of the medium on the position and size of the acoustic potential trap are discussed in detail, providing a more realistic theoretical model of acoustic radiation force for practical experimental applications.

Chapter 2 introduces the study of the acoustic radiation force on microspheres by spherical focusing ultrasound based on the ray acoustics method. Based on the method of geometric acoustics approximation, a model of the acoustic radiation force on microspheres in the spherical focusing sound field is given. The effects of the parameters of the focusing sound field on the acoustic radiation force of spherical particles are analyzed. The results show that conventional spherical focusing sound sources also have three-dimensional capture effects on particle spheres.

Chapter 3 introduces the study of acoustic radiation force on spherical particles by Gaussian traveling waves based on the acoustic scattering method. It discusses the scattering effects of various spherical particles' boundary conditions on the sound field. Using the method of finite series expansion, the Gaussian focusing sound wave is expanded according to spherical functions. A detailed derivation of the acoustic radiation force function of Gaussian sound waves on spherical particles under the acoustic scattering theory is provided, laying the foundation for studying the radiation force of complex sound waveforms. Chapter 4 introduces the study of acoustic radiation force on spherical particles by Gaussian standing waves based on the acoustic scattering method. It presents the acoustic radiation force functions in the scattering theory of Gaussian standing wave fields and quasi-Gaussian standing wave fields. It analyzes the characteristics of the acoustic radiation force experienced by spherical particles in Gaussian standing wave fields and quasi-Gaussian standing wave fields and explains the manipulation of particles by Gaussian standing wave fields. Chapter 5 introduces the solution of the acoustic radiation force on multilayer spherical particles by Gaussian traveling waves based on the acoustic scattering method. It studies the characteristics of the acoustic radiation force on multilayer spherical particles in Gaussian standing wave fields based on the theory of sound wave scattering. Two types of particle models are mainly studied: one is a solid-liquid dual-layer spherical particle model composed of polymer materials for drug delivery; the other is a three-layer spherical particle model composed of viscoelastic materials for the structure and mechanical properties of cells. The focus is on analyzing the impact of material and shell thickness changes in multilayer spherical structures on the acoustic radiation force. Chapter 6 introduces the acoustic radiation force of other sound sources. Based on the method of spherical scattering, it theoretically studies and calculates the acoustic radiation force of a hollow focusing transducer on an axial rigid sphere and calculates the acoustic radiation force of an annular transducer on an axial rigid sphere and a benzene sphere particle. The results show that a pulling force can only occur when the gradient force generated by the sound source on the axial spherical particles is much greater than the scattering force along the direction of the sound wave, which is the physical reason for the generation of acoustic radiation pulling force. This chapter also uses the method of finite series expansion to theoretically calculate the beam factor of the zero-order quasi-Bessel Gaussian wave and uses the spherical scattering method to study the acoustic radiation force of the zero-order quasi-Bessel Gaussian wave on spherical particles on the axis. The results found that there is an acoustic radiation pulling force on the liquid spherical particles on the axis of the zero-order quasi-Bessel Gaussian wave, which proves the feasibility of using non-diffracting waves to achieve single sound source acoustic tweezers. This chapter also introduces the application of Airy Gaussian waves in acoustics, gives its acoustic radiation force expression on cylindrical particles, and shows its acoustic tweezers effect on cylindrical particles. Chapter 7 introduces the influence of boundaries on the acoustic radiation force. It studies the acoustic radiation force experienced by viscoelastic cylindrical particles near the impedance boundary in a plane traveling wave field propagating at any incident angle. It analyzes the acoustic radiation force on viscoelastic cylindrical particles under different particle radii, different boundary distances, different impedance boundaries, and different incident angles. This analysis result can provide effective theoretical support for the micro-manipulation of biological particles in a bounded space. Chapter 8 introduces the acoustic radiation force between multiple particles. Based on the existence of multiple particles in the sound field, the scattering wave caused by one particle will also generate acoustic radiation force on other particles. This chapter starts from the scattering of the sound field, introduces the acoustic radiation force experienced by particles in the Gaussian traveling wave field with multiple elastic particles, derives the expression of the acoustic radiation force on sparsely distributed particles in space, and gives the simulation results. This research helps to use the acoustic radiation force to achieve acoustic manipulation of multiple tiny elastic particles. Chapter 9 introduces the medical applications of acoustic radiation force, including elasticity imaging based on acoustic radiation force, particle manipulation, and neural regulation. Acoustic radiation force elasticity imaging generates acoustic radiation force through an ultrasound beam, which acts on human soft tissues to cause local micro-displacement, subsequently inducing shear waves and propagating around. As a non-invasive quantitative measurement of tissue elasticity modulus, acoustic radiation force imaging has been proven to have great application potential in the diagnosis of various diseases. The principle of sound control mainly uses the reflection, refraction, absorption, etc., of particles in the sound field, so that the momentum carried by the sound field is exchanged between the sound wave and the particles, and the particles are manipulated by the acoustic radiation force to move. The target molecule-specific antibody or ligand is connected to the surface of the ultrasound contrast agent to construct a targeted ultrasound contrast agent. Using the acoustic radiation force, the ultrasound contrast agent actively binds to the target tissue, thereby observing the specific imaging of the target tissue at the molecular or cellular level, reflecting the changes in the diseased tissue at the molecular basis. Ultrasound waves are generated by ultrasound transducers, and their center frequency, energy intensity, pulse change rules, and the size of the irradiated target area can all be adjusted. Based on these advantages, ultrasound neural regulation can be widely applied to the neural regulation fields of cells, nematodes, insects, rodents, non-human primates, and humans at various scales. Special thanks to Prof. Xiufen Gong for her multifaceted guidance and meticulous assistance! I would also like to express my gratitude for the many valuable suggestions and recommendations made by colleagues such as Profs. Jianchun Cheng, Xiaojun Liu and Dong Zhang from the Acoustics Institute of Nanjing University. I am also deeply grateful for the support and assistance from Academician Shining Zhu of the Physics Department of Nanjing University, Prof. De'an Ta of Fudan University, Prof. Xiaofeng Zhang of Shaanxi Normal University, Profs. Fenghua Li, Jun Yang and Xiuming Wang of the Acoustics Research Institute of the Chinese Academy of Sciences. Thanks to the leadership and teachers of the Physics Department of Nanjing University for their encouragement and support! Associate Professor Jiehui Liu of the research group, doctoral students Rongrong Wu, Haibin Wang, Yupei Qiao, Menyang Gong, and master's students Chen Jiang, Sha Gao, Mingxin Hui, Kaixuan Cheng, Teng Liu, and others participated in part of the work, and I would like to express my gratitude here!

This book has received support from National Key Research and Development Program of China (2020YFA0211400, 2017YFA0303702, 2016YFF0203000, 2012CB921504, 2011CB707902), the Key Program of the National Natural Science Foundation of China (11834008), the National Natural Science Foundation of China (12174192, 11774167, 11474160, 11274166, 11074122, 10674066), and State Key Laboratory of Acoustics and Marine Information, Chinese Academy of Sciences. We express our gratitude here.

In the writing process, the authors have strived for meticulousness and rigor, but there are still shortcomings. We warmly welcome criticism and correction from readers.

Nanjing, China
August 2025

Xiaozhou Liu
Hairong Zheng

Contents

Chapter 1
Calculation of Acoustic Radiation Force on Spherical Particles Based on Acoustic Ray Method

Currently, there are two main methods for studying the acoustic radiation force on tiny particles in acoustic tweezers. The acoustic ray method is approximately applicable when the wavelength of the sound wave is relatively small compared to the scattering body, and diffraction is ignored. In this case, only the particle nature of the sound wave is considered, not its wave nature. The strength of the sound field can be represented by the density of the sound rays, i.e., the ray theory of sound waves is used to calculate the acoustic radiation force. The acoustic ray method is simple, fast, and reliable. However, its application range is narrow, limited to cases where the size of the object is much larger than the wavelength of the sound wave.

The second method is the scattering method, which uses sound field theory to calculate the acoustic radiation force. It is based on the analytical theory of sound waves, expands the incident sound wave into a series, uses acoustic boundary conditions to give an analytical expression for the scattering coefficient of sound waves, and calculates the acoustic radiation force on the target particle according to the radiation theory of sound waves. It is applicable when the wavelength of the sound wave is larger than or approximately equal to the size of the scattering body.

1.1 Acoustic Ray Method

The geometric approximation of acoustic rays, commonly known as the acoustic ray method [1], is based on geometric acoustics and is widely used in marine acoustics and architectural acoustics research. The acoustic ray method uses ray or cone tracking to calculate the propagation of sound and can be used in enclosed, semi-enclosed, or completely open spaces. When the frequency of the sound wave is high, i.e., the wavelength of the sound wave is smaller than the reflecting surface, the concept of the reflecting surface in geometric acoustics can be used to view the propagation of sound as sound energy propagating along the sound rays, ignoring the

X. Liu and H. Zheng, *Acoustic Radiation Force*, Advances in Acoustics,
https://doi.org/10.1007/978-981-95-3152-3_1

wave nature of sound. The use of the ray model requires that the size of the reflecting surface is much larger than the wavelength of the sound wave, and the roughness of the reflecting surface is much smaller than the wavelength.

Similarly, under high-frequency ultrasound, when the scale of the reflector is much larger than the wavelength of the sound wave, we can focus on the particle nature of the ultrasound and ignore its wave nature. Therefore, the acoustic ray method can also handle some problems under high-frequency ultrasound. Using the acoustic ray method, we can quickly and effectively analyze the energy transfer of sound waves inside particles, thereby determining the radiation force of sound waves on particles.

Inspired by the research in the theory of optical tweezers in geometric optics [2–5], J. Lee and others applied the method of geometric acoustics to the calculation of acoustic radiation force [6]. The interaction between sound waves and matter involves the exchange of momentum, which manifests as the force of sound waves on objects. In the acoustic ray method, the acoustic radiation force consists of two parts [2, 6], one part is the scattering force along the direction of sound wave propagation, which is caused by the reflection of sound waves and is proportional to the sound intensity; the other part is the gradient force, its direction is the direction of sound intensity gradient, related to the energy gradient of the sound field, pointing to the place with the maximum sound field intensity.

As shown in Fig. 1.1, assume that a sound ray with power P is incident on a spherical particle at an incident angle θ_i, a series of reflections and refractions occur on the particle surface and inside, causing energy changes, PR, PT^2, PT^2R, PT^2R^2, … PT^2R^m, … represent the power of the scattered sound rays that are emitted after m reflections and refractions in the small sphere, where R and T represent the energy reflection and refraction coefficients of sound waves, both of which are less than 1. As marked in Fig. 1.1, taking the counterclockwise direction as positive, the angles between these outgoing sound rays and the positive extension line of the incident sound ray are $\pi + 2\theta_i$, α, $\alpha + \beta$, $\alpha + 2\beta$, …, $\alpha + m\beta$, … Therefore, the force F_Z applied by this beam of sound rays to the particle sphere along the Z direction (i.e., the direction of incidence of this sound ray) is the change per second in this direction produced by this sound ray, therefore:

$$F_Z = \frac{P}{c} - \left[\frac{PR}{c}\cos(\pi + 2\theta_i) + \sum_{m=0}^{\infty}\frac{P}{c}T^2R^m\cos(\alpha + m\beta)\right] \tag{1.1}$$

where: $\frac{P}{c}$ is the momentum incident per second in the Z direction, c is the propagation speed of sound waves in the medium. Similarly, in the Y direction, since the momentum of the incident sound ray is zero, the change in momentum is:

$$F_Y = 0 - \left[\frac{PR}{c}\sin(\pi + 2\theta_i) - \sum_{m=0}^{\infty}\frac{P}{c}T^2R^m\sin(\alpha + m\beta)\right] \tag{1.2}$$

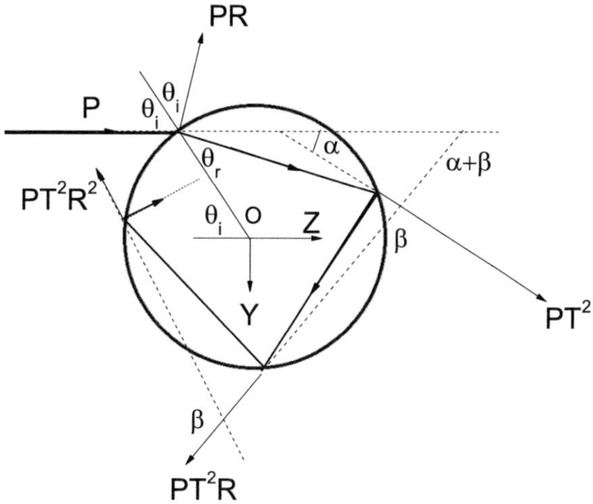

Fig. 1.1 Schematic diagram of calculating acoustic radiation force by acoustic ray method

Therefore, the resultant force exerted on the particle mapped onto the complex plane is:

$$\begin{aligned} F_{tot} &= \frac{P}{c}(1 + R\cos 2\theta_i) + i\frac{P}{c}R\sin 2\theta_i - \frac{P}{c}T^2\sum_{m=0}^{\infty} R^m e^{i(\alpha+m\beta)} \\ &= \frac{P}{c}(1 + R\cos 2\theta_i) + i\frac{P}{c}R\sin 2\theta_i - \frac{P}{c}T^2 e^{i\alpha}\left(\frac{1}{1 - Re^{i\beta}}\right) \end{aligned} \tag{1.3}$$

Using Euler's formula:

$$e^{i\alpha}\left(\frac{1}{1 - Re^{i\beta}}\right) = \frac{[\cos(2\theta_i - 2\theta_r) + R\cos(2\theta_i)] + i[\sin(2\theta_i - 2\theta_r) + R\sin(2\theta_i)]}{1 + R^2 + 2R\cos 2\theta_r} \tag{1.4}$$

As shown in the figure: $\alpha = 2\theta_i - 2\theta_r$, $\beta = \pi - 2\theta_r$

Substituting formula (1.3), we get [2, 18]:

$$\begin{aligned} F_{tot} = \frac{P}{c}\left\{1 + R\cos(2\theta_i) - T^2\frac{\cos(2\theta_i - 2\theta_r) + R\cos(2\theta_i)}{1 + R^2 + 2R\cos 2\theta_r}\right\} + i\frac{P}{c}\{R\sin(2\theta_i) \\ - T^2\frac{\sin(2\theta_i - 2\theta_r) + R\sin(2\theta_i)}{1 + R^2 + 2R\cos 2\theta_r}\} \end{aligned} \tag{1.5}$$

By equating the real and imaginary parts, the components of the scattering force are:

$$F_s = \frac{P}{c}\{1 + R\cos(2\theta_i) - T^2\frac{\cos(2\theta_i - 2\theta_r) + R\cos(2\theta_i)}{1 + R^2 + 2R\cos 2\theta_r}\} \tag{1.6}$$

The components of the gradient force are:

$$F_g = \frac{P}{c}\left\{R\sin(2\theta_i) - T^2\frac{\sin(2\theta_i - 2\theta_r) + R\sin(2\theta_i)}{1 + R^2 + 2R\cos 2\theta_r}\right\} \tag{1.7}$$

where: θ_i and θ_r are the incident and refracted angles respectively; R and T are the energy reflection and refraction coefficients respectively; F_s is the scattering force along the direction of sound wave propagation, F_g is the gradient force pointing towards the maximum intensity of the sound field.

The force exerted on an object by sound wave radiation often manifests as force, hence it is commonly referred to as acoustic radiation force. However, under specific sound field distributions, sound waves can also exert a pulling force on an object, forming a potential well that traps particles. Acoustic tweezers are three-dimensional gradient acoustic potential wells formed by the mechanical effect of momentum transfer between sound waves and matter. When the particle diameter is much larger than the ultrasonic wavelength, if tiny particles are placed within a certain area centered around the focus of the ultrasonic beam forming the acoustic tweezers, the particles will automatically move towards the convergence center of the ultrasonic beam and be stably captured near the focus.

Assuming a series of sound beams with intensity gradients are incident on a particle sphere, when the sound line is incident on the interface between the medium and the particle, momentum is transferred, resulting in acoustic radiation force. As shown in Fig. 1.2, sound line a, b is incident from the medium to the particle, and the path of sound line propagation can be determined by geometric acoustics. The sound line refracts when entering and leaving the particle sphere, and also reflects to some extent on the surface. The reflection action generates a scattering force along the direction of the sound line, causing the particle to move in the direction of the sound beam. We focus on analyzing the gradient force exerted on the small ball by the refraction of sound and the small ball. The incident sound line in the figure a, b propagates along the Z axis, that is, the initial momentum direction is along the Z axis. When the sound line leaves the small ball after two refractions, the direction of sound line propagation changes, that is, the momentum changes.

Due to the conservation of momentum, according to the principle of vector superposition, the sound beam transfers a momentum to the small ball that is equal in magnitude and opposite in direction to this momentum change, thereby generating force F_a, F_b applied to the particle sphere, which is the gradient force, causing the particle to move towards the focus of the sound beam.

When the sound field has an intensity gradient, as shown in Fig. 1.3, the sound line a gives the small ball a larger momentum when it refracts with the small ball, thereby generating a larger F_a, causing the particle to move towards the area with greater sound field intensity.

Only when the refraction of sound waves on the particle surface dominates, that is, the gradient force pulling the particle towards the vicinity of the focus of the sound beam is much greater than the scattering force along the direction of the sound line, can the capture of particles by the sound beam be achieved. Here we can adopt the

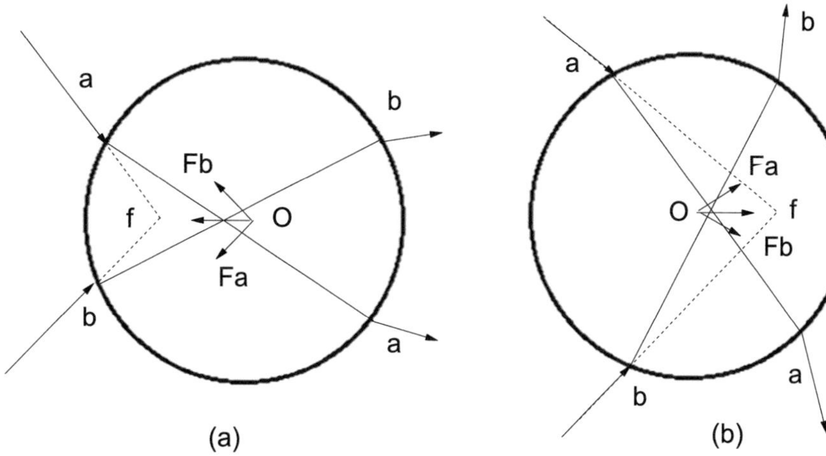

Fig. 1.2 Schematic diagram of the force on a particle in a sound field

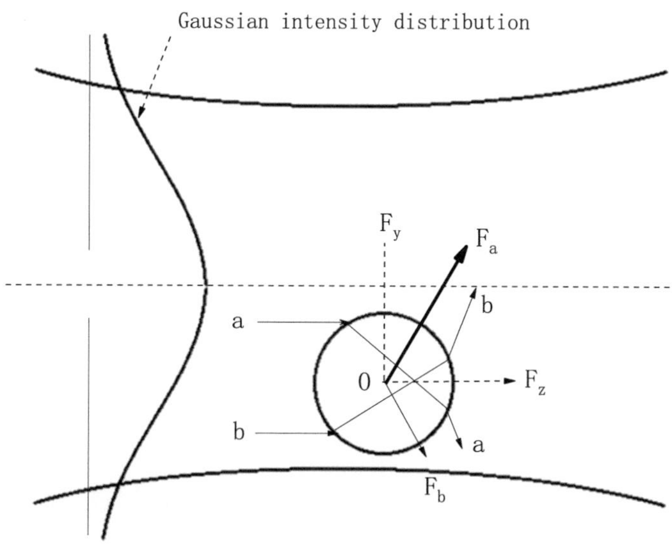

Fig. 1.3 Schematic diagram of the force on a particle in a sound field with an intensity gradient

following methods to achieve this: first, select particles and media with matching acoustic impedance to reduce the reflection of sound lines on the particles. Secondly, the speed of sound in the particles should be slightly lower than that in the medium to avoid total reflection [7].

The ray acoustics approximation has the advantages of intuitive theoretical methods and fast calculation speed, and has been specifically applied in calculating the acoustic radiation force of ultrasound on tiny particles and ray tracing [11–17].

1.2 Acoustic Radiation Force of Gaussian Ultrasound on Microspheres

This section primarily investigates the use of the ray method to solve for the acoustic radiation force of a Gaussian focusing sound beam on spherical particles. It derives the formula for the acoustic radiation force, provides simulation calculations, and analyzes and discusses the simulation results.

1.2.1 Axial Acoustic Radiation Force

The schematic diagram of the Gaussian sound beam is shown in Fig. 1.4, where: y is the radial coordinate, with the center of the sound axis as the reference point, z is the axial coordinate, with the narrowest point of the sound wave on the sound axis (also known as the beam waist) as the reference point, λ is the wavelength, $k = \frac{2\pi}{\lambda}$ is the wave number. The radius of the Gaussian sound beam spot expands according to a hyperbolic rule with the coordinate z. At $z = 0$, $W(z) = W_0$, it corresponds to the thinnest part of the sound wave, that is, the beam waist. It is generally stipulated that the radius of the sound wave cross-section corresponding to when the amplitude decreases to the maximum value of $\frac{1}{e}$ is taken as the nominal sound wave cross-section radius, that is, $W(z)$.

When the spherical particle is exactly on the axis of the Gaussian sound beam, since the acoustic radiation force is symmetrical along the transverse position and cancels each other out, we only consider the axial acoustic radiation force F_z.

As shown in Fig. 1.5, O is the center point of the Gaussian focused ultrasound, θ is the intersection point of the incident sound line and the particle A and the line connecting the center of the sphere C and the angle between the Z axis, θ_i and θ_r are

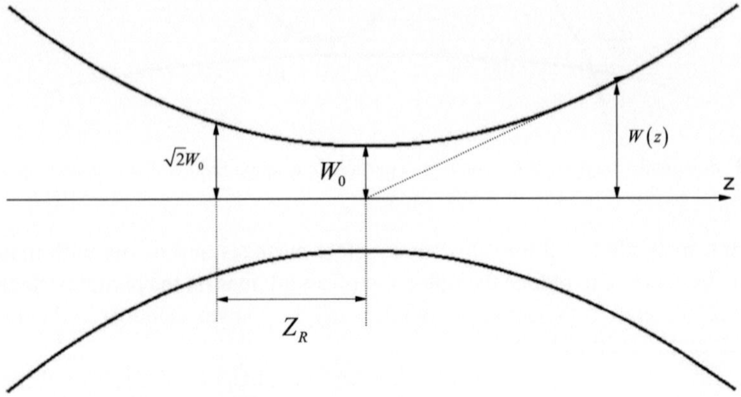

Fig. 1.4 Schematic diagram of Gaussian sound wave

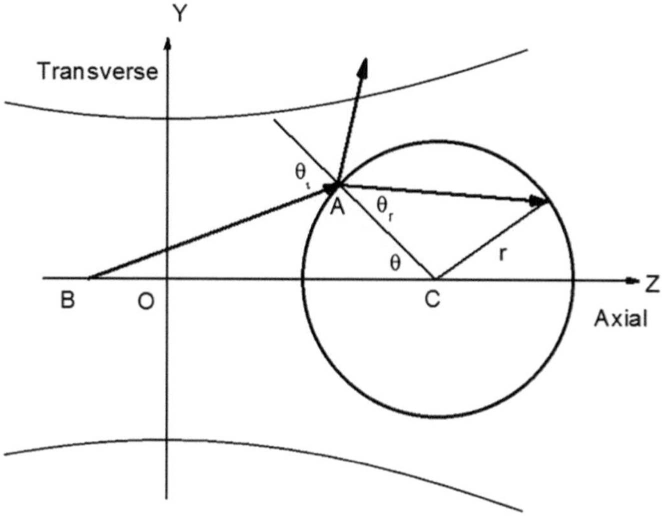

Fig. 1.5 Schematic diagram of the axial acoustic radiation force on a microsphere

the incident angle and the refractive angle, respectively, y and z are the longitudinal and transverse coordinates, respectively.

When the power of the ultrasound is P, incident on a microsphere with a radius of r, the sound field expression of the Gaussian ultrasound is:

$$I = \frac{P}{W^2(z)} \exp\left(\frac{-2y^2}{W^2(z)}\right) \tag{1.8}$$

$$\text{or: } I(r, z) = \frac{|E(r, z)|^2}{2\eta} = I_0\left(\frac{W_0}{W(z)}\right)^2 \exp\left(\frac{-2y^2}{W^2(z)}\right)$$

where: $I_0 = \frac{E_0^2}{2\eta}$ is the sound intensity at the center of the sound beam waist, η is the characteristic impedance in the propagation medium, in air $\eta \approx 337Pa \cdot s/m$.

The radiation force on the spherical particle is equal to the energy of the sound field received per unit time on a unit area, then the axial acoustic radiation force F_z [6, 11]:

$$F_z = \frac{1}{c_w} \oint\limits_s Ids = \frac{P}{c_w} \int \frac{r^2}{W(z)^2} \exp\left[\frac{-2y^2}{W(z)^2}\right] H(\theta) d\theta \tag{1.9}$$

where:

$$H(\theta) = \cos\theta_i \cdot \sin\theta$$

$$\cdot \left\{ \cos(\theta_i - \theta) + R\cos(\theta_i + \theta) - \frac{T^2[\cos(\theta_i + \theta - 2\theta_r) + R\cos(\theta_i + \theta)]}{1 + R^2 + 2R\cos 2\theta_r} \right\} \tag{1.10}$$

c_w is the speed of sound in the medium water.

$$R = \left| \frac{\frac{Z_2}{\cos\theta_r} - \frac{Z_1}{\cos\theta_i}}{\frac{Z_2}{\cos\theta_r} + \frac{Z_1}{\cos\theta_i}} \right|^2 \tag{1.11}$$

$$T = 1 - R \tag{1.12}$$

where: Z_1, A are the acoustic impedances in the medium and the particle, respectively,

1.2.2 *Transverse Acoustic Radiation Force*

When the spherical particle is not symmetrical about the Gaussian ultrasound axis, we need to consider the effect of the transverse radiation force on the particle [11, 12].

As shown in Fig. 1.6, O is the center point of the Gaussian focused ultrasound, θ is the intersection point of the incident sound line and the particle A and the line connecting the center of the sphere C and the angle between the Z axis, α is the line connecting point A and the intersection point of the incident sound line Z axis B and the Z Angle of the axis. θ_i and θ_r are the incident angle and the refracted angle, respectively, y_A, z_A and y_C, z_C are the vertical and horizontal coordinates of points A and C, respectively.

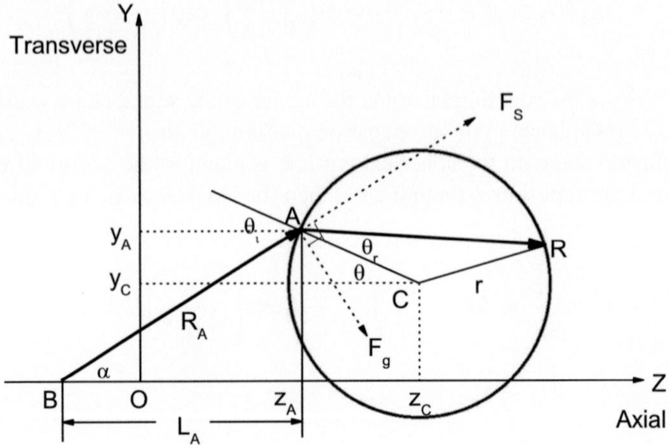

Fig. 1.6 Schematic diagram of the two-dimensional acoustic radiation force on a microsphere

The force of a single beam of sound on a particle is divided into two parts: the scattering force and the gradient force. The scattering force propagates along the direction of the sound beam, while the gradient force is perpendicular to it, pointing towards the direction with a larger sound intensity gradient:

$$d\vec{F} = d\overrightarrow{F_s} + d\overrightarrow{F_g} \tag{1.13}$$

$$d\overrightarrow{F_s} = \frac{IdS\cos\theta_i}{c_\omega}\vec{s} \tag{1.14}$$

$$d\overrightarrow{F_g} = \frac{IdS\cos\theta_i}{c_\omega}\vec{g} \tag{1.15}$$

$$dS = r^2 \sin\theta d\theta \tag{1.16}$$

where: $\vec{s}$, $\vec{g}$ are the unit direction vectors of the scattering force and the gradient force, respectively.

Similar to the one-dimensional case, from Eqs. (1.6) and (1.7), we get:

$$\begin{aligned} Q_s &= 1 + R\cos 2\theta_i - \sum_{n=0}^{\infty} T^2 R^n \cos\{2\theta_i - 2\theta_r + n(\pi - 2\theta_r)\} \\ &= 1 + R\cos 2\theta_i - \frac{T^2[\cos(2\theta_i - 2\theta_r) + R\cos 2\theta_i]}{1 + R^2 + 2R\cos 2\theta_r} \end{aligned} \tag{1.17}$$

$$\begin{aligned} Q_g &= -R\cos 2\theta_i + \sum_{n=0}^{\infty} T^2 R^n \cos\{2\theta_i - 2\theta_r + n(\pi - 2\theta_r)\} \\ &= -R\cos 2\theta_i + \frac{T^2[\cos(2\theta_i - 2\theta_r) + R\cos 2\theta_i]}{1 + R^2 + 2R\cos 2\theta_r} \end{aligned} \tag{1.18}$$

$$\theta_i = \alpha + \theta = \arccos\frac{L_A}{R_A} + \theta \tag{1.19}$$

$$\theta_r = \arcsin\left(\frac{c_2 \cdot \sin\theta_i}{c_1}\right) \tag{1.20}$$

where: c_1, c_2 represent the speed of sound in the medium and in the particle, respectively, Q_s, Q_g define the integral part of the incident sound wave in the propagation direction and the intensity gradient direction on the particle surface.

The scattering force and the gradient force are decomposed in the Y, Z axis direction:

$$d\overrightarrow{F_s} = Q_s\frac{IdS\cos\theta_i}{c_w}\vec{s} = Q_s\frac{IdS\cos\theta_i}{c_w}\left(s_z, s_y\right) = \left(dF_{sz}, dF_{sy}\right) \tag{1.21}$$

$$d\vec{F_g} = Q_g \frac{IdS \cos\theta_i}{c_w} \vec{g} = Q_g \frac{IdS \cos\theta_i}{c_w} (g_z, g_y) = (dF_{gz}, dF_{gy}) \tag{1.22}$$

The geometric vector relationship in Fig. 1.6 gives:

$$\vec{OA} = (z_C - r\cos\theta, y_C + r\sin\theta) \tag{1.23}$$

$$\vec{OB} = (z_C - r\cos\theta - L_A, 0) \tag{1.24}$$

$$\vec{OC} = (z_C, y_C) \tag{1.25}$$

where: z_C, y_C are the coordinates of the particle center.

$$\vec{s} = \frac{\vec{BA}}{\left|\vec{BA}\right|} = \frac{\vec{OA} - \vec{OB}}{R_A} = \left(\frac{L_A}{R_A}, \frac{y_C + r\sin\theta}{R_A}\right) \tag{1.26}$$

$$\vec{g} = \left(\frac{y_C + r\sin\theta}{R_A}, -\frac{L_A}{R_A}\right) \tag{1.27}$$

where: R_A is the radius of curvature of the incident sound beam, L_A is the axial distance between points A, B:

$$R_A = Z_A + \frac{\pi^2\lambda^2}{z_A} \tag{1.28}$$

$$L_A = \pm\sqrt{R_A^2 - (y_C + r\sin\theta)^2} \tag{1.29}$$

Therefore, the sound radiation force produced by a single beam of sound on a particle in the Y, Z axis direction is:

$$dF_y = dF_{sy} + dF_{gy} = \frac{r^2 Id\theta \sin\theta \cos\theta_i}{c_w R_A} \{Q_s(y_C + r\sin\theta) - Q_g L_A\} \tag{1.30}$$

$$dF_z = dF_{sz} + dF_{gz} = \frac{r^2 Id\theta \sin\theta \cos\theta_i}{c_w R_A} \{Q_s L_A + Q_g(y_C + r\sin\theta)\} \tag{1.31}$$

The total force on the particle in the Y, Z axis direction is:

$$F_z = \int_{\theta_{\min}}^{\theta_{\max}} \frac{r^2 I \sin\theta \cos\theta_i}{c_w R_A} \{Q_s L_A + Q(y_C + r\sin\theta)\} d\theta \tag{1.32}$$

$$F_y = \int_{\theta_{min}}^{\theta_{max}} \frac{r^2 I \sin\theta \cos\theta_i}{c_w R_A} \{Q_s(y_C + r\sin\theta) - Q_g L_A\} d\theta \tag{1.33}$$

Therefore:

$$F_y = \frac{2P}{\pi c_w} \int_{\theta_{min}}^{\theta_{max}} \left\{\frac{r}{W(z)}\right\}^2 \exp\left[\frac{-2y^2}{W(z)^2}\right] \frac{\cos\theta_i}{R_A} \{Q_s(y_C + r\sin\theta) - Q_g L_A\} \sin\theta d\theta \tag{1.34}$$

$$F_z = \frac{2P}{\pi c_w} \int_{\theta_{min}}^{\theta_{max}} \left\{\frac{r}{W(z)}\right\}^2 \exp\left[\frac{-2y^2}{W(z)^2}\right] \frac{\cos\theta_i}{R_A} \{Q_s L_A + Q_g(y_C + r\sin\theta)\} \sin\theta d\theta \tag{1.35}$$

Here, θ_{max}, θ_{min} are the upper and lower limits of the angle of the sound radiation force area integral of the sound beam on the particle sphere.

As shown in Fig. 1.7: Let a positive real variable m be such that the ray of mW_0 just touches the particle, with the point of tangency denoted as $A(z_A, y_A)$ [3, 11].

$$mW(z) = y = mW_0\sqrt{1 + \left(\frac{\lambda z}{\pi W_0^2}\right)^2} \tag{1.36}$$

Then, the slope of the tangent at the point of tangency is:

$$y_{A'} = \frac{mW_0 \cdot 2\frac{\lambda z_A}{\pi W_0^2} \cdot \frac{\lambda}{\pi W_0^2}}{2\sqrt{1 + \left(\frac{\lambda z_A}{\pi W_0^2}\right)^2}} = \frac{m^2 \frac{\lambda^2 z_A}{\pi^2 W_0^2}}{mW_0\sqrt{1 + \left(\frac{\lambda z_A}{\pi W_0^2}\right)^2}} = \frac{m^2\lambda^2 z_A}{\pi^2 W_0^2 y_A} \tag{1.37}$$

Therefore, the equation of the tangent line is:

$$y = \left(\frac{m^2\lambda^2 z_A}{\pi^2 W_0^2 y_A}\right)(z - z_A) + y_A = M(z - z_A) + y_A \tag{1.38}$$

where:

$$M = \left(\frac{m^2\lambda^2 z_A}{\pi^2 W_0^2 y_A}\right) \tag{1.39}$$

The line connecting the center of the sphere C and the point of tangency A is perpendicular to this tangent line, therefore:

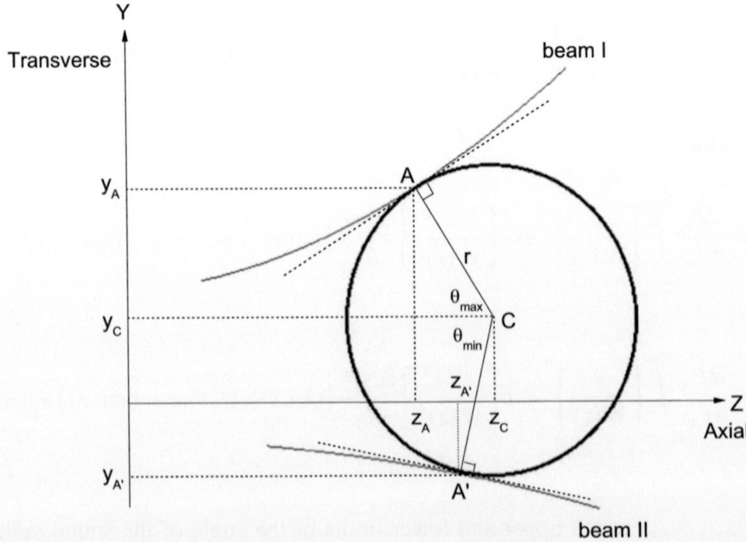

Fig. 1.7 Schematic diagram of the solution, showing the upper and lower limits of the integral $\theta_{\max}, \theta_{\min}$

$$\frac{(y_C - y_A)}{(z_C - z_A)} \cdot M = -1 \tag{1.40}$$

Also, point A lies on the circle, hence:

$$(z_A - z_C)^2 + (y_A - y_C)^2 = r^2 \tag{1.41}$$

From the above four equations, we obtain two solutions for y_A, z_A. Based on these two solutions, we can determine the upper and lower limits of the integrals in Eqs. (1.34) and (1.35):

$$\theta_{\max}, \theta_{\min} = \arctan\left(\frac{y_C - y_A}{z_C - z_A}\right) \tag{1.42}$$

1.2.3 *Considering the Acoustic Radiation Force Under Decay Conditions*

Since the ultrasound used is high-frequency, the impact of decay cannot be ignored [11, 13, 16, 17]. Here, we study the case of ultrasound decay during propagation within particles.

During propagation, the amplitude of the ultrasound decays with the propagation distance z:

$$A_z = A_0 e^{-\mu_0 z} \tag{1.43}$$

Converted into an energy relationship:

$$E_z = E_0 e^{-2\mu_0 z} \tag{1.44}$$

Substituting into Eq. (1.44), we get:

$$E_z = E_0 e^{-\frac{\alpha \cdot z}{10} \ln 10} = E_0 \cdot 10^{\lg e^{-\frac{\alpha \cdot z \ln 10}{10}}} = E_0 \cdot 10^{-\frac{1}{10}\alpha z} \tag{1.45}$$

That is, for every propagation distance z (cm), the energy decays to $10^{-0.1\alpha z}$ of its original value.

As shown in Fig. 1.8, the incident angle is θ_i, and the refracted angle is θ_r. From the geometric relationship, we can deduce:

As long as the particle is a uniform sphere, regardless of the size of the incident and refracted angles, the distance $R_x = 2 \cdot r \cdot \cos\theta_r$ that the sound wave travels in the particle before each reflection, where r is the radius of the particle.

Therefore, each time the sound wave reflects within the particle, the decay factor is $G = 10^{-0.2r\cos\theta_r\alpha}$.

Therefore, apart from the first time the sound wave enters the particle, after each reflection, the reflection coefficient should be multiplied by a decay factor G.

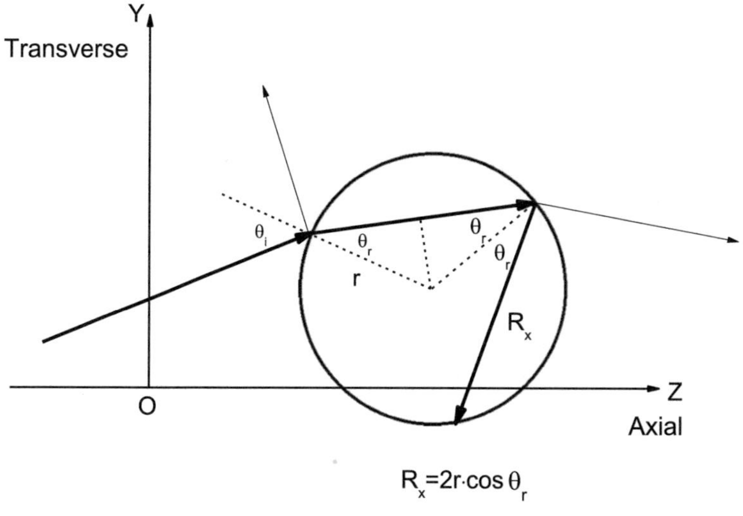

Fig. 1.8 Schematic diagram of the decay process of sound waves within particles

Therefore, the integral part of the two-dimensional acoustic radiation force, Eqs. (1.17) and (1.18), should be corrected to:

$$
\begin{aligned}
Q_s &= 1 + R\cos 2\theta_i - \sum_{n=0}^{\infty} T^2 (RG)^n \cos\{2\theta_i - 2\theta_r + n(\pi - 2\theta_r)\} \\
&= 1 + R\cos 2\theta_i - \frac{T^2[\cos(2\theta_i - 2\theta_r) + RG\cos 2\theta_i]}{1 + (RG)^2 + 2RG\cos 2\theta_r}
\end{aligned} \tag{1.46}
$$

$$
\begin{aligned}
Q_g &= -R\sin 2\theta_i + \sum_{n=0}^{\infty} T^2 (RG)^n \sin\{2\theta_i - 2\theta_r + n(\pi - 2\theta_r)\} \\
&= -R\sin 2\theta_i + \frac{T^2[\sin(2\theta_i - 2\theta_r) + RG\sin 2\theta_i]}{1 + (RG)^2 + 2RG\cos 2\theta_r}
\end{aligned} \tag{1.47}
$$

Therefore, in the two-dimensional case, the acoustic radiation force on the particle in the Y and Z directions is:

$$
F_z = \frac{2P}{\pi c_w} \int_{\theta_{\min}}^{\theta_{\max}} \left\{ \frac{r}{W(z)} \right\}^2 \exp\left[\frac{-2y^2}{W(z)^2} \right] \frac{\cos\theta_i}{R_A} \{ Q_s L_A + Q_g (y_C + r\sin\theta) \} \sin\theta d\theta \tag{1.48}
$$

$$
F_y = \frac{2P}{\pi c_w} \int_{\theta_{\min}}^{\theta_{\max}} \left\{ \frac{r}{W(z)} \right\}^2 \exp\left[\frac{-2y^2}{W(z)^2} \right] \frac{\cos\theta_i}{R_A} \{ Q_s (y_C + r\sin\theta) - Q_g L_A \} \sin\theta d\theta \tag{1.49}
$$

From the above equations, it can be seen that when the decay coefficient α is 0, the equations are consistent with the case where decay is not considered.

1.2.4 Simulation and Discussion

As mentioned earlier, to provide sufficient strength gradient to generate acoustic trapping force for particles, we assume the particles in the simulation are fat cells, surrounded by water. Their acoustic impedances are 1.5MRayls and 1.4MRayls respectively, and the speed of sound in fat is 1450 m/s, while in water it is 1500 m/s. These two media have similar acoustic impedances and sound speeds, which can increase the refraction of sound waves in particles, enhance the gradient force generated by the sound beam on the particles, and thus achieve the capture of particles by sound waves. A 100 MHz Gaussian focused ultrasound is used for the simulation, where the wavelength of the ultrasound λ is 15 μm, the beam width of the ultrasound

ω_0 is 15 μm, and the output acoustic power is 1mW. In the theory of ray approximation, the linearity of the particle should be much larger than the wavelength of the sound wave.

In the simulation, the radius of the particle r ranges from 10 λ to 14 λ, i.e., 150–210 μm.

To calculate the acoustic radiation force on the particle in two dimensions, it is first necessary to calculate how much sound can be incident on the particle surface, i.e., the upper and lower limits of the integral in formulas (1.44) and (1.45) $\theta_{\max}, \theta_{\min}$. Assuming the particle radius $r = 150$ μm, as shown in Figs. 1.9 and 1.10, where Fig. 1.9 represents the size of $\theta_{\max}$, and Fig. 1.10 represents the size of $\theta_{\min}$. The figures list two different initial lateral positions. When $y_{C1} = 100$ μm, the center of the sphere is above the ultrasound axis, i.e., most of the particle is above the axis; when $y_{C2} = -100$ μm, the center of the sphere is below the ultrasound axis, i.e., most of the particle is below the axis. It can be seen that when the center of the particle sphere is to the right of the ultrasound axis, i.e., $z_C > 0$, regardless of the positive or negative angle, y_{C1} under $\theta_{\max}$ is less than y_{C2} under $\theta_{\max}$, $\theta_{\min}$ otherwise, and in both cases, $\theta_{\max}, \theta_{\min}$ are acute angles; when the center of the particle sphere is to the left of the ultrasound axis, i.e., $z_C < 0$, y_{C1} under $\theta_{\max}$ is greater than y_{C2} under $\theta_{\max}$, $\theta_{\min}$ otherwise, and $\theta_{\max}, \theta_{\min}$ are obtuse angles. This is determined by the nature of Gaussian focused ultrasound. When $z_C > 0$, the Gaussian ultrasound diverges outward; when $z_C < 0$, the ultrasound converges inward. When the center of the particle sphere is exactly on the ultrasound beam axis, i.e., $z_C = 0$, both $\theta_{\max}, \theta_{\min}$ are right angles. This is because the wavefront radius of the Gaussian ultrasound at the beam waist position is infinite, and the sound field at $z_C = 0$ is a plane wave. At the same time, when the particle moves along the axis towards the far field, both $\theta_{\max}, \theta_{\min}$ tend to be right angles, which is due to the plane wave nature of the Gaussian ultrasound in the far field. As shown in the figure, the most sound rays can be incident on the particle near both sides of the ultrasound beam axis. Therefore, this is where the particle is most affected by the scattering force and gradient force, and may generate peak acoustic radiation force.

Figure 1.11 reflects the relationship between the axial acoustic radiation force on the particle and the axial position of the particle. As can be seen from the figure, in most cases, the radiation force on the particle is positive, acting on the particle to move it along the positive direction of the ultrasound axis. When it reaches $z_c = 0.5 \times 10^{-3}$ m, near the focus of the ultrasonic beam, the radiation force exerted on the particles becomes negative, pulling the particles towards the negative direction of the ultrasonic axis. This forms a potential well, trapping the particles at the location where the negative radiation force is generated, thus realizing the axial capture of particles by acoustic tweezers. The figure shows the force conditions of particles of different sizes. It can be seen that the larger the radius of the particle, the greater the peak value of the negative radiation force, i.e., the greater the peak value of the acoustic trap force. At the same time, the larger the scale of the particle, the wider the acoustic potential well formed, i.e., the particle can be captured in a wider range. Moreover, as the radius of the particle increases, the potential well formed moves in the positive axis direction. Figure 1.12 It reflects the lateral acoustic radiation force

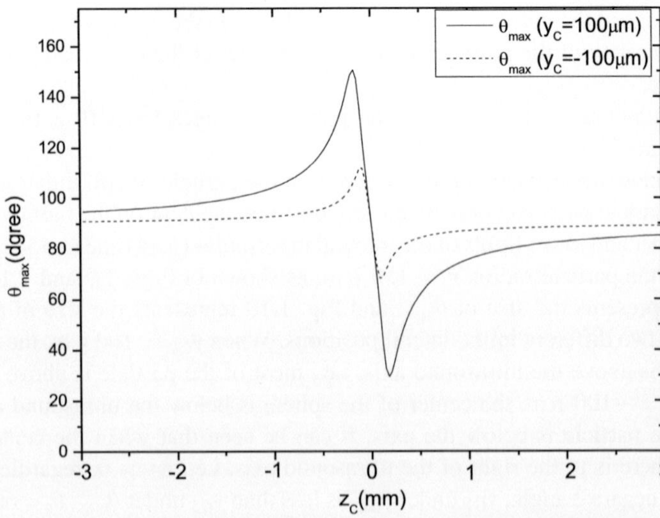

Fig. 1.9 Curve of the upper limit of the integral angle $\theta_{\max}$ with the change of the particle axis position

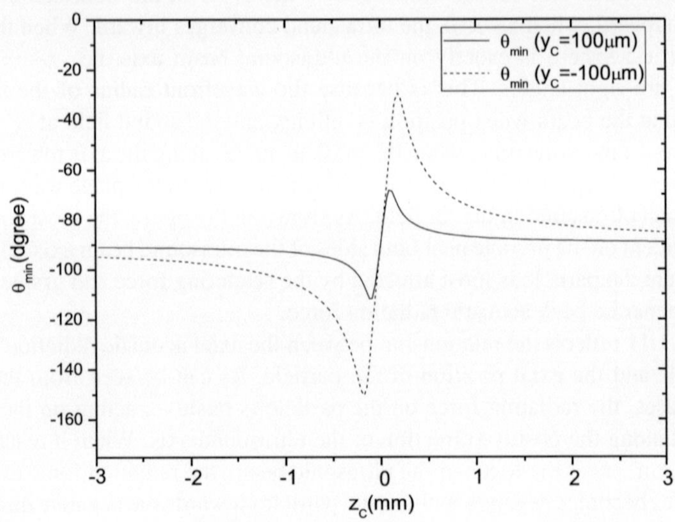

Fig. 1.10 Curve of the lower limit of the integral angle $\theta_{\min}$ with the change of the particle axis position

on the particle in the case of asymmetry about the ultrasonic axis, where the particle's center is above the ultrasonic axis $15\mu m$. It can be seen that in most cases when the particle's center is in the positive or negative direction of the z axis, the radiation force on the particle is negative, causing the particle to move towards the direction of the ultrasonic axis, i.e., towards the place with greater sound field intensity. This is consistent with the previous conclusion that particles will move in the direction of the intensity gradient. However, when the particle is right near the center of the ultrasonic axis, i.e., near $z_c = 0$, the lateral radiation force on the particle becomes positive, pushing the particle away from the axis position. This is because when the particle is near the beam waist, the wavefront is a plane wave, and the reflection part of the sound line on the particle surface relatively increases, i.e., the scattering force generated increases, pushing the particle away from the axis position. The figure shows the force conditions of particles of different sizes. It can be seen that as the radius of the particle gradually increases, although the potential well formed widens, the peak value of the acoustic trap force generated decreases, which is contrary to the nature of the axial radiation force on the particle. At the same time, comparing the two figures, it can be found that the lateral sound capture force on the particle in the Gaussian focused ultrasonic beam is much greater than the axial sound capture force, which is also the main reason why optical tweezers and acoustic tweezers often use lateral capture force to manipulate particles.

Figures 1.13 and 1.14 respectively reflect the relationship between the axial and lateral radiation forces on the particle and the lateral distance of the particle from the ultrasonic axis. Here, the initial axial positions of the particles are selected as $z_c = 0.3$ mm, $z_c = 0.4$ mm, $z_c = 0.5$ mm to analyze the relationship between the radiation force and the initial lateral position of the particle. These three points are exactly the potential well positions in Figs. 1.11 and 1.12, i.e., near the ultrasonic focus. The particle radius is $r = 150$ μm. As shown in Fig. 1.13, when the particle is

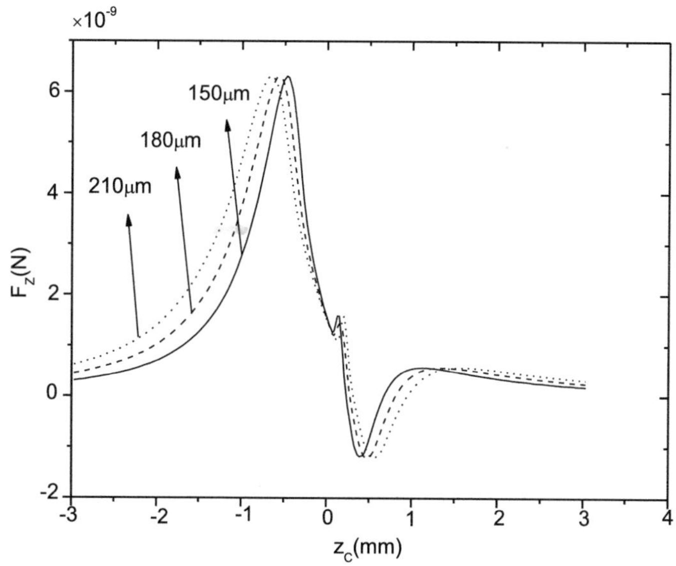

Fig. 1.11 Relationship between axial acoustic radiation force and initial axial position of the particle

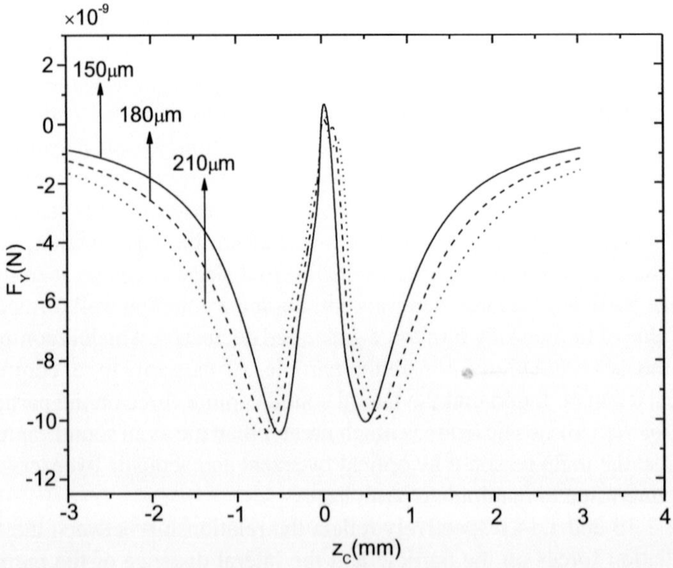

Fig. 1.12 Relationship between lateral acoustic radiation force and initial axial position of the particle

closer to the ultrasonic axis ($y_c < 0.2$ mm), the axial radiation force on the particle is negative, i.e., it is subjected to the action of the acoustic trap force. When the particle's center starts to move laterally away from the ultrasonic axis, the axial acoustic trap force on the particle near the focus first increases, reaches a peak near $y_c = 0.1$ mm, and then gradually decreases. However, when the particle gradually moves away from the ultrasonic axis position, the sound radiation force at the focus increases to a positive value. This indicates that when the particle is far from the ultrasonic beam axis, the particle no longer experiences negative acoustic radiation force, i.e., the ultrasonic beam can no longer capture the particle axially. When the particle is in the lateral far field, the axial radiation force on the particle at the focus gradually approaches zero. Due to the nature of Gaussian focused ultrasound, the intensity gradient and sound intensity of the sound field are very small at this time, and the particle only experiences a slight scattering force, pushing the particle away from the focus position. Figure 1.14 shows the relationship between the lateral radiation force on the particle near the focus position and the initial lateral position of the particle. From the figure, it can be seen that regardless of the lateral distance of the particle from the ultrasonic axis, the lateral acoustic radiation force on the particle at the focus is negative. This is because the gradient force on the particle points in the direction of the sound field intensity gradient. In the Gaussian focused ultrasonic beam, the particle always experiences a lateral radiation force pointing in the direction of the ultrasonic axis. Similar to the situation in Fig. 1.13, when the particle is close to the ultrasonic axis, the lateral acoustic trap force on the particle first gradually increases to a peak, then gradually decreases, and gradually approaches zero in the far field.

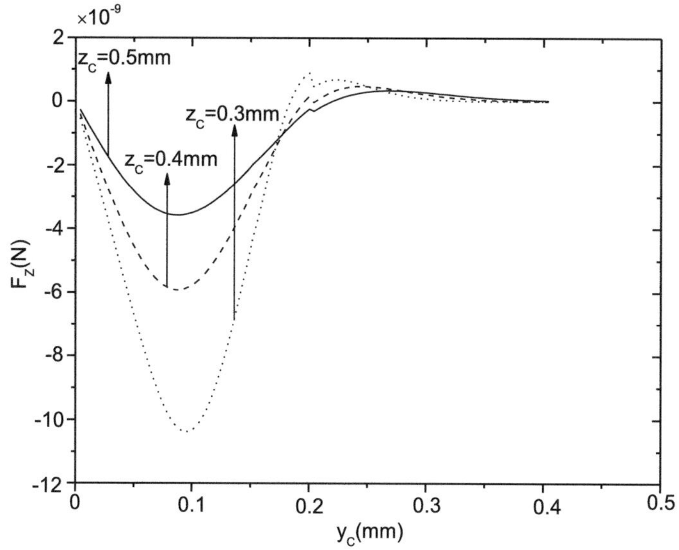

Fig. 1.13 Relationship between axial acoustic radiation force and particle's initial lateral position

This is because at a lateral position far from the ultrasonic beam axis, the intensity gradient of the Gaussian focused ultrasound is very small, and the gradient force on the particle is very small. In summary, when the particle is at $y_c = 0.1$ mm, at the ultrasonic focus, both the axial and transverse acoustic trapping forces reach their peak, resulting in the most noticeable acoustic trapping effect. When the particle is far from the ultrasonic axis, both the axial and transverse acoustic radiation forces exerted by the ultrasonic beam approach zero, making acoustic trapping impossible.

The particles used in this section's simulation are fat particles, which have an attenuation coefficient of 150 dB/cm under 100 MHz ultrasound conditions. As shown in Fig. 1.15, the dashed line describes the axial acoustic radiation force experienced by the particle without considering attenuation, while the solid line represents the force on the particle after considering the attenuation of ultrasound within the particle. It can be seen that when attenuation is considered, the peak of the axial radiation force in the positive direction received by the particle is slightly reduced compared to the situation without attenuation. However, the most important part, the peak of the acoustic trapping force and the width of the acoustic potential well, are almost unchanged compared to the situation without attenuation. This contradicts our assumption that attenuation would lead to energy dissipation, thereby reducing the performance of acoustic tweezers in manipulating and capturing particles. This is because when the ultrasound wave attenuates within the particle, both the scattering force part along the propagation direction of the sound line and the gradient force part along the intensity gradient direction have a certain degree of attenuation. The scattering force pushes the particle away from the focus, and the gradient force pulls the particle towards the focus. The peak of the axial radiation force in the positive

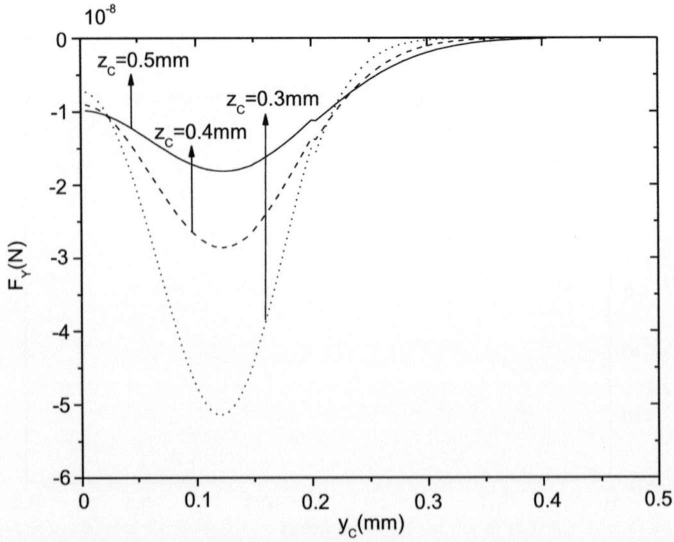

Fig. 1.14 Relationship between lateral acoustic radiation force and particle's initial lateral position

direction received by the particle is determined by the scattering force, so it undergoes a certain degree of attenuation. However, the axial trapping force is mainly due to the combined action of the gradient force and the scattering force, so the attenuation has little effect on the size of the trapping force. However, this phenomenon only exists when the attenuation factor is small. If the attenuation coefficient of the sound wave within the particle is too large, the intensity gradient of the sound wave decays more severely, which affects the gradient force received by the particle and reduces the peak of the trapping force. Figure 1.16 compares the situation of the transverse acoustic radiation force received before and after considering the attenuation of the sound wave within the particle. As can be seen from the figure: after considering attenuation, the positive part of the transverse radiation force received by the particle does not change much compared to the situation without considering attenuation, but the peak of the acoustic trapping force and the width of the acoustic potential well are both slightly reduced, which is exactly the opposite of the situation of the axial force received by the particle. This is because the transverse force received by the particle is determined by the gradient force component of the acoustic radiation force $\vec{g}$ and is irrelevant to the scattering force component $\vec{s}$. Therefore, after considering attenuation, the gradient force part along the intensity gradient direction has a certain degree of attenuation, thereby reducing the peak of the acoustic trapping force received by the particle. Also, as can be seen from Fig. 1.16, when the particle is at the ultrasonic axis, i.e., $z_c = 0$ on both sides, the transverse acoustic radiation force received is negative, causing the particle to move towards the direction of greater ultrasonic intensity. Obviously, the peak of the acoustic trapping force received near $z_c < 0$ is greater than the peak of the acoustic trapping force received near $z_c > 0$.

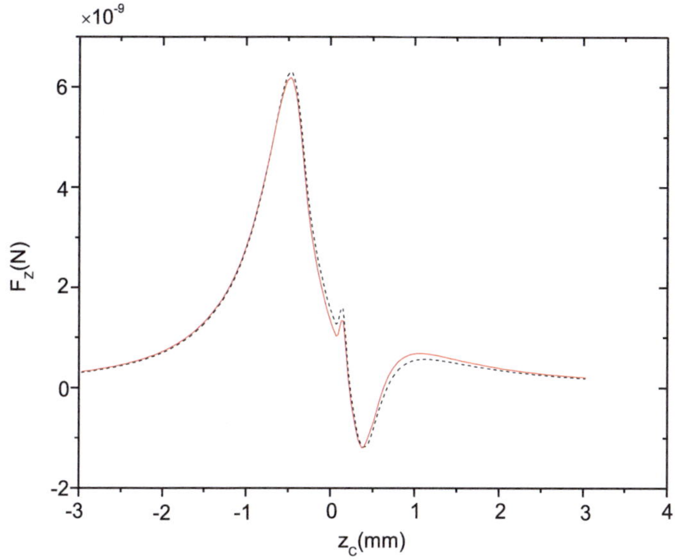

Fig. 1.15 The effect of internal acoustic attenuation in particles on axial acoustic radiation force

This is because when the sphere center is on the left side of the axis, the ultrasonic beam is focused, and when it is on the right side of the axis, the ultrasonic beam is divergent. As shown in Fig. 1.16, the upper and lower limits of integration in the focusing situation are larger than those in the divergent situation, which leads to a greater acoustic trapping force on the particle under the focusing situation of the Gaussian ultrasound. This is also why in practical applications, optical tweezers and acoustic tweezers use the focusing part of the sound beam when moving particles transversely.

1.3 Calculation of Gaussian Acoustic Radiation Force on Double-Layer Spherical Particles Based on Acoustic Ray Method

The previous section introduced the radiation force of the Gaussian sound field on a single-layer sphere, but in the fields of biomedical engineering and materials science, the model of multi-layer spheres has more practical application value: for example, manipulating and separating cell particles for medical research and experiments [19–21], cells are divided into cytoplasm and cell membrane, etc., and the structure and acoustic characteristics of these layers of cells are different, so the radiation force model of multi-layer spheres will be more practical; for example, in the application of drug delivery [22–25], using polystyrene microspheres filled with liquid drugs,

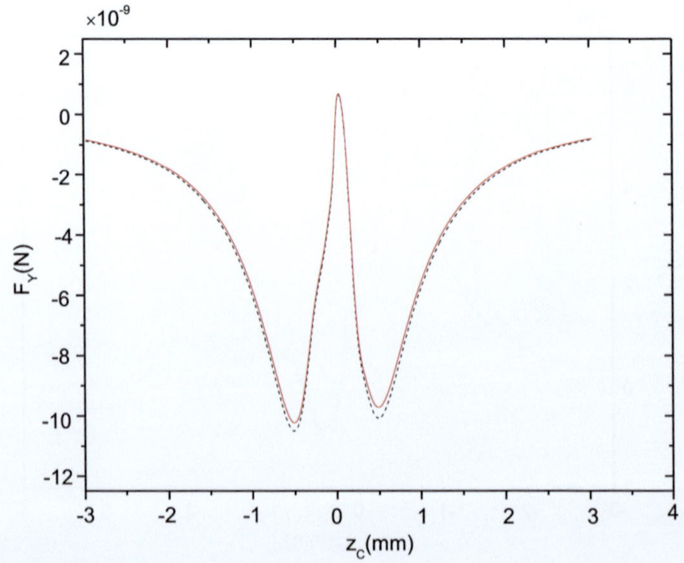

Fig. 1.16 The effect of internal acoustic attenuation in particles on transverse acoustic radiation force

and using acoustic radiation to effectively manipulate the microspheres, so that the shell accurately ruptures at the affected area and medicates.

There are already some acoustic radiation force models of double-layer spheres based on acoustic scattering theory [26–30], which will be introduced in Chap. 5. This chapter will focus on the solution of the radiation force of high-frequency ultrasound field on double-layer spherical particles, and will detail the acoustic radiation force model under the approximation of the acoustic ray method [31]. The results simulate and analyze the effects of shell thickness, acoustic impedance and sound speed of each layer on the final acoustic radiation force and acoustic trapping force. The results provide a theoretical solution for biomedical applications such as cell manipulation, and this conclusion can be extended to multi-layer spheres and more complex structural models.

1.3.1 Acoustic Ray Tracing and Classification in Double-Layer Spherical Particles

Figures 1.17 and 1.18 depict a schematic diagram of a beam of acoustic rays incident on a double-layer symmetrical sphere, where R_c and r_c are the radii of the outer and inner spheres, respectively, z_c is the axial coordinate of the sphere. c_2 and c_3 are the longitudinal sound speeds of ultrasound propagation in the outer and inner spheres, respectively. c_1 is the sound speed of ultrasound propagation in the medium

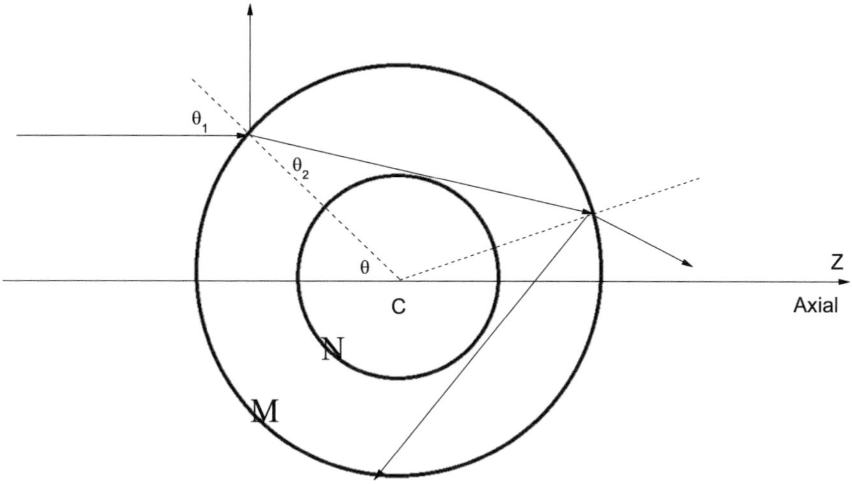

Fig. 1.17 Schematic diagram of acoustic ray incident on a symmetrical double-layer sphere $\left(\theta_1 \geq \arcsin\left[\left(\frac{c_1}{c_2}\right)\left(\frac{r_c}{R_c}\right)\right]\right)$

around the microsphere. We assume that the outer sphere shell is boundary M, and the inner sphere shell is boundary N. θ is the angle between the line connecting the intersection of the incident acoustic ray and the particle with the center of the sphere C and the axis Z, θ_1 and θ_2 are the incident angle and refraction angle of the acoustic ray incident on the outer sphere shell, respectively, y and z are the longitudinal and transverse coordinates, respectively.

From the figure, it is not difficult to find that when a beam of acoustic rays is incident on the microsphere, when $\theta_1 \geq \arcsin\left[\left(\frac{c_1}{c_2}\right)\left(\frac{r_c}{R_c}\right)\right]$, as shown in Fig. 1.17, when the acoustic ray is refracted at boundary M, it will not be incident in the inner sphere. Since the double-layer sphere is completely symmetrical, in this case, the acoustic ray will never be incident on the inner sphere, and will only continuously reflect and refract on the outer sphere shell until all the energy is refracted out of the sphere or decays completely.

We note here that in this case, the contribution of the acoustic wave to the radiation force on the outer part of the double-layer sphere F_{z1}, is similar to the content discussed in the previous section [32].

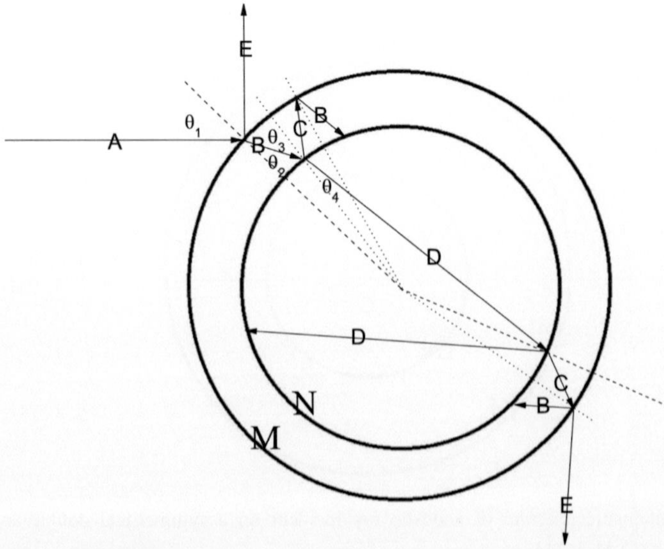

Fig. 1.18 Schematic diagram of sound rays incident on a symmetrical double-layer sphere $\left(\theta_1 < \arcsin\left[\left(\frac{c_1}{c_2}\right)\left(\frac{r_c}{R_c}\right)\right]\right)$

From Eqs. (1.9–1.10) we can get:

$$F_{z1} = \frac{P_0}{c_1} \int_{\theta_{\min}}^{\theta_{\max}} \frac{R_c^2}{W^2(z)} \exp\left[\frac{-2y^2}{W^2(z)}\right] H(\theta) d\theta \tag{1.50}$$

where:

$$H(\theta) = \cos\theta_1 \cdot \sin\theta \cdot \left\{\cos(\theta_1 - \theta) + R_1 \cos(\theta_1 + \theta) - \frac{T_1^2[\cos(\theta_1 + \theta - 2\theta_2) + R_1 G \cos(\theta_1 + \theta)]}{1 + R_1^2 G^2 + 2R_1 G \cos 2\theta_2}\right\} \tag{1.51}$$

$$R_1 = \left| \frac{\frac{Z_2}{\cos\theta_2} - \frac{Z_1}{\cos\theta_1}}{\frac{Z_2}{\cos\theta_2} + \frac{Z_1}{\cos\theta_1}} \right|^2 \tag{1.52}$$

$$T_1 = 1 - R_1 \tag{1.53}$$

$$G = 10^{-0.2 R_c \cos\theta_2 \alpha_1} \tag{1.54}$$

The calculation of $\theta_{\max}$ can be found in Eq. (1.42)

$$\theta_{\min} = \theta 1 - \arcsin\left(\frac{R_c \sin\theta}{z_c\left[1+\left(\frac{\pi W_0^2}{\lambda z_c}\right)^2\right]}\right)$$
$$= \arcsin\left[\left(\frac{c_1}{c_2}\right)\left(\frac{r_c}{R_c}\right)\right] - \arcsin\left(\frac{R_c \sin\theta}{z_c\left[1+\left(\frac{\pi W_0^2}{\lambda z_c}\right)^2\right]}\right) \quad (1.55)$$

where: R_1, T_1 are the energy reflection coefficient and refraction coefficient of the outer sphere shell M, respectively, Z_1, Z_2 are the acoustic impedances of the surrounding medium and the outer sphere, respectively, G is a dimensionless attenuation factor related to the attenuation coefficient of the outer sphere α_1.

When the incident angle satisfies $\theta_1 < \arcsin\left[\left(\frac{c_1}{c_2}\right)\left(\frac{r_c}{R_c}\right)\right]$, after the incident sound ray refracts at interface M and enters the outer sphere, the sound ray will propagate to the surface N of the inner sphere and undergo a series of complex emissions and refractions, as shown in Fig. 1.18.

Here, we define and categorize all sound rays for subsequent classification calculations. As shown in Fig. 1.18, we define all sound rays incident from the medium around the microsphere to boundary M as sound ray A, sound rays from boundary M to boundary N as sound ray B, sound rays reflected from boundary N to boundary M as sound ray C, sound rays propagating within the inner sphere from boundary N to boundary N as sound ray D, and finally, sound rays emitted from boundary M to the surrounding medium as sound ray E. In this way, we find that all sound rays have been classified.

Using the ray tracing method, we have drawn a set of tree diagrams to illustrate the relationship between the sound ray sets, as shown in Fig. 1.19. Considering the case of attenuation, part of the energy is absorbed inside the particle, and the remaining energy will continue to propagate to the subset of sound rays, guiding all energy to propagate out of the sphere through the sound ray set E.

By observing and tracking all the reflections and refractions of sound rays inside the sphere, we can conclude that all sound energy is transmitted to the microsphere through sound ray A, and finally, the energy is transmitted out through sound ray E or attenuated inside the particle. Therefore, the sound rays B, C, and D propagating inside the particle do not contribute to the radiation force received by the particle. We only need to consider the relationship between sound ray sets A and E.

When the sound ray is incident on the outer interface M or the inner interface N of the double-layer sphere, refraction or reflection occurs. At this time, the new refracted or reflected sound ray will change in angle and relative energy compared to the incident sound ray. Table 1.1 lists the changes in angle (positive in the clockwise direction) and energy (relative energy compared to the incident sound ray) of the newly generated sound rays compared to the original sound rays. As shown in Fig. 1.18, θ_3 and θ_4 are the incident and reflection angles at interface N, respectively, R_2 and T_2 are the energy reflection and refraction coefficients of the inner shell N,

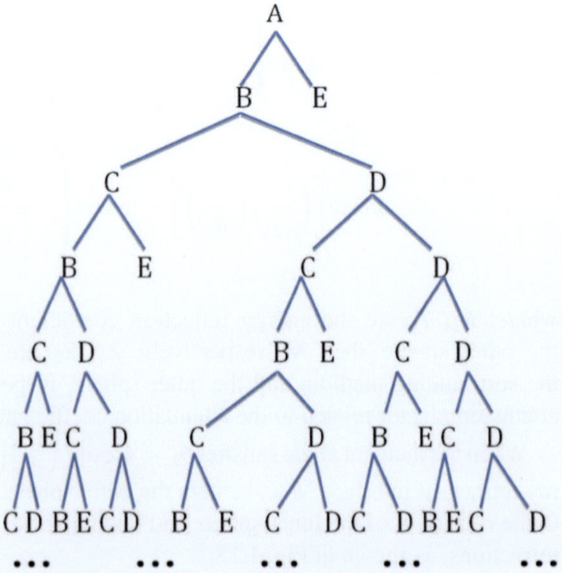

Fig. 1.19 Tree diagram of the relationship between sound ray sets

respectively, G_1 and G_2 These are dimensionless decay factors for an outer sphere and an inner sphere, which will be introduced later. Since we only consider the relationship between sound lines A and E, we reclassify the set of sound lines E here. As shown in Fig. 1.20, sound line E will be emitted from the outer surface M of the particle after several reflections and refractions inside the particle. We assume that sound line E has i reflections on the outer surface of boundary N, j refractions on the outer surface of boundary N, and k reflections on the inner surface of boundary N. We classify the set of sound lines E as P_{ijk}, so all the outgoing sound lines E are classified.

Similarly, we define the relative energy between the outgoing sound line E and the incoming sound line A as τ_{ijk}, and the relative angle as β_{ijk}. By summarizing and generalizing Table 1.1, we conclude:

Table 1.1 Relationship between the angle and energy changes of sound rays incident at the interface

Category	Deflection angle	Relative power
$A \to B$	$\theta_1 - \theta_2$	T_1
$A \to E$	$\pi + 2\theta_1$	R_1
$B \to C$	$-\pi + 2\theta_3$	R_2G_1
$C \to B$	$-\pi - 2\theta_2$	R_1G_1
$C \to E$	$-\theta_1 - \theta_2$	T_1G_1
$B \to D$	$\theta_3 - \theta_4$	T_2G_1
$D \to C$	$-\theta_3 - \theta_4$	T_2G_2
$D \to D$	$-\pi - 2\theta_4$	R_2G_2

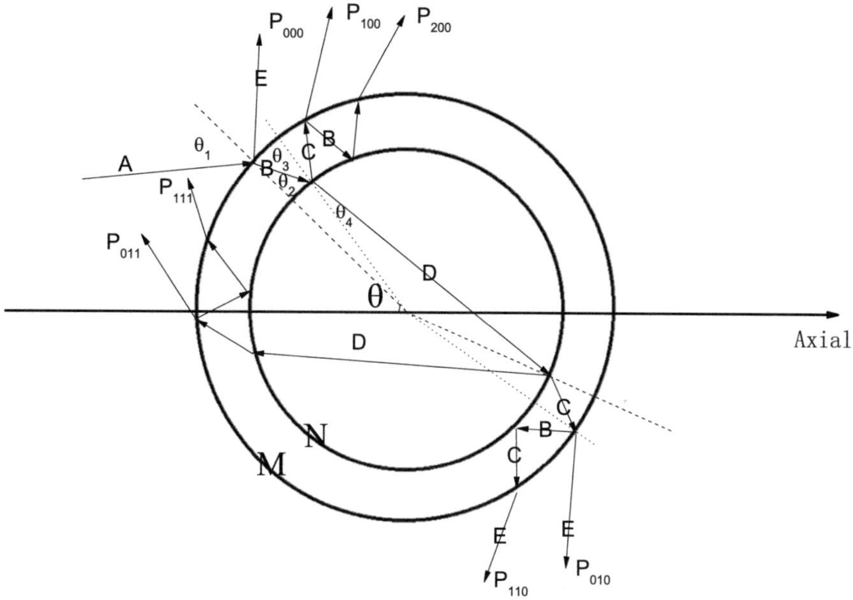

Fig. 1.20 Schematic diagram of sound line classification

$$\left.\begin{array}{l} \tau_{ijk} = T_1 \times (G_1R_2)^i \times (G_1R_1)^{i+j-1} \\ \quad \times \left(T_2^2 G_1 G_2\right)^j \times (R_2G_2)^k \times G_1T_1 \\ \beta_{ijk} = (\theta_1 - \theta_2) + i(\pi + 2\theta_3) + (i+j-1)(\pi - 2\theta_2) \\ \quad + j(2\theta_3 - 2\theta_4) + k(\pi - 2\theta_4) + (\theta_1 - \theta_2) \end{array}\right\} i, j, k = 0, 1, 2, 3\ldots \tag{1.56}$$

where:

$$R_2 = \left| \frac{\frac{Z_3}{\cos\theta_4} - \frac{Z_2}{\cos\theta_3}}{\frac{Z_3}{\cos\theta_4} + \frac{Z_2}{\cos\theta_3}} \right|^2 \tag{1.57}$$

$$T_2 = 1 - R_2 \tag{1.58}$$

$$\theta_2 = \arcsin\left[\left(\frac{c_2}{c_1}\right) \cdot \sin\theta_1\right] \tag{1.59}$$

$$\theta_3 = \arcsin\left[\left(\frac{R_c}{r_c}\right) \cdot \sin\theta_2\right] \tag{1.60}$$

$$\theta_4 = \arcsin\left[\left(\frac{c_3}{c_2}\right) \cdot \sin\theta_3\right] \tag{1.61}$$

Z_3 is the acoustic impedance of the inner sphere.
Where:

$$\begin{cases} \tau_{000} = R_1 \\ \beta_{000} = \pi + 2\theta_1 \end{cases} \tag{1.62}$$

$$\tau_{i0k} = 0(i \neq 0, k \neq 0) \tag{1.63}$$

In addition, when the sound line does not refract on the outside of the interface N, that is, the sound line does not enter the particle sphere, there is no reflection of the sound line on the inside of the interface N.

In summary, the axial sound radiation force part contributed by the inner side of the double-layer sphere F_{z2}:

$$F_{z2} = \frac{P_0}{c_1} \int_0^{\theta_{\min}} \frac{R_c^2}{W(z)^2} \exp\left[\frac{-2y^2}{W(z)^2}\right] H'(\theta) d\theta \tag{1.64}$$

where:

$$H'(\theta) = \cos\theta_1 \cdot \sin\theta \cdot \left\{ \cos(\theta_1 - \theta) - \sum_{i=0}^{\infty}\sum_{j=0}^{\infty}\sum_{k=0}^{\infty}\left[\tau_{ijk} \cdot \cos\left(\beta_{ijk} + \theta - \theta_1\right)\right] \right\} \tag{1.65}$$

1.3.2 Sound Radiation Force Under Decay Conditions

Similar to the previous calculation of decay inside particles, the dimensionless decay factors G_1 and G_2 are related to the decay factors of the outer sphere and the inner sphere α_1 and α_2:

$$G_1 = 10^{-0.1 L_1 \alpha_1} \tag{1.66}$$

$$G_2 = 10^{-0.1 L_2 \alpha_2} \tag{1.67}$$

where:

$$L_1 = R_c \cdot \cos\theta_2 - \sqrt{r_c^2 - (R_c \sin\theta_2)^2} \tag{1.68}$$

$$L_2 = 2r_c \cos\theta_4 \tag{1.69}$$

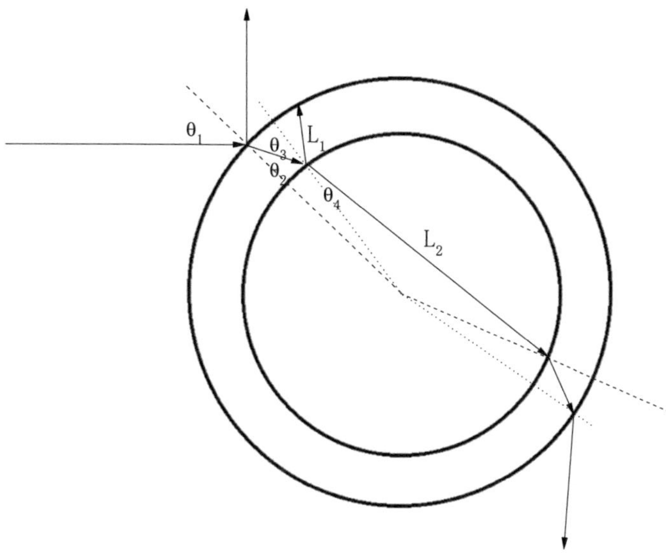

Fig. 1.21 Schematic diagram of decay in particles

L_1 and L_2 represent the distance of sound line propagation in the outer and inner sides of the particle, as shown in Fig. 1.21.

The decay of high-frequency ultrasound in the surrounding medium will further affect the action of the sound radiation force. Therefore, for focused ultrasound, the focus position should be as close to the transducer surface as possible, which requires a smaller F number (the ratio of the focal length to the diameter of the focusing transducer) to provide a closer focus. Therefore, a large-angle focusing transducer is chosen to avoid or reduce the effect of decay. As shown in Fig. 1.22: L_w is the distance from the sound line starting from the transducer surface to the outer surface of the particle, the geometric focal length of the transducer, that is, the distance between the transducer surface and the origin O is l, so the dimensionless decay factor G_w in the surrounding medium (water) is:

$$G_w = 10^{-0.1L\alpha} \tag{1.70}$$

$$\begin{aligned} L_W &= \left|\overrightarrow{BP}\right| \\ &= \sqrt{[(z_c - R_c\cos\theta) - l]^2 + [R_c\sin\theta - (l\tan\varphi + R_c\sin\theta - \tan\varphi(z_c - R_c\cos\theta))]^2} \end{aligned} \tag{1.71}$$

where:

$$\varphi = \angle PAC = \arcsin\left(\frac{R_c\sin\theta}{\eta}\right) \tag{1.72}$$

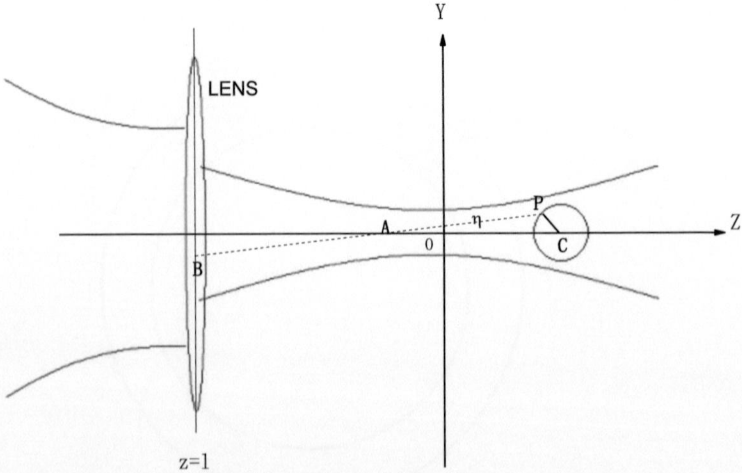

Fig. 1.22 Schematic diagram of decay in the medium around the particle

$$\eta = \left|\overrightarrow{AP}\right| = z_p\left[1 + \left(\frac{\left(\pi W_0^2\right)}{\lambda z_p}\right)^2\right] \tag{1.73}$$

$$z_p = z_c - R_c \cos\theta \tag{1.74}$$

The distance between the transducer surface and the origin O is l.
Therefore:

$$P = P_0 \cdot G_w \tag{1.75}$$

P refers to the power reaching the particle surface after the sound wave has been absorbed and attenuated by the surrounding medium.

In summary, from Eqs. (1.70–1.75), we conclude that the radiation force generated by Gaussian focused ultrasound on a uniformly symmetric double-layer sphere is:

$$F_z = \frac{1}{c_1}\left\{\int\limits_{\theta_{\min}}^{\theta_{\max}} \frac{R_c^2}{\omega(z)^2}\exp\left[\frac{-2y^2}{\omega(z)^2}\right]H(\theta)\cdot Pd\theta + \int\limits_{0}^{\theta_{\min}} \frac{R_c^2}{\omega(z)^2}\exp\left[\frac{-2y^2}{\omega(z)^2}\right]H'(\theta)\cdot Pd\theta\right\} \tag{1.76}$$

1.3.3 Simulation and Discussion

Table 1.2 lists the acoustic linear properties of the materials used in this section's simulation. We use high-frequency focused ultrasound at 100 MHz for the simulation calculations, meaning the wavelength of the ultrasound in water is 15 μm. To satisfy the conditions of the acoustic line method, the scale of the particles should be much larger than the wavelength of the sound wave. Therefore, we choose particles with an outer sphere radius of 150 μm for the simulation. The output power of the ultrasound is 100mW.

First, we assume that the outer layer of the double-layer particle is fat, the inner layer is filled with animal fat, and the surrounding medium of the particle is water. We use this simple model to simulate large cells in human tissue. Figure 1.23 depicts the relationship between the axial radiation force on the particle and the initial axial position of the particle. When $\frac{r_c}{R_c} = 0$, that is, when the inner sphere is zero, the simulation curve is similar to the previous model curve of the axial acoustic radiation force on a single-layer sphere [32] (Fig. 1.11). As the radius of the inner sphere gradually increases from zero, the backward radiation force (i.e., the trapping force at the possible position of the acoustic potential well), or the trapping force, continuously decreases. When the radius of the inner sphere increases to a certain value, the acoustic trapping force even disappears, changing from negative to positive, while at the same time, the forward acoustic radiation force also continuously increases. This indicates that the particle will no longer be axially trapped. The reason is that the contribution of the inner sphere to the radiation force is continuously increasing, resulting in a larger scattering force. As we mentioned in Chap. 2, a larger scattering force will lead to a larger forward radiation force: this will be discussed in depth in Fig. 1.23 later.

However, as the radius of the inner sphere continues to increase, the peak of the backward radiation force begins to appear, and even a larger negative radiation force than the single-layer sphere appears, especially in the simulation diagram of the thin sphere shell $\left(\frac{r_c}{R_c} = 0.99\right)$ This is due to the lower speed of sound in the animal fat

Table 1.2 Acoustic properties of various materials

Material	Acoustic impedance (MRayl)	Sound attenuation coefficient at 100 MHz $\left(\frac{\mathrm{dB}}{\mathrm{cm}}\right)$	Longitudinal sound Speed $\left(\frac{\mathrm{m}}{\mathrm{s}}\right)$
Water	1.5	20	1500
Fat	1.4	150	1450
Animal oil	1.32	120	1400
Low-density polyethylene (LDPE)	1.57	450	1745
Air	0.00043	966	340

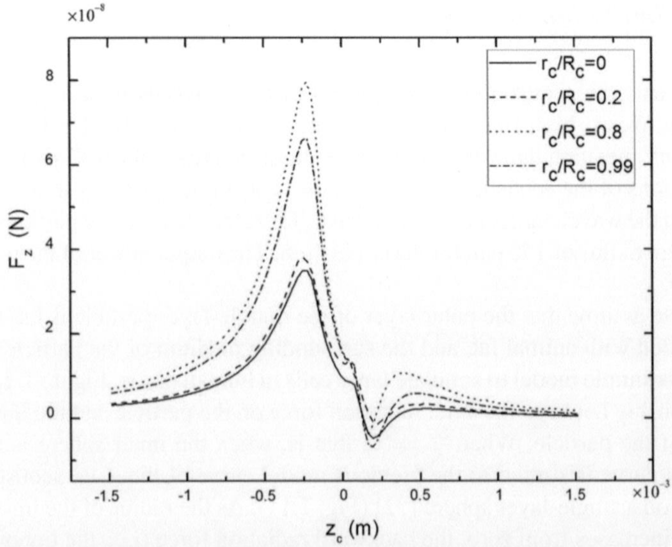

Fig. 1.23 Acoustic radiation force on particles filled with animal fat in a fat sphere shell

of the inner sphere compared to water and fat, resulting in a greater gradient force, causing the particles to experience a greater axial trapping force compared to the fat sphere. At this time, the radius of the inner sphere is close to the radius of the outer sphere, and the contribution of the outer sphere shell to the overall radiation force is minimal, leading to a greater backward radiation force in the case of a thin sphere shell.

We simulated the axial acoustic radiation force on a double-layer sphere filled with water in a fat sphere shell, as shown in Fig. 1.24. Similar to the previous Fig. 1.23, when $\frac{r_c}{R_c} = 0$, that is, when the radius of the inner sphere is zero, the radiation force on the particles is consistent with the single-layer sphere. Then, as the radius of the inner sphere gradually increases, the backward radiation force on the particles decreases. At the same time, when $\frac{r_c}{R_c} = 0.8$, the radiation force-initial position curve of the double-layer sphere shows two peaks, located on the left and right of the focused ultrasound. This is because the inner sphere has a higher speed of sound than the outer sphere, resulting in a higher reflection coefficient of the sound wave on the surface of the inner sphere, thereby providing a greater scattering force on the inner sphere [19], which pushes the double-layer sphere particles in the direction of sound wave propagation and cannot generate an acoustic potential well. As the radius of the inner sphere further increases, the reflection coefficient of the inner sphere gradually decreases (this will be analyzed in Fig. 1.27 later), the radiation force on the double-layer sphere particles will continuously decrease and approach zero.

An important application of acoustic radiation force is drug delivery, which involves filling a polystyrene microsphere shell with liquid drugs and then guiding

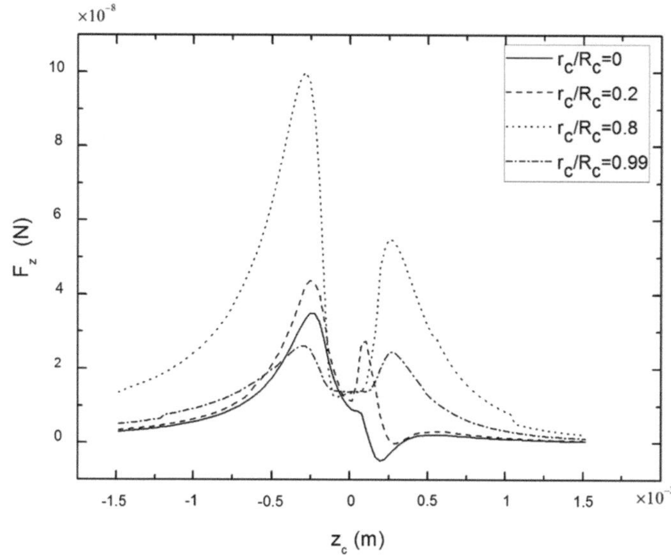

Fig. 1.24 Acoustic radiation force on water-filled fat sphere shell

them to a fixed location through the directional action of sound waves. Here we simulated the acoustic radiation force on double-layer sphere particles filled with water in a low-density polystyrene (LDPE) sphere shell, as shown in Fig. 1.25. Because the LDPE sphere shell has a higher speed of sound than the surrounding medium (water), the backward acoustic radiation force on the particles appears on the left side of the ultrasound focus position, which means that the acoustic potential well generated by the Gaussian focused ultrasound on the LDPE sphere shell particles is located in the divergent part of the Gaussian wave rather than the focus part [32]. At the same time, when $\frac{r_c}{R_c} = 0$, compared with the previous examples, due to the greater acoustic impedance difference between the LDPE sphere shell and the surrounding medium of the particles, the ultrasound generates a greater acoustic radiation force on the particles. As the radius of the inner sphere continues to increase, the backward radiation force on the particles continuously decreases until it disappears. As the radius of the inner sphere continues to increase, approaching the thin sphere shell situation, the forward radiation force also continuously decreases. This is similar to the situation in Fig. 1.24.

In Figs. 1.26 and 1.27, we describe the relationship curves between the maximum value of the backward acoustic radiation force and the thickness of the sphere shell, corresponding to the two double-layer sphere models of the fat sphere shell filled with animal fat and the LDPE sphere shell filled with water. As can be seen from the figure, in both cases, as the sphere shell continues to increase, the absolute value of the maximum backward radiation force on the particles continuously decreases until it becomes positive. At the same time, almost when the radius of the inner sphere increases to half of the radius of the outer sphere, with the further increase

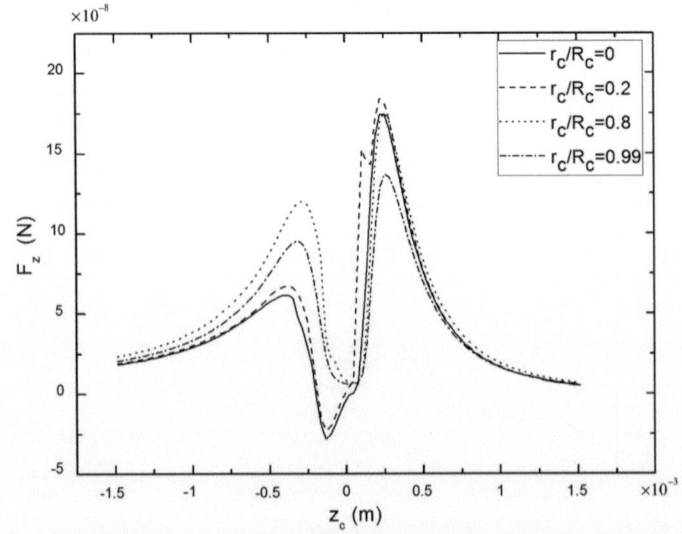

Fig. 1.25 Acoustic radiation force on water-filled low-density polystyrene sphere shell

of the radius of the inner sphere, a negative sound trapping force appears again in Fig. 3.3.4, and in the case of a thin sphere shell, the maximum value of the negative trapping force on the particles is greater than that of a single-layer sphere. Similarly, in Fig. 1.27, the situation of increasing backward radiation force appears again, but the figure shows that in the LDPE model, the backward radiation force on the particles is much smaller than the previous particles, and it does not reach a negative value again.

According to formulas (1.57–1.61), and satisfying $\frac{r_c}{R_c} > \sin\theta_1 \cdot c_1/c_2$ to ensure that the incident sound line can be incident on the inner sphere, we give the relationship between the reflection coefficient of the inner sphere and the radius of the inner sphere, as shown in Fig. 1.28. Here, the incident angle of the sound line on the outer sphere is set to $\theta_1 = \pi/6$, and other parameters and conditions use the previous simulation parameters.

We found that the reflection coefficient of the inner sphere is initially high. In addition, as long as the incident sound line passes through the surface of the outer sphere and is incident on the surface of the inner sphere, that is, when $\frac{r_c}{R_c}$ is around 0.5 in Fig. 1.28, the reflection coefficient of the inner sphere is at its maximum, which means that the inner sphere will be subjected to a greater scattering force. This is why $\frac{r_c}{R_c}$. The reason why double-layer spherical particles experience a greater positive radiation force near 0.5.

Furthermore, as shown in Fig. 1.28, the reflection of the inner sphere decreases continuously as the radius of the inner sphere increases again. This leads to the gradient force beginning to dominate the radiation force, resulting in the reappearance of the negative radiation force shown in Fig. 1.26. At the same time, as the radius of the inner sphere continues to increase, the contribution of the outer sphere to

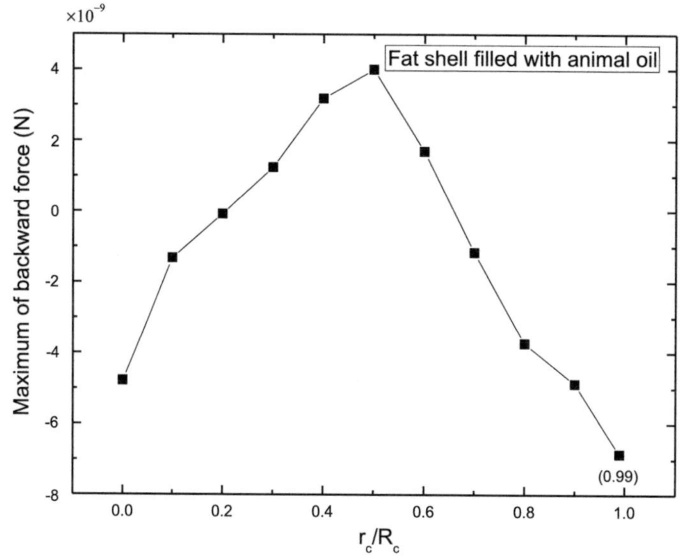

Fig. 1.26 Relationship between maximum backward radiation force and inner sphere radius in fat sphere shell filled with animal fat

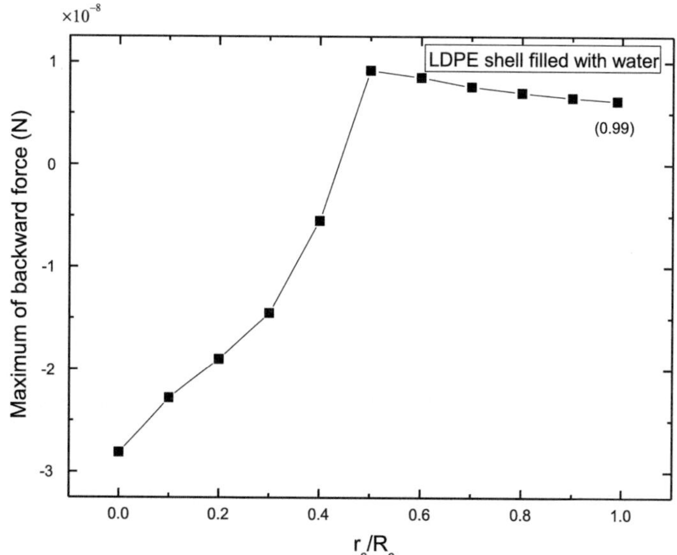

Fig. 1.27 Relationship between maximum backward radiation force and inner sphere radius in LDPE sphere shell filled with water

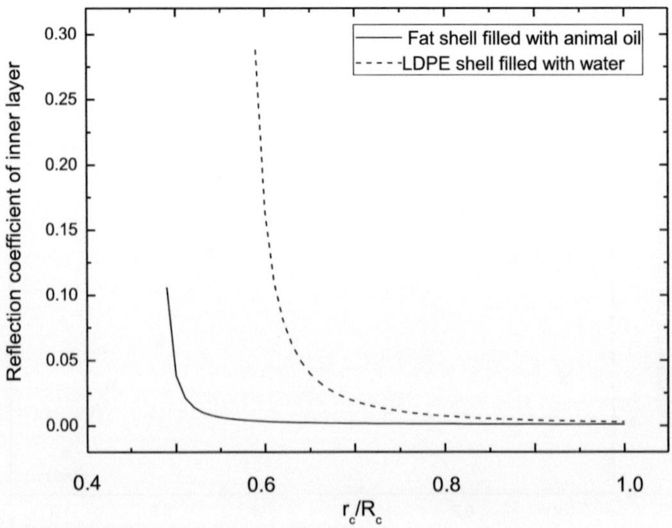

Fig. 1.28 Relationship between reflection coefficient of inner sphere and radius of inner sphere

the radiation force gradually decreases. This results in less gradient force in the double-layer sphere model filled with animal fat in the adipose sphere shell, and more gradient force in the double-layer sphere model filled with water in the LDPE sphere shell. These reasons explain the results of Figs. 1.26 and 1.27.

We also calculated the model of the hollow LDPE sphere shell, where $\frac{r_c}{R_c} = 0.8$, the outer sphere shell is low-density polystyrene, and the inside is filled with air. Figures 1.29 and 1.30 respectively represent the contributions of the outer sphere shell and the inner air to the total acoustic radiation force on the particle. As can be seen from Fig. 1.29, the acoustic radiation force contributed by the outer sphere shell creates an acoustic potential well. Since it is a thin sphere shell, the radiation force and capture force experienced by the outer sphere shell are not much. At the same time, in comparison, the inner air experiences a quite large radiation force, as shown in Fig. 1.30. However, this force is entirely positive, pushing the particles in the direction of sound wave propagation. This is due to the huge acoustic impedance difference between the sphere shell and the internal air, almost no sound energy can transmit through the LDPE-air layer, resulting in a huge scattering force along the direction of sound wave propagation. Especially at the ultrasonic focus position, where the wavefront of the ultrasound is approximately a plane wave, providing the maximum axial thrust. Overall, in the hollow LDPE sphere shell model in water, the radiation force experienced by the particles is large and always positive.

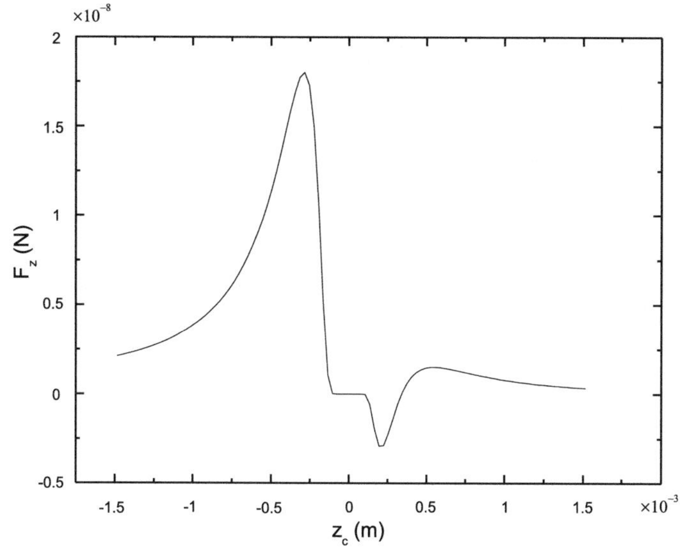

Fig. 1.29 Acoustic radiation force provided by the outer sphere shell in the hollow LDPE model

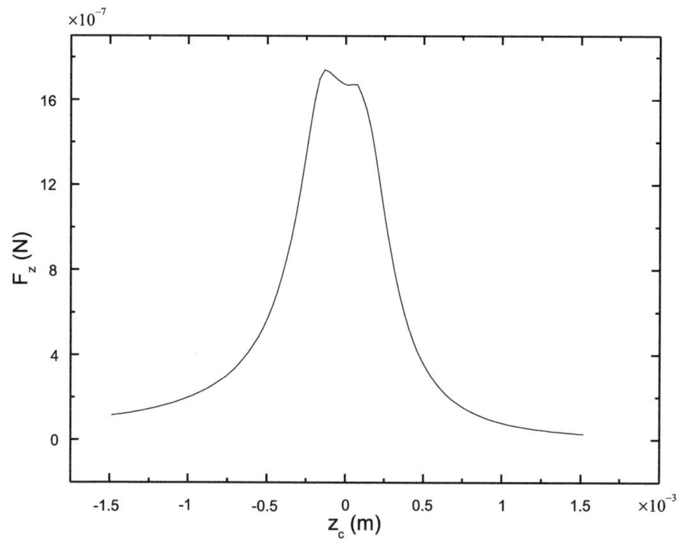

Fig. 1.30 Acoustic radiation force provided by the inner air in the hollow LDPE model

References

1. Gonghuan D, Zhemin Z, Xiufen G (2001) Fundamentals of acoustics, 2nd edn, Nanjing University Press

2. Ashkin A (1992) Forces of a single beam gradient laser trap on a dielectric sphere in the ray optics regime. Biophys J 61:569–582
3. Nemoto S, Togo H (1998) Axial force acting on a dielectric sphere in a focused laser beam. Appl Opt 37:6386–6394
4. Sidick E, Collins SD, Knoesen A (1997) Trapping forces in a multiple beam fiber-optic trap. Appl Opt 36:6423–6433
5. Bakker Schut TC, Hesselink G, de Grooth BG et al (1991) Experimental and theoretical investigations on the validity of the geometrical optics model for calculating the stability of optical traps. Cytometry 12:479–485
6. Lee JW, Ha KL, Shung KK (2005) A theoretical study of the feasibility of acoustical tweezers: ray acoustics approach. J Acoust Soc Am 117:3273–3280
7. Bandres (2004) Ince Gaussian beams. Opt Lett, 29:144–146
8. Garg AK (2012) Classical electromagnetism in a nutshell. Princeton University Press, Princeton, N.J.
9. Wikipedia: Gaussian beam
10. Thompson RB, Lopes EF (1984) The effects of focusing and refraction on gaussian ultrasonic beams. J Nondestr Eval 10:107–123
11. Wu RR, Liu XZ, Gong XF (2013) A study of the acoustical radiation force considering attenuation. Sci China-Phys Mech Astronomy 56:1237–1245
12. Lee JW, Shung KK (2006) Radiation forces exerted on arbitrarily located sphere by acoustic tweezer. J Acoust Soc Am 120:1084–1094
13. Lee JW, Shung KK (2006) Effect of ultrasonic attenuation on the feasibility of acoustical tweezers. Ultrasound Med Biol 32:1575–1583
14. Kang ST, Yeh CK (2013) Trapping of a Mie sphere by acoustic pulses: effects of pulse length. IEEE Trans Ultrason Ferroelectr Freq Control 60:1487–1497
15. Lee J (2014) Numerical analysis for transverse microbead trapping using 30 MHz focused ultrasound in ray acoustics regime. Ultrasonics 54:11–19
16. Wu RR, Liu XZ, Liu JH et al (2014) Calculation of acoustical radiation force on microsphere by spherically-focused source. Ultrasonics 54:1977–1983
17. Wu RR, Cheng KX, Liu XZ et al (2014) Acoustic radiation force acting on double-layer microsphere placed in a Gaussian focused field. Appl Phys 116:144903
18. Roosen G, Imbert C (1976) Optical levitation by means of two horizontal laser beams: a theoretical and experimental study. Phys Lett 59:6–8
19. Lee JW, Lee CY, Shung KK (2019) Calibration of sound forces in acoustic traps. IEEE Trans Ultrason Ferroelectr Freq Control 57:2305–2310
20. Lee JW (2011) Targeted Cell Immobilization by Ultrasound Microbeam. Biotechnol Bioeng 108:1643–1650
21. Meng L, Cai FY, Chen JJ et al (2012) Precise and programmable manipulation of microbubbles by two dimensional standing surface acoustic waves. Appl Phys Lett 100:173701
22. Brannon-Peppas L (1997) Polymers in controlled drug delivery. Med Plast Biomater 97:34–45
23. Hu Y, Qin S, Jiang Q (2004) Characteristics of acoustic scattering from a double-layered micro shell for encapsulated drug delivery. IEEE Trans Ultrason Ferroelectr Freq Control 51:809–821
24. Jeffers R (1995) Activation of anti-cancer drugs with ultrasound. J Acoust Soc Am 98:2380
25. Munshi N, Rapoport N, Pitt WG (1997) Ultrasonic activated drug delivery from Pluronic P-105 micelles. Cancer Lett 118:13–19
26. Hasegawa T, Hino Y, Annou A et al (1993) Acoustic radiation pressure acting on spherical and cylindrical shells. J Acoust Soc A 93:154–161
27. Marston PL (2017) Negative axial radiation forces on solid spheres and shells in a Bessel beam. J Acoust Soc Am 122:3162–3165
28. Marston PL (2008) Scattering of a Bessel beam by a sphere: II. Helicoidal case and spherical shell example. J Acoust Soc Am 124:2905–2910
29. Mitri FG, Lobo TP, Silva GT (2012) Axial acoustic radiation torque of a Bessel vortex beam on spherical shells. Phys Rev E 85:026602

30. Azarpyvand M (2014) Application of acoustic bessel beams for handling of hollow porous spheres. Ultrasound Med Biol 40:422–433
31. Wu RR, Cheng KX, Liu XZ et al (2014) Acoustic radiation force acting on double-layer microsphere placed in a Gaussian focused field. J Appl Phys 116:144903
32. Thompson PB, Lopes EF (1984) The effects of focusing and refraction on gaussian ultrasonic beams. J Nondestr Eval 4:107–123

Chapter 2
Calculation of Acoustic Radiation Force on Microspheres Based on Ray Acoustics Method in Spherical Focusing Ultrasound

Concave spherical focusing transducers [1, 2] have significant applications in the field of biomedical ultrasound, such as the use of high-intensity focused ultrasound (HIFU) for tumor treatment due to its thermal effects [3], the application of phased array focusing transducers [4], and the cavitation effect of sound waves [5]. This chapter uses an ideal spherical focusing transducer as a model, and by using the ray acoustics approximation, calculates the three-dimensional acoustic radiation force on spherical particles in a high-frequency focused sound field in water. This chapter provides a detailed derivation of the three-dimensional radiation force calculation model, studies the impact of various parameters on the radiation force, and provides the location of the three-dimensional acoustic potential well generated by the spherical focusing sound field on particles. This provides a theoretical basis for the application of spherical focusing transducers in acoustic trapping.

2.1 Theoretical Methods

2.1.1 Acoustic Radiation Force on Microspheres in Spherical Focusing Sound Field

Similar to the theoretical part of the ray acoustics approximation introduced in the previous chapter, the acoustic radiation force consists of two parts [6–9], one part is the scattering force along the direction of sound wave propagation, which is caused by the reflection of sound waves and is proportional to the sound intensity; the other part is the gradient force, its direction is the direction of sound intensity gradient, related to the energy gradient of the sound field, pointing to the place where the sound field intensity is the highest.

X. Liu and H. Zheng, *Acoustic Radiation Force*, Advances in Acoustics,
https://doi.org/10.1007/978-981-95-3152-3_2

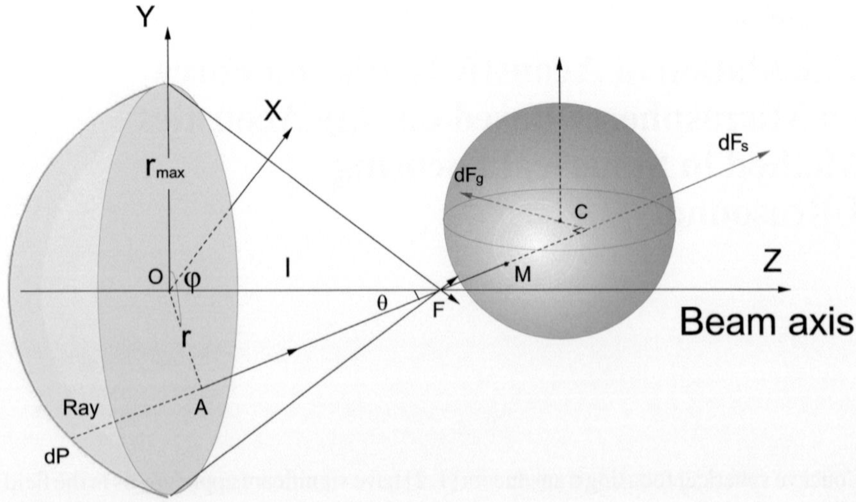

Fig. 2.1 Schematic diagram of the acoustic radiation force of spherical focusing ultrasound on a spherical particle

As an approximation calculation, we assume that the concave spherical focusing sound field is an ideal sound source, that is, the ultrasound focusing focus is a perfect point. As shown in Fig. 2.1, the sound field focuses at point F, the geometric focal length of the transducer is l, the aperture radius of the transducer is r_{max}, and the power of the transducer is P.

It is clear that the sound power of each ray is [9]:

$$dP_0 = \left(\frac{P}{\pi r_{max}^2} r d\varphi dr\right) \tag{2.1}$$

Establish a Cartesian coordinate system as shown in Fig. 2.1, point $A(r\cos\varphi, r\sin\varphi, 0)$ is the intersection of the incident ray from the surface of the spherical focusing transducer with the X–Y plane at z = 0, R_c is the radius of the spherical particle, point $C(a, b, c)$ is the coordinate of the particle center, point $F(0, 0, l)$ is the intersection position of the sound field, $M(x, y, z)$ is the intersection coordinate of the incident ray and the spherical particle, the following relationship can be obtained from the spatial geometric relationship:

$$\begin{cases} \frac{x}{r\cos\varphi} = \frac{y}{r\sin\varphi} = \frac{z-l}{-l} \\ (x-a)^2 + (y-b)^2 + (z-c)^2 = R_c^2 \\ \frac{|ar\sin\varphi + b(l - r\cos\varphi) + cr\sin\varphi - lr\sin\varphi|}{\sqrt{(r\sin\varphi)^2 + (l - r\cos\varphi)^2 + (r\sin\varphi)^2}} < R_c \end{cases} \tag{2.2}$$

By solving Eq. (2.2) for the minimum value z, the intersection coordinates on the spherical surface $M(x_m, y_m, z_m)$ can be obtained.

Due to the use of high-frequency focused ultrasound (100 MHz), the attenuation in water cannot be ignored. According to the derivation in the previous chapter, it is not difficult to obtain the energy of the sound ray incident on the particle surface after attenuation in water:

$$dP = \frac{dP_0}{10^{\frac{\alpha' \cdot l_m}{10}}} \tag{2.3}$$

where:

$$l_m = \left|\overline{AM}\right| = \sqrt{(x_m - r\cos\phi)^2 + (y_m - r\sin\phi)^2 + z_m^2} \tag{2.4}$$

α' is the attenuation coefficient in water, l_m is the propagation distance of each sound ray in water.

Through geometric relationships, we can obtain the acute angle between $\overrightarrow{CM}$ and $\overrightarrow{AF}$, which is the incident angle of the sound ray on the spherical surface, should satisfy the following equation:

$$\cos(\pi - \theta_i) or \cos\theta_i = \frac{\overrightarrow{FA} \times \overrightarrow{CM}}{\left|\overrightarrow{FA}\right| \cdot \left|\overrightarrow{CM}\right|}$$
$$= \frac{r\cos\varphi(x_m - a) + r\sin\varphi(y_m - b) - l(z_m - c)}{\sqrt{r^2\cos^2\varphi + r^2\sin^2\varphi + l^2} \cdot \sqrt{(x_m - a)^2 + (y_m - b)^2 + (z_m - c)^2}} \tag{2.5}$$

2.1.2 *Calculation of Scattering Force and Gradient Force Vectors*

The scattering force of sound waves on particles is along the direction of sound wave propagation, which is $\overrightarrow{AF}$ in Fig. 2.1.

Let $\overrightarrow{dF_s}$ be the scattering force component of the radiation force received by the particle, the direction is the same as the direction of the incident sound ray, as shown in Fig. 2.2, $\overrightarrow{dF_s}$ should satisfy:

$$dF_{sz} = dF_s\cos\theta, dF_{sx} = -dF_s\sin\theta\sin\varphi,$$
$$dF_{sy} = -dF_s\sin\theta\cos\varphi \tag{2.6}$$

where: $\theta = \arctan\frac{r}{l}$.

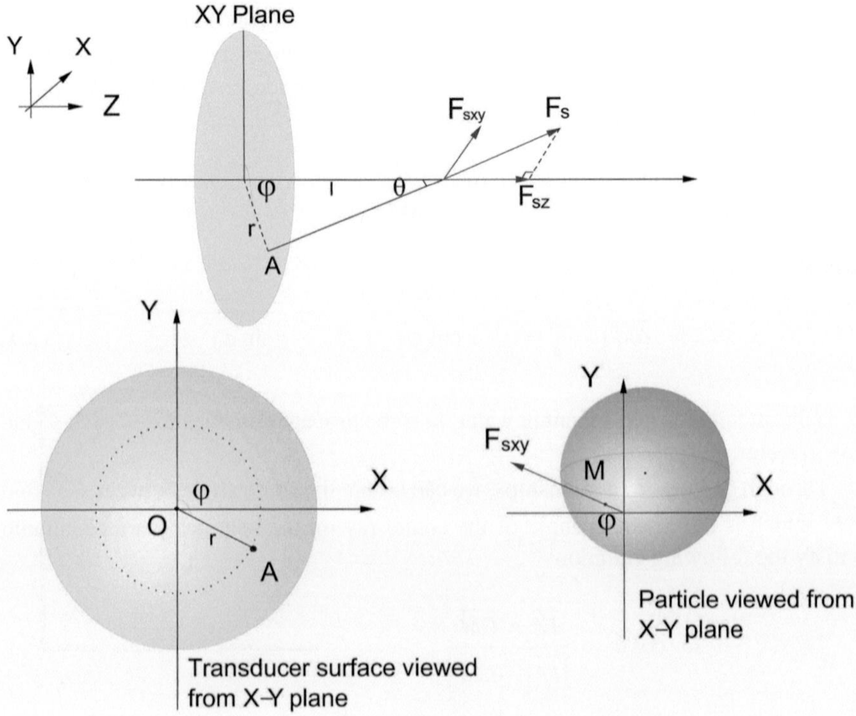

Fig. 2.2 Schematic diagram of the calculation of scattering force components

The direction of the gradient force is the direction of the sound intensity gradient, perpendicular to the direction of sound propagation, and points towards the center of the particle sphere.

Let $\overrightarrow{dF_g}$ be the scattering force component experienced by the particle, as shown in Fig. 2.3. To solve for the scattering force component, we first assume a plane S, perpendicular to the incident sound line $\overrightarrow{AF}$ and passing through the center C of the particle. From spatial geometric relationships, the equation of plane S is:

$$r\cos\varphi(x-a)+r\sin(y-b)-l(z-l)=0 \tag{2.7}$$

The intersection point $B(x_b, y_b, z_b)$ of $\overrightarrow{AF}$ and plane S should satisfy:

$$\begin{cases} r\cos\varphi(x-a)+r\sin\varphi(y-b)-l(z-l)=0 \\ xr\sin\varphi-yr\cos\varphi=0 \\ yl+(z-l)r\sin\varphi=0 \end{cases} \tag{2.8}$$

Meanwhile, the direction cosine of $\overrightarrow{BC}$ with the X-axis is:

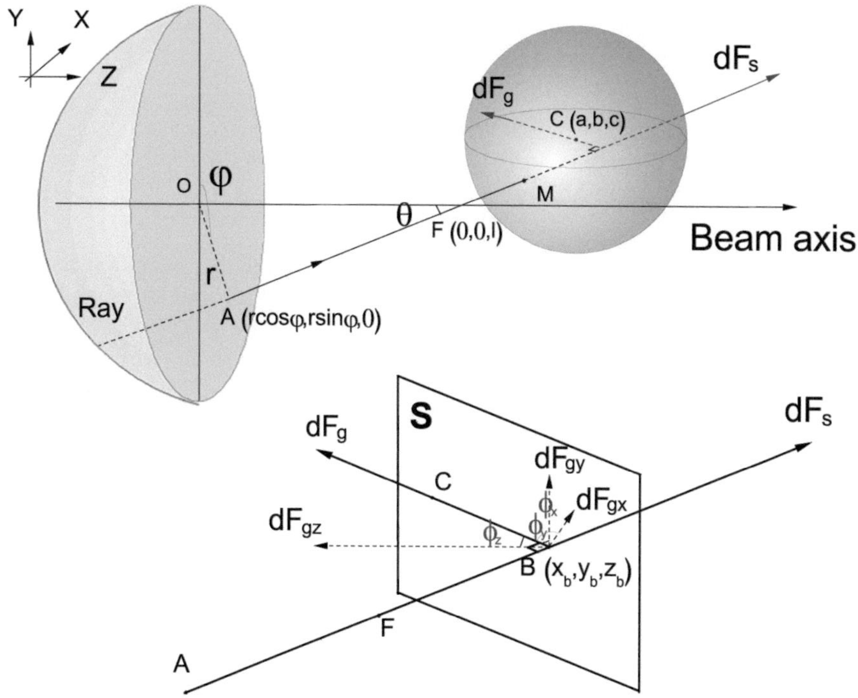

Fig. 2.3 Schematic diagram of the calculation of gradient force components

$$\cos\phi_x = \frac{a - x_b}{\left|\overrightarrow{BC}\right|} = \frac{a - x_b}{\sqrt{(a - x_b)^2 + (b - y_b)^2 + (c - z_b)^2}} \tag{2.9}$$

The direction cosine of $\overrightarrow{BC}$ with the Y-axis is:

$$\cos\phi_y = \frac{b - y_b}{\left|\overrightarrow{BC}\right|} = \frac{b - y_b}{\sqrt{(a - x_b)^2 + (b - y_b)^2 + (c - z_b)^2}} \tag{2.10}$$

The direction cosine of $\overrightarrow{BC}$ with the Z-axis is:

$$\cos\phi_z = \frac{c - z_b}{\left|\overrightarrow{BC}\right|} = \frac{c - z_b}{\sqrt{(a - x_b)^2 + (b - y_b)^2 + (c - z_b)^2}} \tag{2.11}$$

Based on the above, we can obtain:

$$dF_{gz} = dF_g cos\phi_z, dF_{gx} = dF_g cos\phi_x, dF_{gy} = dF_g cos\phi_y \tag{2.12}$$

Therefore, we obtain the three-dimensional radiation force components on the particle sphere from each sound line:

$$\begin{aligned} dF_z &= dF_s \cos\theta + dF_g \cos\phi_z \\ dF_x &= -dF_s \sin\theta \sin\varphi + dF_g \cos\phi_x \\ dF_y &= -dF_s \sin\theta \cos\varphi + dF_g \cos\phi_y \end{aligned} \tag{2.13}$$

Combining Eqs. (2.6), (2.12), and (2.13), we obtain the three-dimensional radiation force on the spherical particle from the spherical focused ultrasound:

$$F_z = \int_0^{2\pi}\int_0^{r_{max}} dF_z,\ F_x = \int_0^{2\pi}\int_0^{r_{max}} dF_x,\ F_y = \int_0^{2\pi}\int_0^{r_{max}} dF_y \tag{2.14}$$

2.1.3 Acoustic Radiation Force of Hollow Focused Sound Beam on Microspheres

In biomedical ultrasound, an opening at the center of the spherical focusing transducer can be used for applications such as HIFU treatment and ultrasound detection. Figure 2.4 depicts the acoustic radiation force of the hollow focused ultrasound on spherical particles.

The aperture radius of the ultrasound transducer is r_{max}, and a circular hole with a radius of r_{min} is opened at its center. It is not difficult to conclude that in order to ensure the same sound power density as the previous spherical focused sound field, the power of each sound line at this time is $dP_0 = \left(P/\pi r_{max}^2\right) r d\varphi dr$.

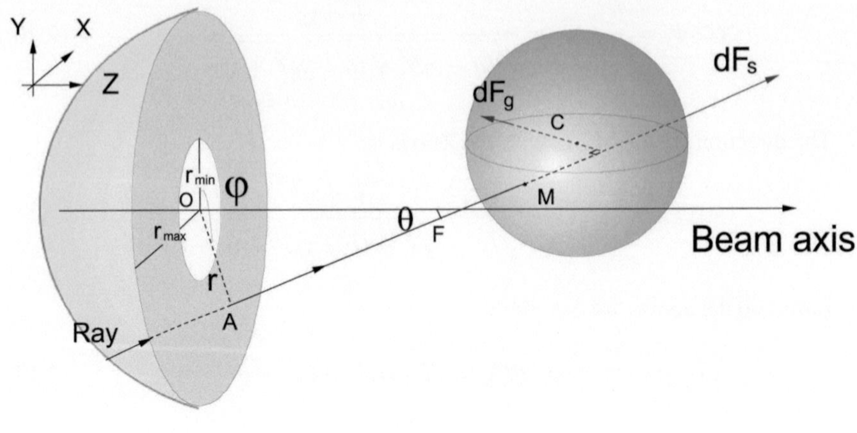

Fig. 2.4 Schematic diagram of hollow focused ultrasound

Therefore, the acoustic radiation force of the hollow focused sound field on the spherical particle is:

$$F_z = \int_0^{2\pi}\int_{r_{min}}^{r_{max}} dF_z, F_x = \int_0^{2\pi}\int_{r_{min}}^{r_{max}} dF_x, F_y = \int_0^{2\pi}\int_{r_{min}}^{r_{max}} dF_y \tag{2.15}$$

2.2 Simulation and Discussion

In order to provide sufficient acoustic energy to generate the required acoustic trapping force, we chose fat microspheres for simulation. Meanwhile, we chose water as the surrounding medium to ensure more acoustic energy refraction rather than reflection. The acoustic impedances of the particle and the surrounding medium are 1.4 and 1.5 MRayls, respectively, and the sound speeds in the particle and water are 1450 and 1500 m/s, respectively. In the simulation, we assumed that the output power of the focusing transducer is 100 mW, the frequency is 100 MHz, which means the wavelength of ultrasound in water is 15 μm. To ensure the validity of the ray approximation theory, we chose the radius of the particle to be more than 10 times the wavelength of the sound wave, that is, 150 μm. The sound attenuation coefficients of 100 MHz ultrasound in water and particles are 20 and 150 dB/cm, respectively. We used the parameters of two 100 MHz transducers for simulation, with aperture sizes of 0.635 cm and focal lengths of 0.508 and 0.635 cm, respectively.

As shown in Figs. 2.5 and 2.6, set the focus position of the focused ultrasound as the origin, and the focal length of the ultrasound is 0.508 cm. The horizontal coordinate Z_c represents the axial distance between the origin and the center of the particle, ranging from -1×10^{-6} to 1×10^{-6} m, at which point the lateral distance from the center of the particle to the Z-axis is 30×10^{-6} m. As can be clearly seen from the two figures, the lateral force exerted on the particles, i.e., F_x and F_y are both negative. This indicates that the lateral force of the ultrasound always pushes the particles towards the ultrasonic axis, and the negative capture force reaches its peak at the focal point, i.e., at z = 0. The axial force of the ultrasound on the particles, F_z, is positive when the particle is on the left side of the ultrasonic focus and negative when it is on the right side. Near the ultrasonic focal point, $F_z = 0$, indicating that the particles will be stably captured near the ultrasonic focus, creating an acoustic potential well. In addition, the lateral radiation force (F_x, F_y) and axial radiation force (F_z) increase with the size of the particles, indicating that larger particles will experience greater acoustic radiation forces and are more likely to be captured. Moreover, since the acoustic capture force is greater than the gravitational force on the particles and the viscous force of the surrounding medium on the particles, the capture and manipulation of particles by the spherical focusing sound field is feasible and effective. Furthermore, as can be seen from the two figures above, the lateral capture force is much greater than the axial capture force, which explains why lateral capture is easier for microparticles and why single-beam acoustic tweezers use lateral capture [10–13].

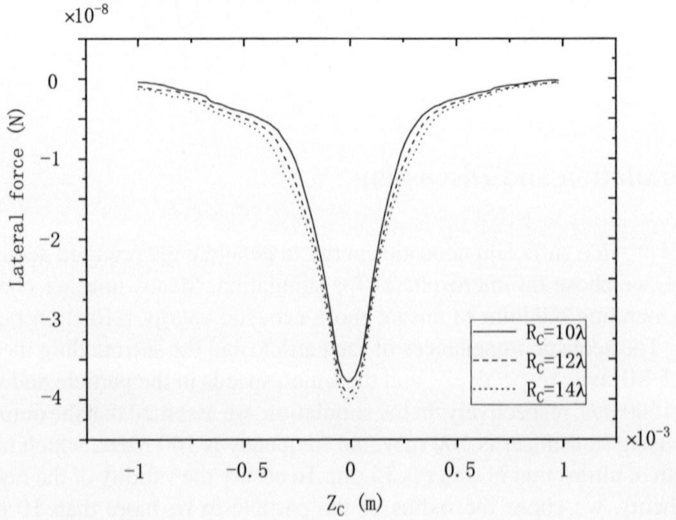

Fig. 2.5 Relationship between lateral acoustic radiation force and initial axial position of the particle under different particle radii

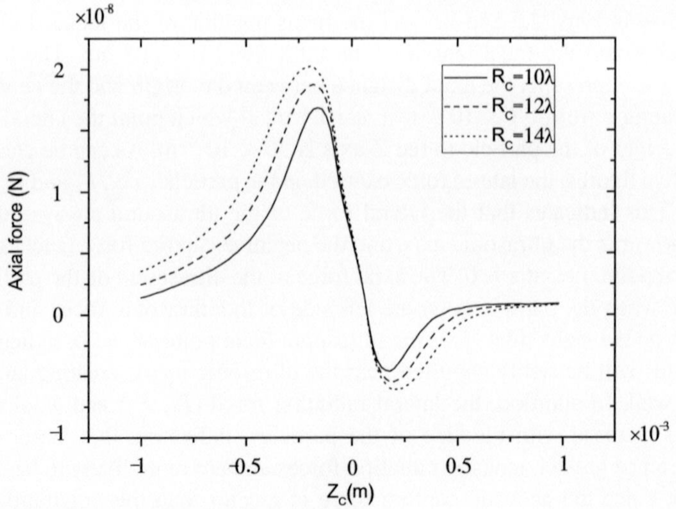

Fig. 2.6 Relationship between axial acoustic radiation force and initial axial position of the particle under different particle radii

As shown in Figs. 2.7 and 2.8, the lateral distance from the center of the particle to the Z-axis is 30×10^{-6} m. When the focal length of the focused ultrasound increases, or in other words, when the ultrasound transducer has a smaller angle, both the lateral and axial acoustic radiation forces decrease significantly. At this time, due to the larger focal length caused by the small angle focusing transducer, the attenuation of sound energy in the medium around the particles is more significant, making it more difficult for ultrasound to capture particles. Therefore, the focus of the ultrasound should be as close as possible to the surface of the transducer, i.e., a focusing transducer with a smaller F-number should be chosen. At the same time, a transducer with a larger angle will provide a larger sound intensity gradient, thereby increasing the refraction of sound waves on the surface of the particles, ultimately providing a larger gradient force to increase the pulling effect on the particles.

Since the lateral capture force and axial capture force decrease with the increase of the ultrasound focal length, in the application of acoustic tweezers, we should choose a large angle focusing transducer to avoid the attenuation of sound waves in the medium around the particles.

Figures 2.9 and 2.10 depict the differences in radiation forces experienced by particles in hollow focused sound fields and spherical focused sound fields. In this case, the aperture of the hollow focused ring transducer is 0.635 cm, the hollow opening in the middle is 0.0635 cm, the focal lengths of the two transducers are both 0.508 cm, and the sound field power density in both cases is the same. The lateral distance from the center of the particle to the Z-axis is 30×10^{-6} m.

When the particle is in the hollow focused sound field, it can be seen that both the axial and lateral sound capture forces are larger than those in the spherical focused sound field under the same conditions. This is because the sound lines near the Z-axis

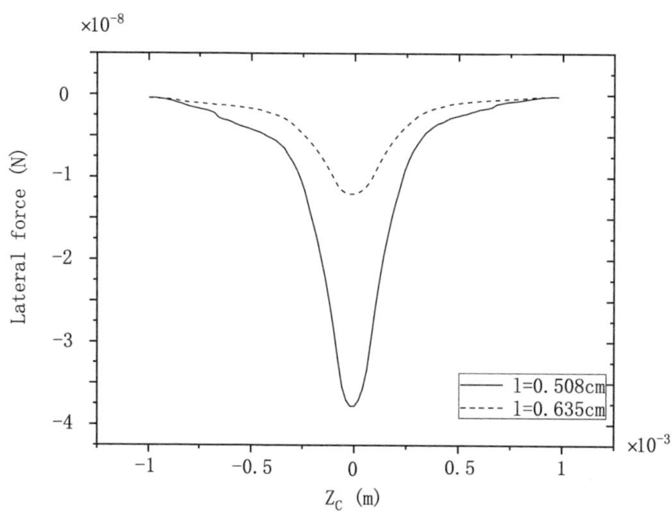

Fig. 2.7 Relationship between lateral acoustic radiation force and initial axial position of particles under different focused ultrasound

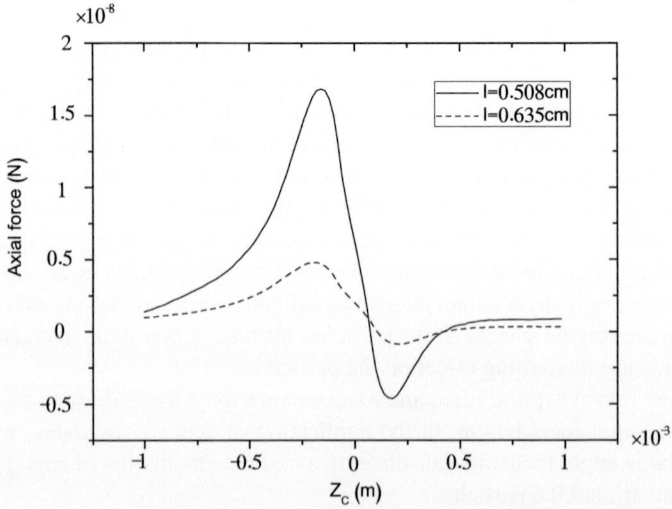

Fig. 2.8 Relationship between axial acoustic radiation force and initial axial position of particles under different focused ultrasound

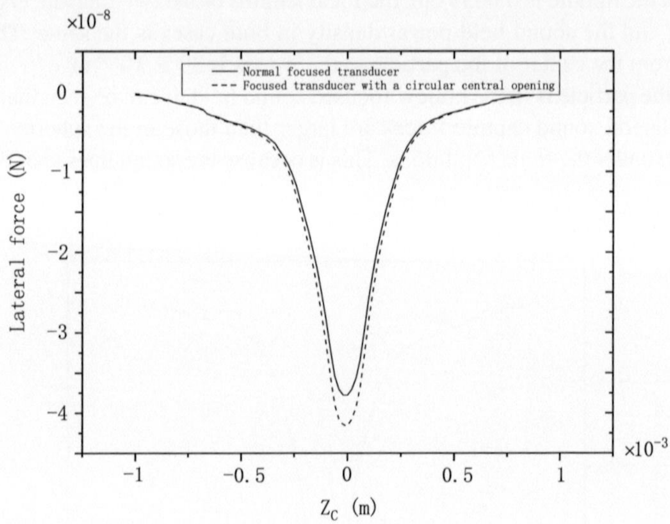

Fig. 2.9 The effect of hollow focused sound field on lateral acoustic radiation force

can be approximated as propagating along the Z-axis direction, where there is not enough intensity gradient to generate gradient forces, thus the dominant scattering force component will push the particles away from the focal point. In contrast, in the case of the hollow focused sound field, the scattering force component near the axis

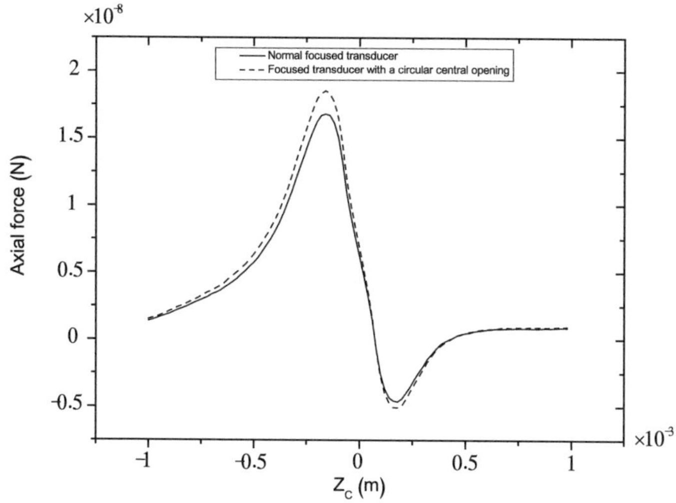

Fig. 2.10 The effect of hollow focused sound field on axial acoustic radiation force

is reduced, thereby increasing the lateral capture force, axial capture force, and the width of the acoustic potential well.

Figures 2.11 and 2.12 depict the spatial distribution of the acoustic radiation force on a particle moving along the Y–Z plane on the X-axis in a spherical focusing sound field. At this time, the focal length of the transducer is 0.508 cm. Figure 2.11 clearly shows the acoustic potential trap in the lateral direction. When the center of the particle is on the z-axis, the lateral radiation force on the particle is zero, indicating that the three-dimensional model degenerates into a one-dimensional model. Since the particle is symmetrical about the z-axis, it is not subject to lateral force, and the particle will stabilize on the ultrasonic axis. However, when the center of the particle moves in the direction away from the z-axis, the lateral acoustic radiation force gradually increases to its maximum value, and then gradually decreases. When the particle is far from the z-axis, the lateral radiation force is very small, approximately zero, so the particle will no longer be able to be captured back to the position of the ultrasonic axis when it is in the far field of the ultrasonic axis. In addition, we can see that when the particle is outside the range of the focused ultrasonic beam, the lateral force on the particle is also zero and is no longer subject to the radiation force of the sound field. From Fig. 2.12, we can clearly see the phenomenon of axial sound capture. At the same time, when the center of the particle is exactly on the ultrasonic axis, the axial radiation force on the particle is the maximum. And when the center of the particle gradually moves away from the ultrasonic axis in the Y direction, the sound radiation force on the particle continuously decreases. Like the lateral radiation force, when the particle leaves the range of the ultrasonic beam, the axial radiation force also becomes zero and cannot be captured again.

Assuming that the center of the particle is stable on the X-axis and moves along the Y–Z plane, Fig. 2.13 provides a schematic diagram of the size and direction of the

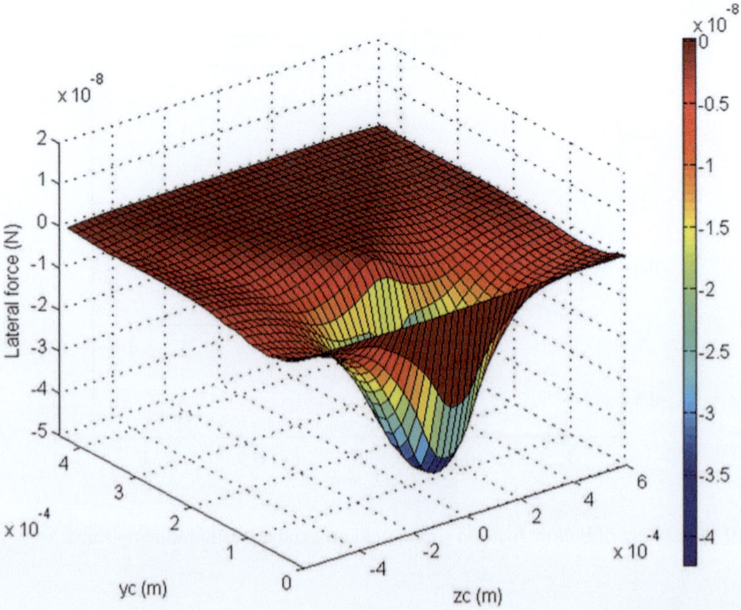

Fig. 2.11 Spatial distribution of lateral acoustic radiation force on particles moving along the Y–Z plane on the X-axis

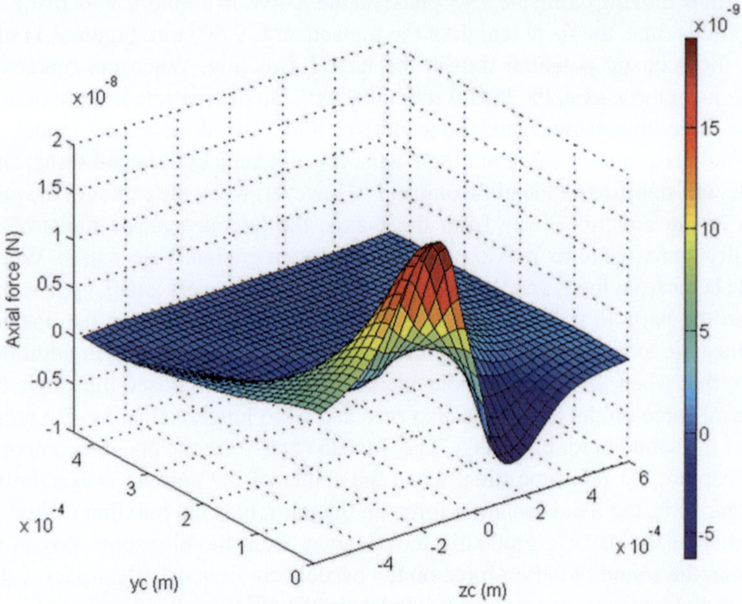

Fig. 2.12 Spatial distribution of axial acoustic radiation force on particles moving along the Y–Z plane on the X-axis

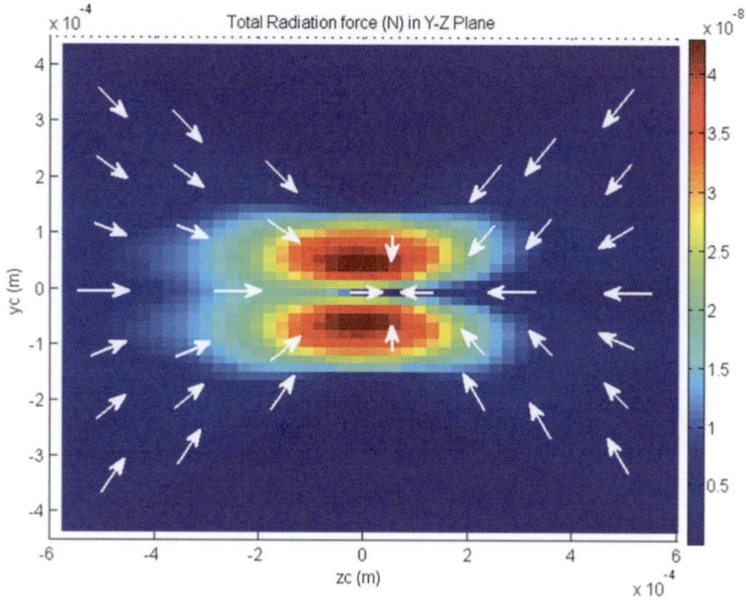

Fig. 2.13 The size and direction of the acoustic radiation force on the particle moving along the Y–Z plane on the X-axis

radiation force on the particle in the Y–Z space plane. The color represents the force size, and the arrow represents the force direction. From the figure, we can see that as long as the particle is within the range of the focused ultrasonic beam, the particle will be subjected to the action of the acoustic radiation force, causing the particle to move towards the position of the ultrasonic focus. When the particle moves near the focus position, the force on the particle is zero, so the particle will stabilize at this position. At the same time, we can see that when the center of the particle is in the focusing area of focused ultrasound, the radiation force received is much greater than when the particle is in the divergent area of ultrasound. Moreover, when the center of the particle is near the lateral plane of the ultrasonic focus, the radiation force received is greater than other positions, which also explains the importance of lateral capture by single-beam acoustic tweezers. Finally, we can see that when the particle is far from the range of the ultrasonic beam, the particle will no longer be affected by the acoustic radiation force.

References

1. Ruo F (2001) Ultrasound handbook. Nanjing University Press
2. Wan M, Bian Z, Cheng J (1992) Medical ultrasound—principles and techniques. Xi'an Jiaotong University Press

3. Kennedy JE (2005) High-intensity focused ultrasound in the treatment of solid tumours. Nat Rev Cancer 5:321–327
4. Wang HZ, Qian JP (1986) Ultrasound dynamically focused phased array for local hyperthermia in cancer therapy. Acoust Lett 10:5–10
5. Suslick KS, Price GJ (1995) Applications of ultrasound to materials chemistry. Annu Rev Mater Res 26:295–326
6. Ashkin A (1992) Forces of a single beam gradient laser trap on a dielectric sphere in the ray optics regime. Biophys J 61:569–582
7. Lee JW, Ha KL, Shung KK (2005) A theoretical study of the feasibility of acoustical tweezers: ray acoustics approach. J Acoust Soc Am 117:3273–3280
8. Wu RR, Liu XZ, Gong XF (2013) A study of the acoustical radiation force considering attenuation. Sci China-Phys Mechan Astron 56:1237–1245
9. Wu JR (1995) Calculation of acoustic radiation force generated by focused beams using the ray acoustics approach. J Acoust Soc Am 97:2747–2750
10. Lee JW, Lee CY, Shung KK (2010) Calibration of sound forces in acoustic traps. IEEE Trans Ultrason Ferroelectr Freq Control 57:2305–2310
11. Lee JW (2001) Targeted cell immobilization by ultrasound microbeam. Biotechnol Bioeng 108:1643–1650
12. Ying L, Shung KK (2013) A simple method for evaluating the trapping performance of acoustic tweezers. Appl Phys Lett 102:084102
13. Wu RR, Liu XZ, Liu JH et al (2014) Calculation of acoustical radiation force on microsphere by spherically-focused source. Ultrasonics 54:1977–1983

Chapter 3
Solving the Acoustic Radiation Force of Plane Waves on Spheres Based on Acoustic Scattering Method

When sound waves propagating in a fluid medium encounter obstacles, refraction and scattering occur on the surface of the obstacles. Therefore, the interaction between objects in the sound field and sound waves is the mutual interaction between the object and the incident sound waves, refracted sound waves, and scattered sound waves. By calculating the scattering coefficient of the object in the sound field, the acoustic radiation force can be solved according to the theory of sound wave radiation.

Mie scattering, proposed by Mie in 1908, is used to calculate the scattering of objects to sound waves [1]. In 1934, King and others first proposed the theory of acoustic radiation force based on Mie scattering method, providing an analytical expression for the acoustic radiation force on a rigid sphere in an ideal fluid [2]. In 1955, Yosioka and others studied the acoustic radiation force of plane waves on compressible spheres based on this [3]. In 1969, Hasegawa and others considered the acoustic radiation force of elastic spherical particles and verified the correctness of the theory through experiments [4]. Marston, Mitri and others extended the study of radiation force to Bessel beams, providing the relationship between the acoustic radiation force and frequency, semi-cone angle on the spherical shell and sphere in the Bessel beam sound field [5–9]. Zhang and others used finite series expansion to expand Gaussian sound waves according to spherical function expressions, deriving the acoustic radiation force of spherical or cylindrical particles in the Gaussian focusing sound field [10, 11].

This chapter will start with the analytical theory of sound waves, focusing on the scattering effect of various boundary conditions on sound waves, introducing the analytical expression of the scattering coefficient of sound waves using acoustic boundary conditions; at the same time, the incident Gaussian sound wave will be expanded in series, and the acoustic radiation force of the Gaussian wave on the sphere under various boundary conditions will be obtained according to the nonlinear acoustic theory, providing a basis for the theory in the following chapters.

X. Liu and H. Zheng, *Acoustic Radiation Force*, Advances in Acoustics,
https://doi.org/10.1007/978-981-95-3152-3_3

3.1 Acoustic Radiation Force of Plane Waves on Spherical Particles

3.1.1 Acoustic Scattering of Plane Waves on Spheres

As shown in Fig. 3.1, in an ideal situation, assuming that the plane wave with an angular frequency of ω propagates in the positive direction of the Z-axis in water, when the sound wave encounters a spherical particle, scattering and refraction will occur, thus generating scattered waves outside the sphere and refracted waves inside the sphere.

The incident plane wave can be expressed as:

$$p_i = p_0 \sum_{n=0}^{\infty} (2n+1)(i)^n j_n(k_0 r) P_n(\cos\theta) \tag{3.1}$$

where: p_0 is the amplitude of the incident sound pressure, k_0 is the wave number of the sound wave in the medium, $j_n(\cdot)$ is the first kind of n-order spherical Bessel function, $P_n(\cdot)$ is the n-order Legendre function.

The expression of the scattered wave can be easily obtained [12, 13]:

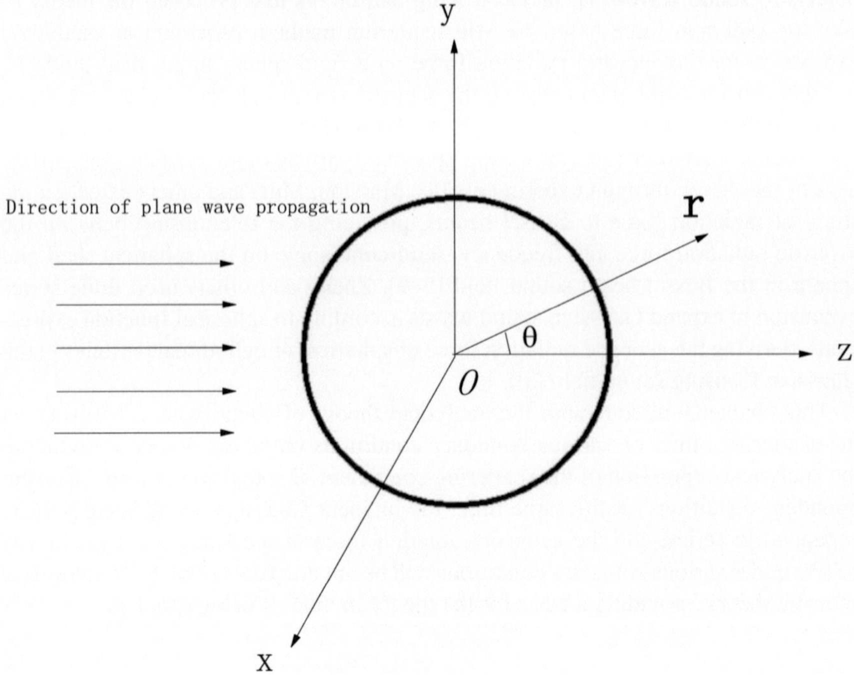

Fig. 3.1 Schematic diagram of the acoustic scattering of plane waves on spherical particles in water

$$p_s = p_0 \sum_{n=0}^{\infty} A_n (2n+1)(i)^n h_n^{(1)}(k_0 r) P_n(\cos\theta) \tag{3.2}$$

where, $A_{n,s}$ is the expansion coefficient of the scattered wave, $h_n(\cdot)$ is the n-order spherical Hankel function.

The acoustic scattering coefficients under various boundary conditions will be introduced next [14].

When a plane wave propagating in a fluid medium encounters a rigid sphere, the sound wave cannot propagate inside the sphere, and there is no sound wave inside the sphere. Therefore, the vibration speed of the point on the boundary of the sphere is 0, then:

$$\left(v_{s,n} + v_{i,n}\right)\big|_{r=a} = 0 \tag{3.3}$$

That is:

$$\frac{\partial}{\partial r}(p_i + p_s)\bigg|_{r=a} = 0 \tag{3.4}$$

where: a is the radius of the sphere.

Substituting Eqs. (3.1) and (3.2) into (3.4), the scattering coefficient $A_{n,s}$ can be obtained:

$$A_{n,s} = -\frac{j_n'(ka)}{h_n'(ka)} \tag{3.5}$$

When a plane wave propagating in a fluid medium encounters a liquid sphere, the sound field outside the sphere is the superposition of the incident plane wave and the scattered wave, and the inside of the sphere is the refracted wave. The refracted wave inside the sphere can also be expressed as:

$$p_r = p_0 \sum_{n=0}^{\infty} A_{n,r} (2n+1)(i)^n h_n^{(1)}(k_1 r) P_n(\cos\theta) \tag{3.6}$$

where: $A_{n,r}$ is the expansion coefficient of the refracted wave, k_1 For the wave number of sound waves in the medium inside the sphere.

At the boundary r = a, the following two acoustic boundary conditions should be satisfied:

The sound pressure is continuous on the surface of the sphere:

$$p_i + p_s = p_t \tag{3.7}$$

The radial direction of the particle velocity is continuous:

$$v_i + v_s = v_t \tag{3.8}$$

By substituting Eqs. (3.1), (3.2), and (3.6) into the above two equations and simplifying, we get:

$$A_{n,s} = -\frac{\left[\frac{\rho_0 c_0}{\rho_1 c_1}\right]\left[j_n(k_1 a)\right]'\left[j_n(k_0 a)\right] - \left[j_n(k_0 a)\right]'\left[j_n(k_1 a)\right]}{\left[\frac{\rho_0 c_0}{\rho_1 c_1}\right]\left[j_n(k_1 a)\right]'\left[h_n(k_0 a)\right] - \left[h_n(k_0 a)\right]'\left[j_n(k_1 a)\right]} \tag{3.9}$$

where: ρ_0, c_0 represent the density and sound speed of the medium in the water around the particle sphere, respectively, ρ_1, c_1 represent the density and sound speed of the liquid sphere, respectively.

When a plane wave propagating in a fluid medium encounters an elastic sphere, there are both transverse and longitudinal waves in the sphere. At this time, the displacement of the medium particles satisfies the following sound wave equation:

$$\nabla^2 u + \frac{\lambda + \mu}{\mu}\frac{\partial}{\partial t}\nabla(\nabla \cdot u) = \frac{1}{c_s^2}\frac{\partial^2 u}{\partial t^2} \tag{3.10}$$

where, λ and μ are Lamé coefficients, $c_s = \sqrt{\frac{\mu}{\rho_0}}$ is the speed of transverse waves.

In the spherical coordinate system:

Equation (3.10) can be written as:

$$\nabla_{r\theta}^2 u_r + \frac{2u_r}{r^2} - \frac{2}{r^2} - \frac{\partial u_\theta \cot\theta}{r^2} + \frac{\lambda + \mu}{\mu} \cdot \frac{\partial \Delta}{\partial r} + k_s^2 u_r = 0 \tag{3.11}$$

$$\nabla_{r\theta}^2 u_\theta + \frac{2}{r^2}\frac{\partial u_r}{\partial \theta} - \frac{u_\theta}{r^2 \sin^2\theta} + \frac{\lambda + \mu}{\mu} \cdot \frac{1}{r} + k_s^2 u_\theta = 0 \tag{3.12}$$

where: $k_s = \frac{\omega}{c_s} = \omega\sqrt{\frac{\mu}{\rho_0}}$ is the wave number of transverse waves.

Introduce two scalar velocity potentials Φ_1 and Φ_2, which satisfy the Helmholtz equation:

$$\left(\nabla_{r\theta}^2 + k_d^2\right)\Phi_1 = 0 \tag{3.13}$$

$$\left(\nabla_{r\theta}^2 + k_s^2\right)\Phi_2 = 0 \tag{3.14}$$

where: $k_d = \frac{\omega}{c_d}$ is the wave number of longitudinal waves, $c_d = \sqrt{\frac{\lambda + 2\mu}{\rho_1}}$ is the speed of transverse waves.

In the sphere, the velocity potentials Φ_1 and Φ_2 are expressed as:

$$\Phi_1(r,\theta,t) = \phi_0 \sum_{n=0}^{\infty} (2n+1)(i)^n P_n(\cos\theta) B_n j_n(k_d r) \tag{3.15}$$

$$\Phi_2(r,\theta,t) = \phi_0 \sum_{n=0}^{\infty} (2n+1)(i)^n P_n(\cos\theta) C_n j_n(k_s r) \tag{3.16}$$

The expression for stress is:

$$\sigma_{rr} = -\lambda k_d^2 \Phi_1 + 2\mu \left[\frac{\partial^2}{\partial r^2} \left(\Phi_1 + \frac{\partial}{\partial r}(r\Phi_2) \right) + k_s^2 \left(\frac{\partial}{\partial r}(r\Phi_2) \right) \right] \tag{3.17}$$

$$\sigma_{r\theta} = \mu \left[\frac{2\partial}{\partial} \left[\frac{1}{r} \frac{\partial}{\partial \theta} \left(\Phi_1 + \frac{\partial}{\partial r}(r\Phi_2) \right) \right] + k_s^2 \frac{\partial}{\partial r}(r\Phi_2) \right] \tag{3.18}$$

$$\sigma_{\varphi\varphi} = -\lambda k_d^2 + 2\mu \left[\left(\frac{1}{r}\frac{\partial}{\partial r} + \frac{\cot\theta}{r^2}\frac{\partial}{\partial \theta} \right) \left(\Phi_1 + \frac{\partial}{\partial r}(r\Phi_2) \right) + k_s^2 \frac{\partial}{\partial r}(r\Phi_2) \right] \tag{3.19}$$

$$\sigma_{\varphi\varphi} = -\lambda k_d^2 + 2\mu \left[\left(\frac{1}{r}\frac{\partial}{\partial r} + \frac{\cot\theta}{r^2}\frac{\partial}{\partial \theta} \right) \left(\Phi_1 + \frac{\partial}{\partial r}(r\Phi_2) \right) + k_s^2 \frac{\partial}{\partial r}(r\Phi_2) \right] \tag{3.20}$$

At the boundary r = a, the boundary conditions should satisfy displacement continuity, normal stress continuity, and tangential stress continuity:

$$u_{1r} = u_{2r}, \sigma_{rr} = -p, \sigma_{r\theta} = 0 \tag{3.21}$$

At this time, the expression for the total sound pressure p outside the sphere is:

$$p = p_0 \sum_{n=0}^{\infty} (2n+1)(i)^n P_n(\cos\theta) \left[j_n(k_0 r) + A_{n,s} h_n^1(k_0 r) \right] \tag{3.22}$$

The particle displacement can be expressed as:

$$u_{1r} = -\frac{1}{\rho\omega^2}\frac{\partial p}{\partial r} \tag{3.23}$$

By substituting Eqs. (3.22) and (3.23) into the boundary conditions, and combining with Eqs. (3.17) and (3.18), a relationship matrix about the coefficient $A_{n,s}$, B_n, C_n can be obtained, and the scattering coefficient $A_{n,s}$ can be finally obtained:

$$A_{n,s} = \frac{\begin{vmatrix} \frac{\rho_0}{\rho} x_2^2 j_n(x) & \left[2n(n+1) - x_2^2\right] j_n(x_1) - 4x_1 j_n'(x_1) & 2n(n+1)\left[x_2 j_n'(x_2) - j_n(x_2)\right] \\ x j_n'(x) & x_1 j_n'(x_1) & n(n+1) j_n(x_2) \\ 0 & 2\left[j_n(x_1) - x_1 j_n'(x_1)\right] & 2x_2 j_n'(x_2) + \left[x_2^2 - 2n(n+1) + 2\right] j_n(x_2) \end{vmatrix}}{\begin{vmatrix} \frac{\rho_0}{\rho} x_2^2 h_n(x) & \left[2n(n+1) - x_2^2\right] j_n(x_1) - 4x_1 j_n'(x_1) & 2n(n+1)\left[x_2 j_n'(x_2) - j_n(x_2)\right] \\ x h_n'(x) & x_1 j_n'(x_1) & n(n+1) j_n(x_2) \\ 0 & 2\left[j_n(x_1) - x_1 j_n'(x_1)\right] & 2x_2 j_n'(x_2) + \left[x_2^2 - 2n(n+1) + 2\right] j_n(x_2) \end{vmatrix}} \tag{3.24}$$

where: $x = k_0 a$, $x_1 = x\frac{c_0}{c_d}$, $x_2 = x\frac{c_0}{c_s}$

For viscoelastic materials [15], specifically reflected in the normalized absorption coefficients of longitudinal waves γ_1 and transverse wave absorption coefficients γ_2, just use $\tilde{x}_l, \tilde{x}_s$: $\tilde{x}_l = k_l a(1 - j\gamma_1)$, $\tilde{x}_s = k_s a(1 - j\gamma_s)$ to replace $x_1 = k_l a$, $x_2 = k_s a$ in the above elastic boundary conditions.

3.1.2 Derivation of the Acoustic Radiation Force Formula

As shown in Fig. 3.1, assume that the isotropic liquid medium is under equilibrium pressure. P_0, with a density of ρ_0, during the transmission of sound waves, the pressure becomes P, and the density becomes ρ. The velocity of the particle in Cartesian coordinates v has a vector of x, y, z: u, q, w, and the superimposed velocity potential of the medium outside the sphere is Φ. According to the motion equation of sound waves [13] and the relationship between velocity and velocity potential, we can derive:

$$\begin{gathered} \frac{Du}{Dt} = -\frac{1}{\rho}\frac{\partial p}{\partial x}, \frac{Dq}{Dt} = -\frac{1}{\rho}\frac{\partial p}{\partial y}, \frac{Dw}{Dt} = -\frac{1}{\rho}\frac{\partial p}{\partial z} \\ u = -\frac{\partial \Phi}{\partial x}, v = -\frac{\partial \Phi}{\partial y}, w = -\frac{\partial \Phi}{\partial z} \end{gathered} \tag{3.25}$$

Since velocity is a function of both time and space, therefore:

$$\begin{aligned} \frac{D}{Dt} &= \frac{\partial}{\partial t} + \frac{\partial}{\partial x}\frac{\partial x}{\partial t} + \frac{\partial}{\partial y}\frac{\partial y}{\partial t} + \frac{\partial}{\partial z}\frac{\partial z}{\partial t} \\ &= \frac{\partial}{\partial t} + u\frac{\partial}{\partial x} + v\frac{\partial}{\partial y} + w\frac{\partial}{\partial z} \\ \frac{Du}{Dt} &= \frac{\partial u}{\partial t} + u\frac{\partial u}{\partial x} = \frac{\partial}{\partial t}\left(-\frac{\partial \Phi}{\partial x}\right) + u\frac{\partial u}{\partial x} \\ \frac{Dq}{Dt} &= \frac{\partial q}{\partial t} + q\frac{\partial q}{\partial y} = \frac{\partial}{\partial t}\left(-\frac{\partial \Phi}{\partial y}\right) + q\frac{\partial q}{\partial y} \\ \frac{Dw}{Dt} &= \frac{\partial w}{\partial t} + w\frac{\partial w}{\partial z} = \frac{\partial}{\partial t}\left(-\frac{\partial \Phi}{\partial z}\right) + w\frac{\partial w}{\partial z} \end{aligned} \tag{3.26}$$

We set $\overline{w} = \int \frac{dP}{\rho}$, then we have:

$$\frac{Du}{Dt} = \frac{-\partial\overline{w}}{\partial x}, \frac{Dq}{Dt} = \frac{-\partial\overline{w}}{\partial y}, \frac{Dw}{Dt} = \frac{-\partial\overline{w}}{\partial z}$$

$$\frac{\partial}{\partial t}\left(-\frac{\partial\Phi}{\partial x}\right) + u\frac{\partial u}{\partial x} = -\frac{\partial\overline{w}}{\partial x}$$

$$\frac{\partial}{\partial t}\left(-\frac{\partial\Phi}{\partial y}\right) + q\frac{\partial q}{\partial y} = -\frac{\partial\overline{w}}{\partial y}$$

$$\frac{\partial}{\partial t}\left(-\frac{\partial\Phi}{\partial z}\right) + w\frac{\partial w}{\partial z} = -\frac{\partial\overline{w}}{\partial z} \tag{3.27}$$

When $\overline{w} = \frac{\partial\Phi}{\partial t} = -\frac{1}{2}u^2$,

$$-\frac{\partial\overline{w}}{\partial x} = -\frac{\partial}{\partial x}\left(\frac{\partial\Phi}{\partial t} - \frac{1}{2}u^2\right) = -\frac{\partial}{\partial x}\left(\frac{\partial\Phi}{\partial t}\right) + u\frac{\partial u}{\partial x} = -\frac{\partial}{\partial t}\left(\frac{\partial\Phi}{\partial x}\right) + u\frac{\partial u}{\partial x} \tag{3.28}$$

Considering the velocity components in three directions together, we get:

$$\overline{w} = \int \frac{dP}{\rho} = \frac{\partial\Phi}{\partial t} - \frac{1}{2}\left(u^2 + q^2 + w^2\right) = \frac{\partial\Phi}{\partial t} - \frac{1}{2}\left(v^2\right) \tag{3.29}$$

where: v represents the velocity of the particle.

Also, since pressure is a function of density, we can obtain:

$$P = f(\rho), c^2 = \frac{dP}{d\rho} = f\prime(\rho) \tag{3.30}$$

We set $s = \frac{\rho-\rho_0}{\rho_0}$, at the equilibrium state of the liquid medium $\rho = \rho_0, P = P_0, s = 0$, we expand the sound pressure P as a function of s:

$$\begin{aligned} P &= f(\rho) = f(\rho_0 + s\rho_0) = f(\rho_0) \\ &\quad + s\rho_0 f'(\rho_0) + \frac{1}{2!}s^2\rho_0^2 f''(\rho_0) + \cdots \\ dP &= \left(\rho_0 f'(\rho_0) + s\rho_0^2 f''(\rho_0) + \frac{1}{2!}s^2\rho_0^3 f'''(\rho_0)\ldots\right)ds \end{aligned} \tag{3.31}$$

According to the relationship between ρ and s:

$$\frac{1}{\rho} = \frac{1}{\rho + s\rho_0} = \frac{1}{\rho_0}\left(1 - s + s^2 - s^3 + s^4\ldots\right) \tag{3.32}$$

After processing the above formulas, we can get:

$$\overline{w}=\int\frac{dP}{\rho}=\int\frac{1}{\rho}dP=\int\left(\frac{1}{\rho}(1-s+s^2-s^3+s^4\ldots)\right)$$
$$\left(\rho_0 f'(\rho_0)+s\rho_0^2 f''(\rho_0)+\frac{1}{2!}s^2\rho_0^3 f'''(\rho_0)\ldots\right)ds$$
$$=\int\begin{bmatrix}((f'(\rho_0)-sf'(\rho_0)+s^2f'(\rho_0)-s^3f'(\rho_0)+s^4f'(\rho_0)+\cdots)\\ +(\rho_0 sf''(\rho_0)-\rho_0 s^2f''(\rho_0)+\rho_0 s^3f''(\rho_0)-\rho_0 s^4f''(\rho_0)+\cdots)\\ +(\rho_0^2 s^2\frac{1}{2!}f'''(\rho_0)-\rho_0^2 s^3\frac{1}{2!}f'''(\rho_0)+\rho_0^2 s^4\frac{1}{2!}f'''(\rho_0)+\cdots)+\cdots)\end{bmatrix}ds$$
$$=\begin{bmatrix}sf'(\rho_0)+\frac{1}{2}s^2(\rho_0 f''(\rho_0)-f'(\rho_0))\\ +\frac{1}{3}s^3(f'(\rho_0)-\rho_0 f''(\rho_0)+\rho_0^2\frac{1}{2!}f'''(\rho_0))+\cdots\end{bmatrix} \tag{3.33}$$

According to [2]:

$$s=\frac{\overline{w}}{f'(\rho_0)}-\frac{1}{2}\frac{\rho_0 f''(\rho_0)-f'(\rho_0)}{f'(\rho_0)}\left(\frac{\overline{w}}{f'(\rho_0)}\right)^2+\ldots \tag{3.34}$$

Thus, we can obtain:

$$P-P_0=-\rho_0 f'(\rho_0)\left(\frac{\overline{w}}{f'(\rho_0)}-\frac{1}{2}\frac{\rho_0 f''(\rho_0)-f'(\rho_0)}{f'(\rho_0)}\left(\frac{\overline{w}}{f'(\rho_0)}\right)^2\right)$$
$$+\frac{1}{2}\rho_0^2\frac{f''(\rho_0)}{(f'(\rho^2))^2}(\overline{w})^2+\cdots$$
$$=\rho_0\overline{w}+\frac{1}{2}\frac{\rho_0}{f'(\rho_0)}(\overline{w})^2+\cdots=\rho_0\overline{w}+\frac{1}{2}\frac{\rho_0}{c^2}(\overline{w})^2+\cdots \tag{3.35}$$

Therefore

$$\delta P=P-P_0=\rho_0\frac{\partial\Phi}{\partial t}-\frac{1}{2}\rho_0 v^2$$
$$+\frac{1}{2}\frac{\rho_0}{c^2}\left(\frac{\partial\Phi}{\partial t}\right)^2-\frac{1}{2}\frac{\rho_0}{c^2}\frac{\partial\Phi}{\partial t}v^2+\frac{1}{4}v^4+\ldots \tag{3.36}$$

The sound radiation force can be obtained by integrating the sound pressure $(p-p_0)$ over any surface surrounding the scatterer [3, 4, 16]:

$$\langle F\rangle=-\iint\limits_{s(t)}\langle P-P_0\rangle\vec{n}ds \tag{3.37}$$

Retaining the second-order small quantity, we can get:

$$\langle F\rangle=\left\langle-\iint\rho_0\frac{\partial\Phi}{\partial t}\vec{n}ds\right\rangle+\left\langle\iint\frac{1}{2}\rho_0 v^2\vec{n}ds\right\rangle-\left\langle\iint\frac{1}{2}\frac{\rho_0}{c^2}\left(\frac{\partial\Phi}{\partial t}\right)^2\vec{n}ds\right\rangle$$

$$= -\left\langle \iint \rho_0 \left(v_n \vec{n} + v_t \vec{t} \right) v_n ds \right\rangle + \left\langle \iint \frac{1}{2} \rho_0 v^2 \vec{n} ds \right\rangle$$
$$- \left\langle \iint \frac{1}{2} \frac{\rho_0}{c^2} \left(\frac{\partial \Phi}{\partial t} \right)^2 \vec{n} ds \right\rangle \tag{3.38}$$

3.1.3 Sound Radiation Force Function of Plane Wave on Sphere

From formula (3.38), when a sound wave is incident on a spherical object, we define the sound radiation force as the pressure passing through the surface of S within a period of time T:

$$\boldsymbol{F} = -\iint_{S(t)} \langle P - P_0 \rangle \boldsymbol{n} dS = -\left\langle \iint \rho (v_n \boldsymbol{n} + v_t \boldsymbol{t}) v_n dS \right\rangle$$
$$+ \left\langle \iint \frac{1}{2} \rho v^2 \boldsymbol{n} dS \right\rangle - \left\langle \iint \frac{1}{2} \frac{\rho}{c^2} \left(\frac{\partial \Phi_t}{\partial t} \right)^2 \boldsymbol{n} dS \right\rangle$$

Where: S_t is the cross-section surrounding the scatterer. $v_n|_{r=a} = -\partial \Phi_t / \partial r$, $v_t|_{r=a} = -\partial \Phi_t / (r \partial r)$ represent the normal and tangential components of the velocity at the particle boundary, respectively. $\boldsymbol{n}$ is the normal vector of the surface dS when $\boldsymbol{t}$ is the tangential component of the surface dS, $dS = 2\pi r^2 \sin\theta d\theta$.

In the spherical coordinate system, as shown in Fig. 3.1, when a plane wave is incident on a spherical object, the velocity potential of the incident wave in the liquid is:

$$\Phi_i = \phi_0 e^{-i\omega t} \sum_{n=0}^{\infty} (2n+1)(i)^n P_n(\cos\theta) j_n(kr) \tag{3.39}$$

ϕ_0 represents the amplitude of the velocity potential.

Meanwhile, it is not difficult to derive the velocity potential of the scattered wave:

$$\Phi_s = \phi_0 e^{-i\omega t} \sum_{n=0}^{\infty} (2n+1)(i)^n A_{n,s} P_n(\cos\theta) h_n^1(kr) \tag{3.40}$$

Therefore, the total velocity potential in the liquid is:

$$\Phi_t = \phi_0 e^{-i\omega t} \sum_{n=0}^{\infty} (2n+1)(i)^n (U_n + iV_n) P_n(\cos\theta) \tag{3.41}$$

where:

$$(U_n + iV_n) = j_n(kr) + A_{n,s}h_n^{(1)}(kr) = j_n(kr)(1 + \alpha_n) \\ - \beta_n y_n(kr) + i\left[\beta_n j_n(kr) + \alpha_n y_n(kr)\right] \tag{3.42}$$

Let α_n, β_n be the real and imaginary parts of the scattering coefficient $A_{n,s}$ respectively.

Then:

$$\begin{cases} U_n = (1 + \alpha_n)j_n(kr) - \beta_n n_n(kr) \\ V_n = \beta_n j_n(kr) + \alpha_n n_n(kr) \end{cases} \tag{3.43}$$

Projecting the radiation force formula (3.38) along the z-axis in the direction of sound wave propagation, we can obtain:

The projection of $-\iint\limits_{S_0} \rho v_n^2 \boldsymbol{n} dS$ in the z-axis direction is:

$$-\iint\limits_{S_0} \rho v_n^2 \boldsymbol{n} dS = -\iint\limits_{S_0} \rho v_n^2 \cos\theta dS = -2\pi a^2 \rho \int\limits_0^\pi \left(\frac{\partial \Phi_t}{\partial r}\right)_{r=a}^2 \sin\theta \cos\theta d\theta \tag{3.44}$$

The projection of $-\iint\limits_{S_0} \rho v_n \cdot v_t \boldsymbol{t} dS$ in the z-axis direction is:

$$\begin{aligned} -\iint\limits_{S_0} \rho v_n \cdot v_t \boldsymbol{t} dS &= \iint\limits_{S_0} \rho v_n \cdot v_t \cos\left(\frac{\pi}{2} - \theta\right) dS \\ &= 2\pi a \rho \int\limits_0^\pi \left(\frac{\partial \Phi_t}{\partial r}\right)_{r=a} \left(\frac{\partial \Phi_t}{\partial \theta}\right)_{r=a} \sin^2\theta d\theta \end{aligned} \tag{3.45}$$

The projection of $\iint\limits_{S_0} \rho v^2 \boldsymbol{n} dS$ in the z-axis direction is:

$$\begin{aligned} \iint\limits_{S_0} \rho v^2 \boldsymbol{n} dS &= \iint\limits_{S_0} \rho\left(v_n^2 + v_t^2\right) \cos\theta dS \\ &= \pi a^2 \rho \int\limits_0^\pi \left(\frac{\partial \Phi_t}{\partial r}\right)_{r=a}^2 \sin\theta \cos\theta d\theta \\ &\quad + \pi \rho \int\limits_0^\pi \left(\frac{\partial \Phi_t}{\partial \theta}\right)_{r=a}^2 \sin\theta \cos\theta d\theta \end{aligned} \tag{3.46}$$

The projection of $-\iint\limits_{S_0} \frac{1}{2}\frac{\rho}{c^2}\left(\frac{\partial \Phi_t}{\partial t}\right)^2 \boldsymbol{n} dS$ in the z-axis direction is:

$$
\begin{aligned}
-\iint\limits_{S_0} \frac{1}{2}\frac{\rho}{c^2}\left(\frac{\partial \Phi_t}{\partial t}\right)^2 \boldsymbol{n} dS &= -\iint\limits_{S_0} \frac{1}{2}\frac{\rho}{c^2}\left(\frac{\partial \Phi_t}{\partial t}\right)^2 \cos\theta dS \\
&= -\frac{\pi a^2 \rho}{c^2}\int_0^{\pi}\left(\frac{\partial \Phi_t}{\partial t}\right)^2_{r=a} \sin^2\theta\cos\theta d\theta
\end{aligned}
\tag{3.47}
$$

By rearranging formulas (3.44)--(3.47), we can obtain:

$$\langle F_z \rangle = \langle F_r \rangle + \langle F_\theta \rangle + \langle F_{r,\theta} \rangle + \langle F_t \rangle \tag{3.48}$$

$$\langle F_r \rangle = \left\langle -\pi a^2 \rho \int_0^{\pi}\left(\frac{\partial \Phi_t}{\partial r}\right)^2_{r=a} \sin\theta\cos\theta d\theta \right\rangle \tag{3.49}$$

$$\langle F_\theta \rangle = \left\langle \pi \rho \int_0^{\pi}\left(\frac{\partial \Phi_t}{\partial \theta}\right)^2_{r=a} \sin\theta\cos\theta d\theta \right\rangle \tag{3.50}$$

$$\langle F_{r,\theta} \rangle = \left\langle 2\pi a \rho \int_0^{\pi}\left(\frac{\partial \Phi_t}{\partial r}\right)_{r=a}\left(\frac{\partial \Phi_t}{\partial \theta}\right)_{r=a} \sin^2\theta d\theta \right\rangle \tag{3.51}$$

$$\langle F_t \rangle = \left\langle -\frac{\pi a^2 \rho}{c^2} \int_0^{\pi}\left(\frac{\partial \Phi_t}{\partial t}\right)^2_{r=a} \sin^2\theta\cos\theta d\theta \right\rangle \tag{3.52}$$

From formulas (3.41)–(3.43), it is known that the velocity potential Φ_t is a complex number. The physically meaningful part is its real part, which we denote as:

$$R_n = \mathrm{Re}\left[i^n (U_n + iV_n)(\cos\omega t - i\sin\omega t)\right] \tag{3.53}$$

$$\Phi_t = \sum_{n=0}^{\infty} (2n+1) R_n P_n(\cos\theta) \tag{3.54}$$

Then:

$$
\begin{aligned}
R_n \cdot R_{n+1} = \mathrm{Re}\big\{&\left[i^n (U_n + iV_n)(\cos\omega t - i\sin\omega t)\right] \\
&\cdot\left[i^{n+1} (U_{n+1} + iV_{n+1})(\cos\omega t - i\sin\omega t)\right]\big\}
\end{aligned}
\tag{3.55}
$$

Also, because the $(\cos\omega t - i\sin\omega t)^2$ term averages to zero over time due to the trigonometric terms.

Through algebraic calculations, it is not difficult to derive:

$$\langle R_n \cdot R_{n+1}\rangle = \frac{1}{2}(U_n V_{n+1} - V_n U_{n+1}) \tag{3.56}$$

Similarly, we can calculate:

$$\left\langle R_n' \cdot R_{n+1}\right\rangle = \frac{1}{2}\left(U_n' V_{n+1} - V_n' U_{n+1}\right) \tag{3.57}$$

$$\left\langle R_n \cdot R_{n+1}'\right\rangle = \frac{1}{2}\left(U_n V_{n+1}' - V_n U_{n+1}'\right) \tag{3.58}$$

$$\left\langle R_n' \cdot R_{n+1}'\right\rangle = \frac{1}{2}\left(U_n' V_{n+1}' - V_n' U_{n+1}'\right) \tag{3.59}$$

Due to the orthogonality of the Legendre functions [12, 17]:

$$\int_0^\pi P_n(\cos\theta)P_m(\cos\theta)\cos\theta\sin\theta d\theta = \frac{2(n+1)}{(n+1)(2n+3)}\cdots m = n+1 \tag{3.60}$$

$$\int_0^\pi P_n(\cos\theta)P_m(\cos\theta)\cos\theta\sin\theta d\theta = 0\cdots m > n+1 \tag{3.61}$$

We can obtain:

$$\langle F_r\rangle = -4\pi a^2\rho\sum_{n=0}^{\infty}(n+1)\left\langle \frac{\partial R_n}{\partial r}\cdot\frac{\partial R_{n+1}}{\partial r}\right\rangle\Bigg|_{r=a} \tag{3.62}$$

$$\langle F_\theta\rangle = 4\pi\rho\sum_{n=0}^{\infty} n(n+1)(n+2)\langle R_n \cdot R_{n+1}\rangle|_{r=a} \tag{3.63}$$

$$\left\langle F_{r,\theta}\right\rangle = 4\pi a\rho\left[\sum_{n=0}^{\infty} n(n+1)\left\langle R_n \cdot \frac{\partial R_{n+1}}{\partial r}\right\rangle\Bigg|_{r=a} - (n+1)(n+2)\left\langle\frac{\partial R_n}{\partial r}\cdot R_{n+1}\right\rangle\Bigg|_{r=a}\right] \tag{3.64}$$

$$\langle F_t\rangle = -\frac{4\pi a^2\rho}{c^2}\sum_{n=0}^{\infty}(n+1)\left\langle \frac{\partial R_n}{\partial t}\cdot\frac{\partial R_{n+1}}{\partial t}\right\rangle\Bigg|_{r=a} \tag{3.65}$$

Finally, using the recursion formula of the spherical Bessel functions [12, 17]:

$$xh_n'(x) = nh_n(x) - xh_{n+1}(x) \tag{3.66}$$

$$xh_{n+1}'(x) = xh_n(x) - (n+1)h_{n+1}(x) \tag{3.67}$$

$$x^2\left[j_{n+1}(x)n_n(x+1) - j_n(x)n_{n+1}(x+1)\right] = 1 \tag{3.68}$$

Based on the above formulas, we can finally simplify and obtain the formula for the axial acoustic radiation force on a sphere due to a plane wave:

$$F_z = \pi a^2 EY_p \tag{3.69}$$

where: $E = \frac{1}{2}\rho k^2 A^2$ is the energy density of the incident wave, Y_p The function for acoustic radiation force is expressed as follows [3, 4]:

$$Y_p = -\frac{4}{(ka)^2}\sum_{n=0}^{\infty}(n+1)\left\{\begin{array}{l}\mathrm{Re}(\alpha_n + \alpha_{n+1} + 2\alpha_n\alpha_{n+1} + 2\beta_n\beta_{n+1}) \\ +Im\left[\beta_{n+1}(1+2\alpha_n) - \beta_n(1+2\alpha_n)\right]\end{array}\right\} \tag{3.70}$$

3.2 Gaussian Wave Acoustic Radiation Force on Spherical Particles

As mentioned in Chap. 2, Gaussian beams play a significant role in the application of biomedical ultrasound. This section will start with the method of function series expansion, approximately expanding the incident Gaussian beam into a superposition form of spherical waves containing beam factors. Then, based on the corresponding theory of acoustic radiation force, we derive the analytical expression of the acoustic radiation force of Gaussian sound waves acting on spherical particles in water.

When the Gaussian beam propagates in space, we approximate the wavefront near its waist as a plane. That is, the Gaussian beam can be approximately regarded as a Gaussian amplitude distribution of sound waves in the waist area. Therefore, when dealing with the scattering of sound waves by spherical particles in liquid, as shown in Fig. 3.2,

The radius of the small sphere is a, a column of Gaussian sound waves propagates along the z-axis in a uniform fluid medium, the real number A is the velocity potential amplitude of the incident sound wave, the medium density is ρ, and the speed of sound in the medium is c. The spherical particle is located at the focal point of the Gaussian beam, and the center of the sphere O is set as the origin. At this time, the velocity potential of the incident Gaussian beam can be expressed as:

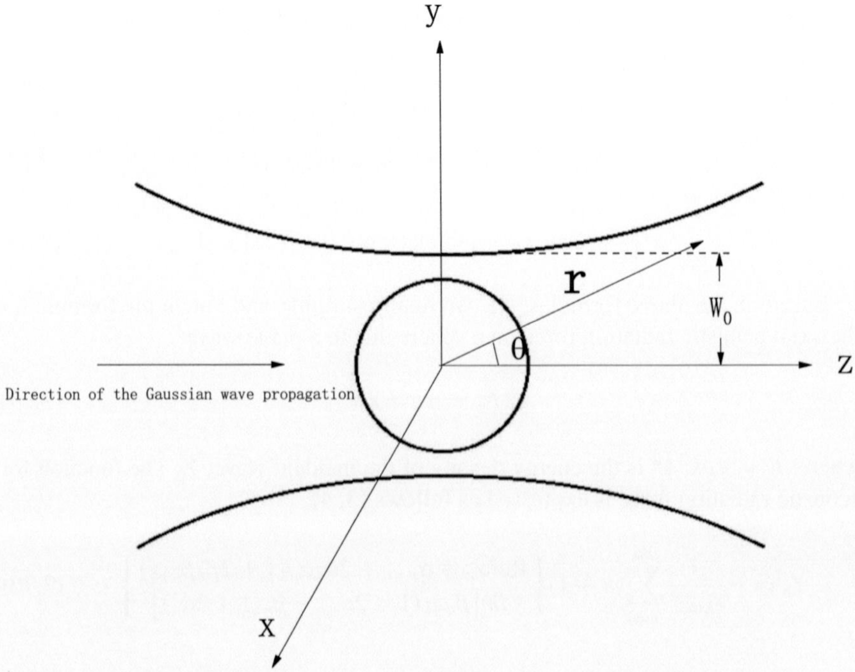

Fig. 3.2 Schematic diagram of the scattering of a Gaussian beam by a spherical particle

$$\Phi_i = Ae^{-\left(\frac{x^2+y^2}{W_0^2}\right)}e^{-ikz}e^{-i\omega t} \tag{3.71}$$

k is the wave number of the sound wave in the medium, k is the wave number, W_0 is the wave number width of the Gaussian wave, ω is the angular frequency of the sound wave.

With the center of the sphere as the origin, a spherical coordinate system is established, and the coordinates of any point (x, y, z) in space can be expressed as:

$$\begin{cases} x = r\sin\theta\cos\varphi \\ y = r\sin\theta\sin\varphi \\ z = r\cos\theta \end{cases} \tag{3.72}$$

where: θ is the angle between r and the z-axis.

We set:

$$\begin{cases} R = \sqrt{x^2 + y^2} = r\sin\theta \\ \delta = \frac{1}{kW_0} = \frac{\lambda}{2\pi W_0} \\ \gamma = kr \end{cases} \tag{3.73}$$

The incident wave is expanded according to the spherical function, and we get:

$$e^{-\left(\frac{x^2+y^2}{W_0^2}\right)}e^{-ikz} = e^{-\delta^2\gamma^2\sin^2\theta}e^{i\gamma\cos\theta}$$
$$= \sum_{n=0}^{\infty} G_n(2n+1)\cdot i^n j_n(\gamma)P_n(\cos\theta) \tag{3.74}$$

where, $P_n(\cdot)$ is the nth order Legendre function, $j_n(\cdot)$ is the nth order Bessel function of the first kind, G_n is the beam factor of the incident Gaussian wave.

We refer to the method of calculating the beam factor G_n in the electromagnetic field [18, 19].

Take $\theta = \frac{\pi}{2}$, Eq. (3.74) can be expressed as:

$$e^{-\delta^2\gamma^2} = \sum_{n=0}^{\infty} G_n(2n+1)\cdot i^n j_n(\gamma)P_n(0) \tag{3.75}$$

Using the formula:

$$\begin{cases} P_{2n}(0) = \frac{(-1)^n\Gamma\left(n+\frac{1}{2}\right)}{\sqrt{\pi}\Gamma(n+1)} \\ P_{2n+1}(0) = 0 \end{cases} \tag{3.76}$$

We get:

$$e^{-\delta^2\gamma^2} = \sum_{n=even} (2n+1)G_n j_n(\gamma)\frac{\Gamma\left(\frac{n}{2}+\frac{1}{2}\right)}{\sqrt{\pi}\Gamma\left(\frac{n}{2}+1\right)} \tag{3.77}$$

There is also the formula:

$$j_n(\gamma) = \sqrt{\frac{\pi}{2\gamma}}J_{n+1}(\gamma) \tag{3.78}$$

Then:

$$e^{-\delta^2\gamma^2} = \sum_{n=0}^{\infty} (2n+1)G_n J_n(\gamma)\frac{\Gamma\left(\frac{n}{2}+\frac{1}{2}\right)}{\sqrt{\pi}\Gamma\left(\frac{n}{2}+1\right)} \tag{3.79}$$

In the above equations, $j_n(\cdot)$ is the nth order Bessel function of the first kind, $J_n(\cdot)$ is the nth order Bessel function of the first kind, $\Gamma(\cdot)$ is the gamma function.

There is also the Neumann series:

$$
\begin{aligned}
z^{\nu} f(z) &= \sum_{n=0}^{\infty} a_n J_{\nu+n}(z) \\
f(z) &= \sum_{n=0}^{\infty} b_n z^n
\end{aligned}
\tag{3.80}
$$

Comparing Eqs. (3.79) and (3.80), we get:

$$
a_n = (2n+1)\frac{\Gamma\left(\frac{n}{2}+\frac{1}{2}\right)}{\sqrt{\pi}\,\Gamma\left(\frac{n}{2}+1\right)} G_n,\ \nu = \frac{1}{2} \tag{3.81}
$$

$$
e^{-\delta^2\gamma^2} = \sum_{n=0}^{\infty} (2n+1) G_n J_n(\gamma) \frac{\Gamma\left(\frac{n}{2}+\frac{1}{2}\right)}{\sqrt{\pi}\,\Gamma\left(\frac{n}{2}+1\right)} = \sum_{n=0}^{\infty} b_n z^n \tag{3.82}
$$

It is not difficult to derive the coefficient:

$$
b_n = \frac{\left(-\delta^2\right)^{\frac{n}{2}}}{\left(\frac{n}{2}\right)!} \tag{3.83}
$$

Finally, according to the formula:

$$
a_n = (\nu+n)\sum 2^{\nu+n-2m}\frac{\Gamma(\nu+n-m)}{m!} b_{n-2m} \tag{3.84}
$$

After simplifying the calculations, we can finally obtain the expression for the beam factor when n is even [10, 20]:

$$
G_{2p} = \frac{\Gamma(p+1)}{\Gamma\left(p+\frac{1}{2}\right)} \sum_{j=0}^{p} \frac{\Gamma\left(p+j+\frac{1}{2}\right)}{\Gamma(p-j)!j!} \times \left(-4\left(\frac{1}{kW_0}\right)^2\right)^j \tag{3.85}
$$

Similarly, as before, using the formula:

$$
\begin{cases}
P'_{2n+1}(0) = \frac{(-1)^n \Gamma\left(n+\frac{3}{2}\right)}{\sqrt{\pi}\,\Gamma(n+1)} \\
P'_{2n}(0) = 0
\end{cases}
\tag{3.86}
$$

We can obtain the expression for the wave number factor when n is odd:

$$
G_{2p+1} = \frac{\Gamma(p+1)}{\Gamma\left(p+\frac{3}{2}\right)} \sum_{j=0}^{p} \frac{\Gamma\left(p+j+\frac{3}{2}\right)}{\Gamma(p-j)!j!} \times \left(-4\left(\frac{1}{kW_0}\right)^2\right)^j \tag{3.87}
$$

The above method uses a finite series to expand the wave number factor. According to the expressions for the wave number factor (3.85) and (3.87), we can see that the

Table 3.1 Convergence of Gaussian beam factor at different beam waist widths

n	G_n		
	$W_0 = \lambda$	$W_0 = 2\lambda$	$W_0 = 3\lambda$
1	1.0000	1.0000	1.0000
2	1.0000	1.0000	1.0000
3	0.8480	0.9620	0.9831
4	0.7467	0.9367	0.9719
5	0.4524	0.8790	0.9448
10	0.0976	0.5639	0.7780
15	0.0040	0.2619	0.5517
16	0.0001	0.2153	0.5078
20		0.0889	0.3414
25		0.0235	0.1841
30		0.0044	0.0859
35		0.0007	0.0356
39		0.0001	0.0159
45			0.0041
50			0.0011
55			0.0003
58			0.0001

wave number factor G_n is related to the parameter $\frac{1}{kW_0}$. Table 3.1 shows the relationship between different beam waist widths and wave number factors in Gaussian waves. When the calculated value $G_n \le 0.0001$, we consider the function to have converged at this point.

As can be seen from the calculations in Table 3.1, the larger the ratio between the beam waist radius W_0 and the acoustic wavelength λ, the more terms the function converges to. Moreover, when $W_0 \gg \lambda$, $G_n \to 1$, indicating that when the waist width of the Gaussian wave is very large, it can be approximated as a plane wave.

At the same time, to verify the correctness of the finite series method for expanding the Gaussian beam, we have simulated the incident waveforms represented by the analytical solution formula (3.71) and the finite series expansion formula (3.74).

The simulation results are shown in Figs. 3.2, 3.3, 3.4, 3.5 and 3.6. From the simulation results in the figures, it can be seen that when the beam width of the Gaussian wave is larger than the wavelength of the sound wave, the spatial distribution of the Gaussian wave simulated by the finite series method fits well with the analytical results. Therefore, when the beam width of the Gaussian wave is large, the above method can be used for approximate solution [21, 22].

Assume that the incident Gaussian wave is symmetric about the z-axis, as shown in Fig. 3.2. From formula (3.74), we can obtain the total velocity potential of the incident wave in the spherical coordinate system:

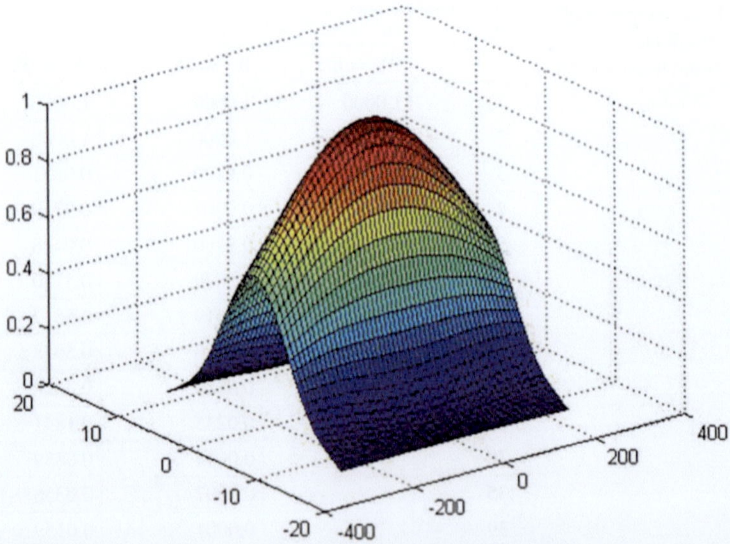

Fig. 3.3 The three-dimensional diagram of the analytical expression of the incident Gaussian waveform at $W_0 = 5\lambda$

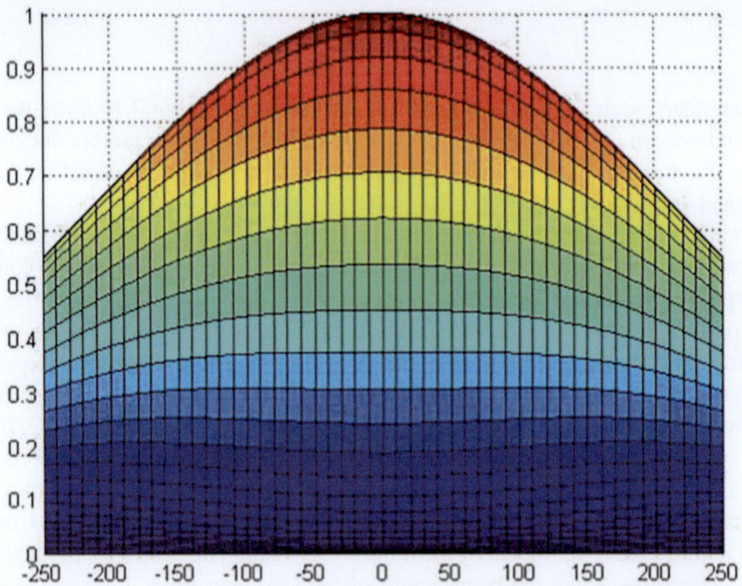

Fig. 3.4 The X–Z view of the analytical expression of the incident Gaussian waveform at $W_0 = 5\lambda$

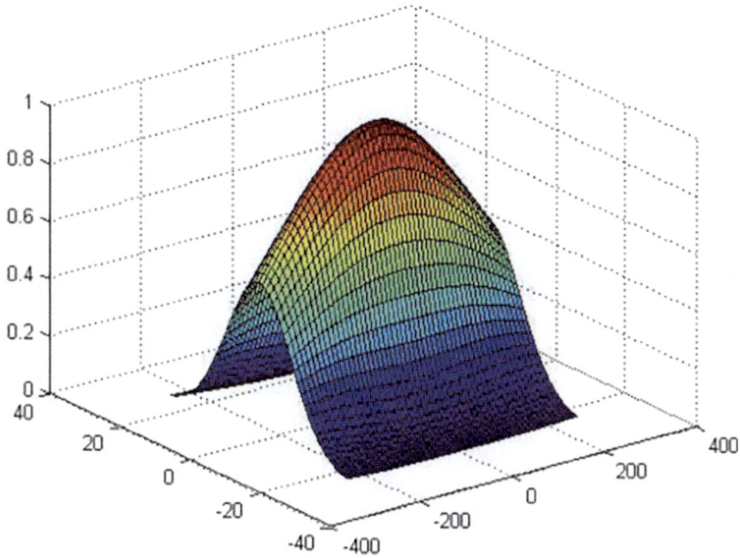

Fig. 3.5 The three-dimensional diagram of the finite series expansion method of the incident Gaussian waveform at $W_0 = 5\lambda$

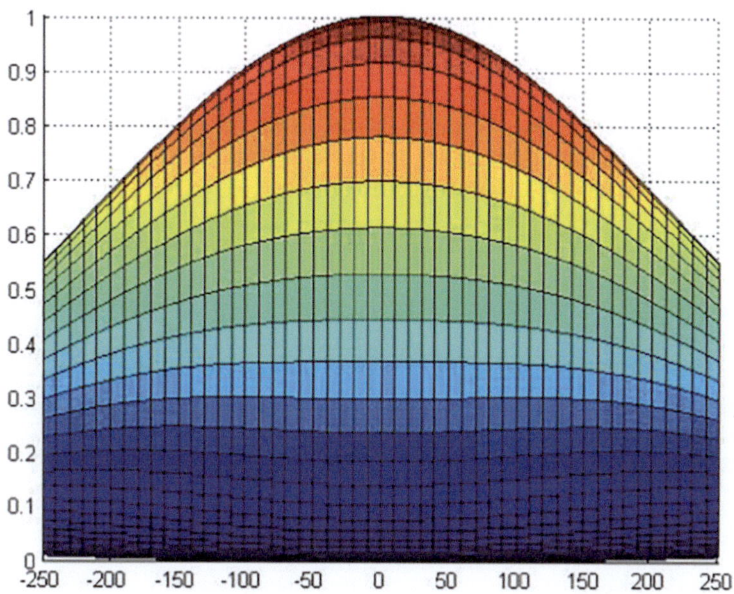

Fig. 3.6 The X–Z view of the finite series expansion method of the incident Gaussian waveform at $W_0 = 5\lambda$

$$\Phi_t = \phi_0 e^{-i\omega t} \sum_{n=0}^{\infty} G_n(2n+1) i^n (U_n + iV_n) P_n(\cos\theta) \tag{3.88}$$

where: α_n, β_n are the real and imaginary parts of the scattering coefficient $A_{n,s}$ respectively.

$$\begin{cases} U_n = (1+\alpha_n) j_n(kr) - \beta_n n_n(kr) \\ V_n = \beta_n j_n(kr) + \alpha_n n_n(kr) \end{cases} \tag{3.89}$$

Following the method of solving the acoustic radiation force function of plane waves in Sect. 3.1, we can obtain the expression of the axial acoustic radiation force of the Gaussian wave on the spherical particle as follows [10]:

$$Y_G = -\frac{4}{(ka)^2} \sum_{n=0}^{\infty} (n+1) \left\{ \begin{array}{l} \mathrm{Re}\left[G_n G_{n+1}^*\right]\left[\alpha_n + \alpha_{n+1} + 2\alpha_n\alpha_{n+1} + 2\beta_n\beta_{n+1}\right] \\ +Im\left[G_n G_{n+1}^*\right]\left[\beta_{n+1}(1+2\alpha_n) - \beta_n(1+2\alpha_{n+1})\right] \end{array} \right\} \tag{3.90}$$

where: $*$ represents the complex conjugate.

$$F_z = \pi a^2 E Y_G \tag{3.91}$$

where: $E = \frac{1}{2}\rho k^2 A^2$ is the energy density of the incident wave, Y_G is the acoustic radiation force function of the Gaussian wave.

From formula (3.90), it can be seen that the radiation force function of the Gaussian wave Y_G is a dimensionless number, which represents the change in unit energy density within the unit cross-sectional area. Its size is related to the wave number k, the radius a of the particle, and the scattering coefficient $A_{n,s}$.

References

1. Beitrage M (2008) On the optics of turbid media, especially colloidal metal solutions. Ann Phys 25:377–445
2. King LV (1934) On the acoustic radiation pressure on a sphere. Proc R Soc A 147:212–240
3. Yosioka K (1995) Acoustic radiation pressure on a compressible sphere. Acta Acust Acust 5:167–173
4. Hasegawal T, Yosioka K (1996) Acoustic radiation force on a solid elastic sphere. J Acoust Soc Am 46:1139
5. Marston PL (2006) Axial radiation force of a Bessel beam on a sphere and direction reversal of the force. J Acoust Soc Am 120:3518–3524
6. Mitri FG (2008) Acoustic scattering of a high-order Bessel beam by an elastic sphere. Ann Phys 323:2840–2850
7. Marston PL (2007) Negative axial radiation forces on solid spheres and shells in a Bessel beam. J Acoust Soc Am 122:3162–3165
8. Mitri FG (2009) Negative axial radiation force on a fluid and elastic spheres illuminated by a high-order Bessel beam of progressive waves. J Phys A-Math Theor 42:1947–1957

9. Azarpeyvand M (2014) Prediction of negative radiation forces due to a Bessel beam. J Acoust Soc Am 136:547–555
10. Zhang XF, Zhang GB (2012) Acoustic radiation force of a Gaussian beam incident on spherical particles in water. Ultrasound Med Biol 38:2007–2017
11. Zhang XF, Song ZG, Chen DM et al (2015) Finite series expansion of a Gaussian beam for the acoustic radiation force calculation of cylindrical particles in water. J Acoust Soc Am 137:1826–1833
12. Liang K (1998) Mathematical physics methods, 2nd edn. Higher Education Press
13. Du G (2001) Fundamentals of acoustics. Nanjing University Press
14. Ayres VM, Gaunaurd GC (1986) Acoustic resonance scattering by viscoelastic objects. J Acoust Soc Am 81:301–311
15. Hartmann B, Jarzynski J (1972) Ultrasonic hysteresis absorption in polymers. J Appl Phys 43:4304
16. Song Z (2013) Study on the acoustic radiation force of micro cylindrical particles by a single beam. Master's Thesis of Shaanxi Normal University
17. Liu S, Liu S (2002) Special functions, 2nd edn. Meteorological Press
18. Love A (1906) A treatise on the mathematical theory of elasticity. Nature 9:1385
19. Gouesbet G (1998) Scattering of a Gaussian beam by a sphere using a Bromwich formulation: case of an arbitrary location. Opt Part Siz 27–42
20. Wu RR, Cheng KX, Liu XZ (2016) Study of axial acoustic radiation force on a sphere in a Gaussian quasi-standing field. Wave Motion 62:63–74
21. Hinders MK (1993) Elastic-wave scattering from an elastic cylinder. IL Nuovo Cimento B 108:285–301
22. Jiang YF (2010) Radiation force of highly focused cosine-Gaussian beam on a particle in the Rayleigh scattering regime. Acta Physica Sinica 6:3959–3964

Chapter 4
Study of Acoustic Radiation Force on Spherical Particles Based on Acoustic Scattering Method

A standing wave refers to two types of sound waves with the same frequency and opposite transmission directions, forming a certain distribution state along the transmission direction. One of these waves is generally the reflection of the other. At the point where the sound pressures of the two add up, a wave crest is formed, and at the point where their sound pressures subtract, a wave trough is formed. On the waveform, the positions of the wave troughs and wave crests are always constant, but their instantaneous values change over time. If the amplitudes of these two waves are equal, then the amplitude of the wave trough is zero. The manipulation of particles and cells in a standing wave field has rapidly developed in recent years [1–4], and the calculation of radiation forces in the standing wave field has gradually expanded to areas such as Bessel standing wave fields [6–8].

This chapter will derive the axial acoustic radiation force function on a small sphere in a Gaussian standing wave field and a quasi-Gaussian standing wave field based on the acoustic scattering method, and solve the boundary conditions of various types of small spheres, including rigid spheres, fluid spheres, elastic spheres, and viscoelastic spheres. This chapter will also analyze and discuss the impact of various parameters in the sound field on the radiation force function.

4.1 Theoretical Methods

Wu and others first achieved the trapping effect of sound waves on tiny particles in 1991 [9], using a focused Gaussian standing wave beam. Two Gaussian focused sound beams were placed opposite each other to form a standing wave field, trapping particles at the wave troughs of the sound field. Similar to plane standing waves [6] and Bessel standing waves [7, 8], a Gaussian standing wave field or a quasi-Gaussian standing wave field refers to two Gaussian sound waves with the same or different

X. Liu and H. Zheng, *Acoustic Radiation Force*, Advances in Acoustics,
https://doi.org/10.1007/978-981-95-3152-3_4

amplitudes propagating in opposite directions along the same axis, with the same frequency.

4.1.1 Radiation Force of Gaussian Standing Wave on Sphere

For the sake of calculation accuracy, we assume that the Gaussian standing wave field here is a weakly focused sound field, that is, it has a larger beam waist width W_0, which makes the propagation phase of the wavefront consistent with the plane wave. At this time, we can regard the waveform at this time as a plane standing wave field with a Gaussian distribution.

As shown in Fig. 4.1, in the uniform medium of water, two columns of Gaussian focused sound waves with a velocity potential amplitude of A propagate in opposite directions along the same axis z. The beam waist positions of the two Gaussian waves are 2h apart. A Cartesian coordinate system is established with the center position of the standing wave field as the origin. A spherical particle is placed between the two sound waves. Assuming that the speed of sound in the medium is c and the density is ρ, the incident wave velocity potential can be expressed as follows:

$$\Phi_i = Ae^{-\left(\frac{x^2+y^2}{W_0^2}\right)}\left[e^{ik(z+h)} + e^{-ik(z+h)}\right]e^{-i\omega t} \tag{4.1}$$

where k is the wave number, ω is the angular frequency of the sound wave.

It can be seen that Eq. (4.1) is not an exact solution to the Helmholtz equation, so the traditional scattering method cannot be used to solve it. According to the spherical function expansion of the Helmholtz equation [10, 11]:

First, expand the incident wave into a finite series:

$$\Phi_i = \sum_{n=0}^{\infty}(2n+1)i^n G_n \delta_n j_n(kr)P_n(\cos\theta)e^{-i\omega t} \tag{4.2}$$

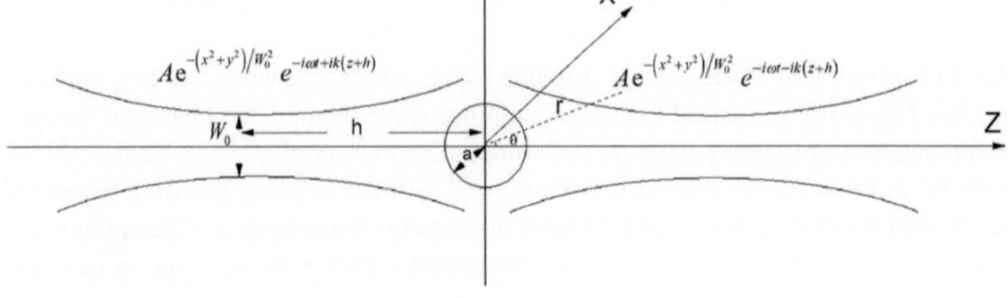

Fig. 4.1 Schematic diagram of the Gaussian standing wave field

where:

$$\delta_n = A\left[e^{ikh} + (-1)^n e^{-ikh}\right] \tag{4.3}$$

G_n is the expansion factor of the Gaussian beam:

$$\begin{cases} G_{2p} = \frac{\Gamma(p+1)}{\Gamma\left(p+\frac{1}{2}\right)} \sum_{j=0}^{p} \frac{\Gamma\left(p+j+\frac{1}{2}\right)}{\Gamma(p-j)!j!} \times \left(-4\left(\frac{1}{kW_0}\right)^2\right)^j \\ G_{2p+1} = \frac{\Gamma(p+1)}{\Gamma\left(p+\frac{3}{2}\right)} \sum_{j=0}^{p} \frac{\Gamma\left(p+j+\frac{3}{2}\right)}{\Gamma(p-j)!j!} \times \left(-4\left(\frac{1}{kW_0}\right)^2\right)^j \end{cases} \tag{4.4}$$

$j_n(\cdot)$ is the n order spherical Bessel function, $P_n(\cdot)$ is the n order Legendre function, $\Gamma(\cdot)$ is the gamma function.

At this time, the velocity potential of the scattered wave can be expressed as:

$$\Phi_s = \sum_{n=0}^{\infty} (2n+1) i^n A_n G_n \delta_n h_n(kr) P_n(\cos\theta) e^{-i\omega t} \tag{4.5}$$

where $A_n = \alpha_n + i\beta_n$ represents the scattering factor, which is determined by the boundary conditions of the particle sphere, α_n, β_n respectively represent the real and imaginary parts of the scattering coefficient; $h_n(\cdot)$ is the first kind n-order spherical Hankel function.

Therefore, the total velocity potential should include the incident wave and the scattered wave, and the meaningful part is its real part, which can be expressed as:

$$\Phi_t = \mathrm{Re}\left[\sum_{n=0}^{\infty} (2n+1) i^n G_n \delta_n (U_n + iV_n) P_n(\cos\theta) e^{-i\omega t}\right] \tag{4.6}$$

where:

$$\begin{cases} U_n = (1+\alpha_n) j_n(kr) - \beta_n n_n(kr) \\ V_n = \beta_n j_n(kr) + \alpha_n n_n(kr) \end{cases} \tag{4.7}$$

$n_n(\cdot)$ is the n-th order spherical Neumann function.

Therefore, the total velocity potential can be expressed as:

$$\Phi_t = \sum_{n=0}^{\infty} (2n+1) R_n P_n(\cos\theta) \tag{4.8}$$

where,

$$R_n = \mathrm{Re}\left[G_n \delta_n i^n (U_n + iV_n)(\cos\omega t - i\sin\omega t)\right] \tag{4.9}$$

Then:

$$\begin{aligned} R_n \cdot R_{n+1} = \mathrm{Re}\big(\big[i^n G_n \delta_n (U_n + iV_n)(\cos\omega t - i\sin\omega t)\big] \\ \cdot\big[i^{n+1} G_{n+1}\delta_{n+1}(U_{n+1} + iV_{n+1})(\cos\omega t - i\sin\omega t)\big]\big\} \end{aligned} \tag{4.10}$$

Also, because:

$$\delta_n \cdot \delta_{n+1} = A^2\big[e^{ikh} + (-1)^n e^{-ikh}\big]\cdot\big[e^{ikh} + (-1)^{n+1}e^{-ikh}\big] = A^2(2i\sin 2kh) \tag{4.11}$$

Through algebraic operations and simplification, we can conclude:

$$\begin{aligned} \langle R_n \cdot R_{n+1}\rangle = (-1)^{n+1}A^2\big[\big(g_n^i g_{n+1}^i + g_n^r g_{n+1}^r\big)(U_n U_{n+1} + V_n V_{n+1}) \\ +\big(g_n^i g_{n+1}^r - g_n^r g_{n+1}^i\big)(V_n U_{n+1} - U_n V_{n+1})\big]\sin(2kh) \end{aligned} \tag{4.12}$$

$$\begin{aligned} \langle R_n' \cdot R_{n+1}'\rangle = (-1)^{n+1}A^2\big[\big(g_n^i g_{n+1}^i + g_n^r g_{n+1}^r\big)\big(U_n' U_{n+1}' + V_n' V_{n+1}'\big) \\ +\big(g_n^i g_{n+1}^r - g_n^r g_{n+1}^i\big)\big(V_n' U_{n+1}' - U_n' V_{n+1}'\big)\big]\sin(2kh) \end{aligned} \tag{4.13}$$

$$\begin{aligned} \langle R_n \cdot R_{n+1}'\rangle = (-1)^{n+1}A^2\big[\big(g_n^i g_{n+1}^i + g_n^r g_{n+1}^r\big)\big(U_n U_{n+1}' + V_n V_{n+1}'\big) \\ +\big(g_n^i g_{n+1}^r - g_n^r g_{n+1}^i\big)\big(V_n U_{n+1}' - U_n V_{n+1}'\big)\big]\sin(2kh) \end{aligned} \tag{4.14}$$

$$\begin{aligned} \langle R_n' \cdot R_{n+1}\rangle = (-1)^{n+1}A^2\big[\big(g_n^i g_{n+1}^i + g_n^r g_{n+1}^r\big)\big(U_n' U_{n+1} + V_n' V_{n+1}\big) \\ +\big(g_n^i g_{n+1}^r - g_n^r g_{n+1}^i\big)\big(V_n' U_{n+1} - U_n' V_{n+1}\big)\big]\sin(2kh) \end{aligned} \tag{4.15}$$

$$\left\langle \frac{\partial R_n}{\partial t}\cdot\frac{\partial R_{n+1}}{\partial t}\right\rangle = \omega^2\langle R_n \cdot R_{n+1}\rangle \tag{4.16}$$

where $g_n^r = \mathrm{Re}[G_n]$, $g_n^i = \mathrm{Im}[G_n]$ represents the real and imaginary parts of the Gaussian wave expansion factor G_n respectively.

The formula for the sound radiation force in the direction of the sound wave axis can be expressed as:

$$\langle F_z\rangle = \langle F_r\rangle + \langle F_\theta\rangle + \langle F_{r,\theta}\rangle + \langle F_t\rangle \tag{4.17}$$

$$\langle F_r\rangle = \left\langle -\pi a^2\rho\int_0^\pi\left(\frac{\partial\Phi_t}{\partial r}\right)^2_{r=a}\sin\theta\cos\theta d\theta\right\rangle \tag{4.18}$$

$$\langle F_\theta\rangle = \left\langle \pi\rho\int_0^\pi\left(\frac{\partial\Phi_t}{\partial r}\right)^2_{r=a}\sin\theta\cos\theta d\theta\right\rangle \tag{4.19}$$

$$\langle F_{r,\theta}\rangle = \left\langle 2\pi a\rho \int_0^{\pi} \left(\frac{\partial \Phi_t}{\partial r}\right)_{r=a} \left(\frac{\partial \Phi_t}{\partial \theta}\right)_{r=a} \sin^2\theta d\theta \right\rangle \tag{4.20}$$

$$\langle F_t\rangle = \left\langle -\frac{\pi a^2 \rho}{c^2} \int_0^{\pi} \left(\frac{\partial \Phi_t}{\partial t}\right)^2_{r=a} \sin^2\theta \cos\theta d\theta \right\rangle \tag{4.21}$$

Due to the orthogonality of the Legendre function, the recursion formula of the spherical Bessel function, etc., we can finally conclude that the axial (z-axis) sound radiation force on the particle sphere in the Gaussian standing wave field is:

$$F_{st} = \pi a^2 E Y_{st} \cdot \sin(2kh) \tag{4.22}$$

where $E = \frac{1}{2}\rho k^2 A^2$ represents the sound energy density, Y_{st} is the sound radiation force function [12]:

$$Y_{st} = \frac{8}{(ka)^2} \sum_{n=0}^{\infty} (n+1)(-1)^{n+1} \left\{ \begin{array}{l} \left(g_n^r g_{n+1}^r + g_n^i g_{n+1}^i\right)\left[\beta_{n+1}(1+2\alpha_n) - \beta_n(1+2\alpha_n)\right] \\ -\left(g_n^i g_{n+1}^r - g_n^r g_{n+1}^i\right)(\alpha_n + \alpha_{n+1} + 2\alpha_n\alpha_{n+1} + 2\beta_n\beta_{n+1}) \end{array} \right\} \tag{4.23}$$

4.1.2 Radiation Force of Quasi-Gaussian Standing Wave on Sphere

When two Gaussian sound waves of the same frequency and different amplitudes propagate along the same axis, a quasi-Gaussian standing wave field will be formed.

As shown in Fig. 4.2, the velocity potential of the incident quasi-Gaussian standing wave can be expressed as:

$$\Phi_i = e^{-(x^2+y^2)/W_0^2}\left[A e^{ik(z+h)} + B e^{-ik(z+h)}\right] e^{-i\omega t} \tag{4.24}$$

The first and second terms in the above formula represent the Gaussian traveling waves propagating in the positive and negative directions along the z-axis, and their velocity potential amplitudes are real numbers A and B (let $A \geq B$).

$$\text{Define:} \delta_n' = A e^{ikh} + (-1)^n B e^{-ikh} \tag{4.25}$$

Then the velocity potential of the incident wave can be expressed as:

$$\Phi_i = \sum_{n=0}^{\infty} (2n+1) i^n G_n \delta_n' j_n(kr) P_n(\cos\theta) e^{-i\omega t} \tag{4.26}$$

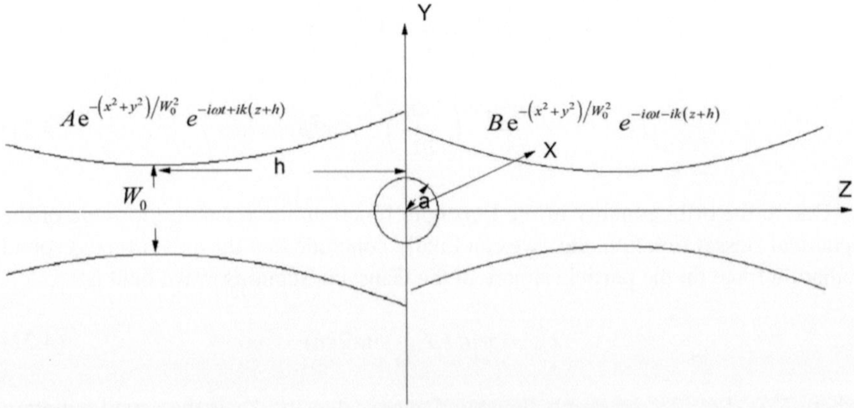

Fig. 4.2 Schematic diagram of the quasi-Gaussian standing wave field

$$R_n = \mathrm{Re}\left[G_n\delta_n' i^n (U_n + iV_n)(\cos\omega t - i\sin\omega t)\right] \tag{4.27}$$

Following the simplification method of the previous sections, we can get:

$$\begin{aligned}\langle R_n \cdot R_{n+1}\rangle = &(-1)^{n+1}AB\left[\left(g_n^i g_{n+1}^i + g_n^r g_{n+1}^r\right)(U_nU_{n+1} + V_nV_{n+1})\right.\\ &\left.+\left(g_n^i g_{n+1}^r - g_n^r g_{n+1}^i\right)(V_nU_{n+1} - U_nV_{n+1})\right]\\ &\sin(2kh) + \frac{A^2 - B^2}{2}\left[\left(g_n^i g_{n+1}^r - g_n^r g_{n+1}^i\right)(U_nU_{n+1} + V_nV_{n+1})\right.\\ &\left.+\left(g_n^i g_{n+1}^i + g_n^r g_{n+1}^r\right)(U_nV_{n+1} - V_nU_{n+1})\right]\end{aligned} \tag{4.28}$$

$$\begin{aligned}\left\langle R_n' \cdot R_{n+1}'\right\rangle = &(-1)^{n+1}AB\left[\left(g_n^i g_{n+1}^i + g_n^r g_{n+1}^r\right)\left(U_n'U_{n+1}' + V_n'V_{n+1}'\right)\right.\\ &\left.+\left(g_n^i g_{n+1}^r - g_n^r g_{n+1}^i\right)\left(V_n'U_{n+1}' - U_n'V_{n+1}'\right)\right]\\ &\sin(2kh) + \frac{A^2 - B^2}{2}\left[\left(g_n^i g_{n+1}^r - g_n^r g_{n+1}^i\right)\left(U_n'U_{n+1}' + V_n'V_{n+1}'\right)\right.\\ &\left.+\left(g_n^i g_{n+1}^i + g_n^r g_{n+1}^r\right)\left(U_n'V_{n+1}' - V_n'U_{n+1}'\right)\right]\end{aligned} \tag{4.29}$$

$$\begin{aligned}\left\langle R_n \cdot R_{n+1}'\right\rangle = &(-1)^{n+1}AB\left[\left(g_n^i g_{n+1}^i + g_n^r g_{n+1}^r\right)\left(U_nU_{n+1}' + V_nV_{n+1}'\right)\right.\\ &\left.+\left(g_n^i g_{n+1}^r - g_n^r g_{n+1}^i\right)\left(V_nU_{n+1}' - U_nV_{n+1}'\right)\right]\\ &\sin(2kh) + \frac{A^2 - B^2}{2}\left[\left(g_n^i g_{n+1}^r - g_n^r g_{n+1}^i\right)\left(U_nU_{n+1}' + V_nV_{n+1}'\right)\right.\\ &\left.+\left(g_n^i g_{n+1}^i + g_n^r g_{n+1}^r\right)\left(U_nV_{n+1}' - V_nU_{n+1}'\right)\right]\end{aligned} \tag{4.30}$$

$$\left\langle R_n' \cdot R_{n+1}\right\rangle = (-1)^{n+1}AB\left[\left(g_n^i g_{n+1}^i + g_n^r g_{n+1}^r\right)\left(U_n'U_{n+1} + V_n'V_{n+1}\right)\right.$$

$$+\left(g_n^i g_{n+1}^r - g_n^r g_{n+1}^i\right)\left(V_n' U_{n+1} - U_n' V_{n+1}\right)\big]$$
$$\sin(2kh) + \frac{A^2 - B^2}{2}\big[\left(g_n^i g_{n+1}^r - g_n^r g_{n+1}^i\right)\left(U_n' U_{n+1} + V_n' V_{n+1}\right)$$
$$+\left(g_n^i g_{n+1}^i + g_n^r g_{n+1}^r\right)\left(U_n' V_{n+1} - V_n' U_{n+1}\right)\big] \tag{4.31}$$

$$\left\langle \frac{\partial R_n}{\partial t} \cdot \frac{\partial R_{n+1}}{\partial t} \right\rangle = \omega^2 \langle R_n \cdot R_{n+1} \rangle \tag{4.32}$$

where: $g_n^r = \mathrm{Re}[G_n]$, $g_n^i = \mathrm{Im}[G_n]$ represents the real and imaginary parts of the Gaussian wave expansion factor G_n respectively.

Thus, the radiation force of the quasi-Gaussian standing wave field on the spherical particle F_{qst} can be obtained:

$$F_{qst} = \pi a^2 E Y_{qst} \tag{4.33}$$

The radiation force function Y_{qst} is [12]:

$$Y_{qst} = \frac{8}{(ka)^2} \sum_{n=0}^{\infty} (n+1)$$
$$\left\{ \begin{array}{l} (-1)^{n+1} \frac{B}{A} \left[\begin{array}{l} \left(g_n^r g_{n+1}^r + g_n^i g_{n+1}^i\right)\left[\beta_{n+1}(1+2\alpha_n) - \beta_n(1+2\alpha_n)\right] \\ -\left(g_n^i g_{n+1}^r - g_n^r g_{n+1}^i\right)(\alpha_n + \alpha_{n+1} + 2\alpha_n\alpha_{n+1} + 2\beta_n\beta_{n+1}) \end{array} \right] \sin(2kh) \\ -\frac{A^2 - B^2}{2A^2} \left[\begin{array}{l} \left(g_n^i g_{n+1}^r - g_n^r g_{n+1}^i\right)\left[\beta_{n+1}(1+2\alpha_n) - \beta_n(1+2\alpha_n)\right] \\ +\left(g_n^r g_{n+1}^r + g_n^i g_{n+1}^i\right)(\alpha_n + \alpha_{n+1} + 2\alpha_n\alpha_{n+1} + 2\beta_n\beta_{n+1}) \end{array} \right] \end{array} \right\} \tag{4.34}$$

By comparing the radiation force function Y_G in the Gaussian traveling wave obtained in Chap. 3 and the radiation force function Y_{st} in the Gaussian standing wave obtained in this chapter, with formulas (3.90) and (4.23), we can conclude that the radiation force function Y_{qst} of the quasi-Gaussian standing wave field on spherical particles can be expanded into the relationship between Y_G and Y_{st}.

$$Y_{qst} = \left(1 - \frac{B^2}{A^2}\right) \cdot Y_G + \frac{B}{A} \cdot Y_{st} \cdot \sin(2kh) \tag{4.35}$$

We can also derive the relationship between the radiation forces of various sound waves:

$$F_{qst} = F_p\left[\left(1 - \frac{B^2}{A^2}\right) + \frac{B}{A} \cdot R_{ps} \cdot \sin(2kh)\right]$$
$$R_{ps} = \frac{Y_{st}}{Y_G} \tag{4.36}$$

$$F_G = \pi a^2 E Y_G \tag{4.37}$$

From formula (4.36), it can be seen that when $A = B$, the radiation force F_{qst} of the quasi-Gaussian standing wave field on the particle sphere is consistent with the radiation force F_{st} of the Gaussian standing wave field on the particle sphere. When $B = 0$, the radiation force F_{qst} of the quasi-Gaussian standing wave field on the particle sphere is consistent with the radiation force F_G of the Gaussian traveling wave on the particle sphere. This is in line with the theory and formulas.

4.1.3 Radiation Force of Gaussian Standing Wave Field on a Sphere Deviating from the Focus Center

As shown in Fig. 4.3, when the position of the particle sphere deviates from the center point h_0 of the Gaussian standing wave field, according to the previous calculation derivation method, the acoustic radiation force function can be simplified into the following formula [12], and the derivation process is not repeated here.

$$F_{st} = \pi a^2 E Y_{st} \cdot \sin(2kh) \tag{4.38}$$

$$Y_{st} = -\frac{8}{(ka)^2}\sum_{n=0}^{\infty}(-1)^n \left\{ \begin{array}{l} \mathrm{Re}\left[G_n G_{n+1}^* \cdot (\cos(kh_0) - i\sin(kh_0))\right] \cdot (\beta_{n+1}(1+2\alpha_n) - \beta_n(1+2\alpha_n)) \\ -\mathrm{Im}\left[G_n G_{n+1}^* \cdot (\cos(kh_0) - i\sin(kh_0))\right] \cdot (\alpha_n + \alpha_{n+1} + 2\alpha_n\alpha_{n+1} + 2\beta_n\beta_{n+1}) \end{array} \right\} \tag{4.39}$$

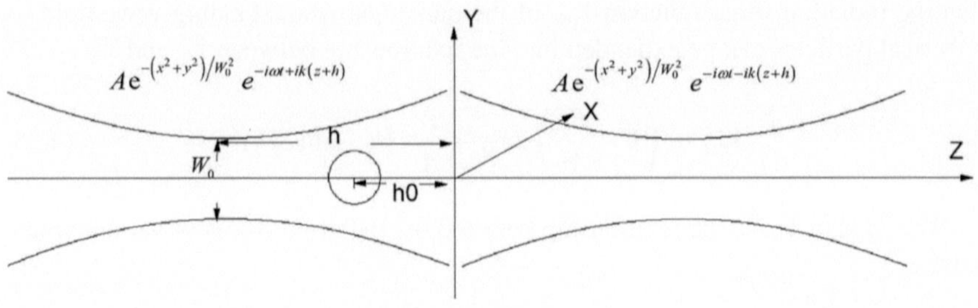

Fig. 4.3 Acoustic radiation force of Gaussian standing wave field on a sphere deviating from the center

4.2 Simulation and Discussion

In this section, we calculate and analyze several theoretical models of acoustic radiation force functions in Gaussian standing wave fields, and discuss the radiation force properties of two types of fluid spheres, two types of elastic spheres, and one type of viscoelastic sphere. Table 4.1 lists the properties of all materials used in this section [12]. The simulation calculation uses the method of finite series expansion, as introduced in Chap. 3. To improve the accuracy of the theory, the beam step rate $\Delta(ka) = 10^{-4}$ used in the following simulation figures, and the truncation number n in the radiation force function formulas (4.23), (4.24) and (4.39) to be solved is $G_n < 0.0001$.

4.2.1 *Acoustic Radiation Force of Gaussian Standing Wave on Liquid Spheres in Air*

In the application of acoustic radiation force fluid dynamics, manipulating liquid particles in air is an important application, especially for simulating the acoustic field environment under low gravity conditions. Under these conditions, the liquid medium will maintain a spherical shape. Since the density of water is about 813 times that of air, and the acoustic impedance of water and air does not match, water droplets in the air medium can be approximated as rigid objects in water [7]. Through computational simulation, the acoustic radiation force function of the rigid sphere in water under the Gaussian standing wave field is exactly the same as the acoustic radiation force function of the water droplet in air. Figure 4.4 shows the variation curve of the acoustic radiation force function Y_{st} in the range of the horizontal coordinate $0 \leq ka \leq 10$.

Table 4.1 Acoustic properties of different materials

Material	Density $(10^3\ \text{k/m}^3)$	Longitudinal wave speed (m/s)	Shear wave speed (m/s)	Normalized longitudinal wave attenuation coefficient γ_1	Normalized shear wave attenuation coefficient γ_2
Air	0.00123	340	–	–	–
Water	1	1500	–	–	–
Oleic acid	0.938	1450	–	–	–
Animal fat	0.8	1400	–	–	–
Poly(methyl methacrylate) (PMMA)	1.19	2680	1380	0.0119	0.0257
Stainless steel	7.9	5240	2978	–	–
Brass	8.1	3830	2050	–	–

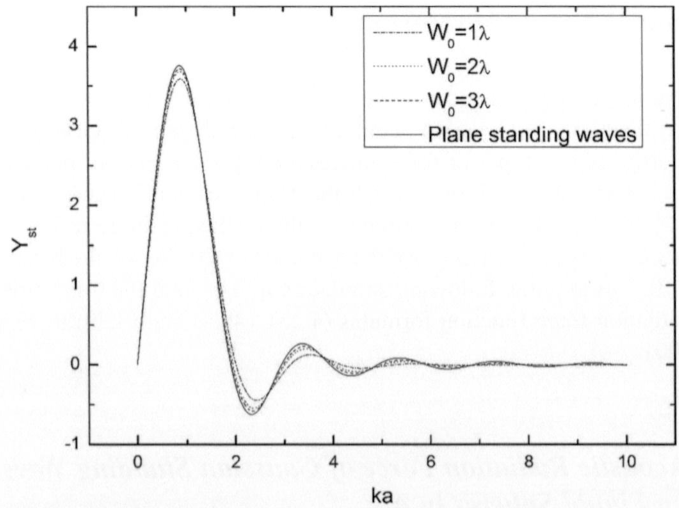

Fig. 4.4 Acoustic radiation force function of Gaussian standing wave field on a water droplet in air (or a rigid sphere in water)

From the figure, it can be seen that when $ka < 1$, the acoustic radiation force increases rapidly with the increase of ka. The peak at low ka values is determined by the characteristic properties of the interaction between the standing wave and the rigid sphere. When $ka > 1$, the acoustic radiation force drops sharply to 0, then $Y_{st} - ka$ curve shows a series of obvious peaks and troughs, which are determined by the resonance characteristics of the rigid sphere. As ka further increases, these peak values and trough values gradually decrease, and this result is similar to the radiation force function result of the plane standing wave field on the sphere [6].

At the same time, Fig. 4.4 also shows the acoustic radiation force function of the Gaussian standing wave under different beam waist widths. From the figure, it can be seen that the beam waist width of the Gaussian standing wave also has a certain impact on the acoustic radiation force function. As the beam waist width W_0 increases, the peak values and trough values on the curve are also increasing. When the beam waist width is much larger than the wavelength of the incident wave in the medium ($W_0 > 3\lambda$), the acoustic radiation force function of the Gaussian standing wave tends to the form of a plane standing wave, and this result is also consistent with previous research on Gaussian traveling waves [11, 13].

4.2.2 Acoustic Radiation Force of Gaussian Standing Wave on Liquid Spheres in Water

In bioengineering applications, the use of sound waves to manipulate cells or particles is also gaining more and more attention. Here, we use oleic acid and animal fat to

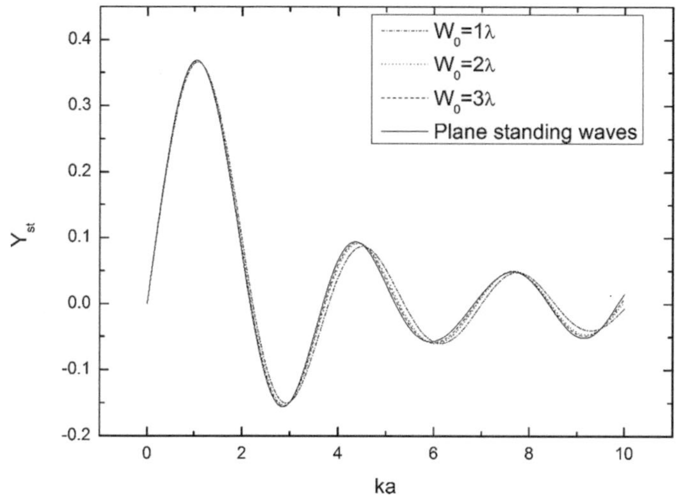

Fig. 4.5 Acoustic radiation force function of an oleic acid sphere in water under Gaussian standing wave field

simulate and calculate fat particles. Figure 4.5 shows the acoustic radiation function produced by the Gaussian standing wave field on oleic acid spherical particles under different beam waist widths. The dimensionless frequency range is $0 \leq ka \leq 10$. From the figure, it can be seen that, $Y_{st} - ka$ curve reaches a peak near $ka = 1$, and then similar to the rigid sphere, as ka increases, a series of peaks and troughs also appear. At the same time, by comparison, these peaks and troughs are more pronounced than those of the previous rigid sphere's acoustic radiation force function. Similarly, as the waist width of the Gaussian standing wave increases, these peaks and troughs also increase.

Figure 4.6 shows the radiation force function curve of the Gaussian standing wave field acting on a fat sphere. By comparing Figs. 4.5 and 4.6, it can be seen that due to the different acoustic properties of the two types of spherical particles, the acoustic radiation force function curve is different from that of oleic acid, especially in terms of the amplitude of the acoustic radiation force: the acoustic radiation force of the fat sphere is relatively larger than that of the oleic acid sphere, but the acoustic radiation forces of both liquid spheres are much smaller than the acoustic radiation force of the previous rigid sphere (Fig. 4.4).

4.2.3 Acoustic Radiation Force of Viscoelastic Spheres in Water Under Gaussian Standing Wave Field

Polymer materials also play a significant role in the application of acoustic radiation force, and polymer particles are widely used in drug delivery technology.

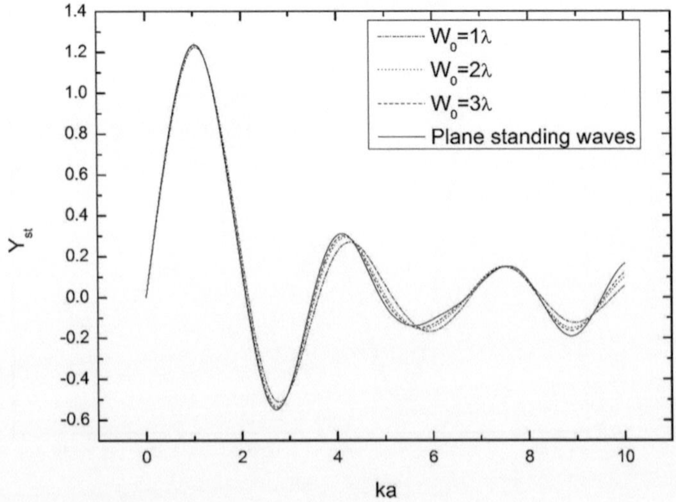

Fig. 4.6 Acoustic radiation force function of a fat sphere in water under Gaussian standing wave field

In the sound field, polymer viscoelastic particles will absorb the energy of sound waves. Referring to the solution method of the elastic sphere boundary condition in Sect. 3.1.1, the absorption of sound waves by particles should be considered in the viscoelastic boundary condition, and the wave number model should be reconstructed [13]. For viscoelastic materials, this is specifically reflected in the normalized absorption coefficients of longitudinal waves γ_1 and transverse waves γ_2, which are replaced by $\tilde{x}_l$ and $\tilde{x}_s$ respectively in the elastic boundary conditions: $\tilde{x}_l = k_l a(1 - j\gamma_1), \tilde{x}_s = k_s a(1 - j\gamma_s)$ replaces $x_l = k_l a$ and $x_s = k_s a$.

Figure 4.7 shows the acoustic radiation force function curve of polymethyl methacrylate (PMMA) microspheres in water under a Gaussian standing wave field. As can be seen from the figure, a series of peaks and troughs also appear, but compared with the previous curves, the peaks and troughs here are more pronounced and sharp, indicating that the compressibility of PMMA microspheres makes the resonance characteristics more pronounced. In addition, if the absorption of sound waves by viscoelastic materials is considered, the peaks and troughs of the acoustic radiation force curve decrease at all positions except for $ka \sim 8$. Therefore, the acoustic radiation force is considered to be frequency-dependent when describing the characteristics of polymer materials.

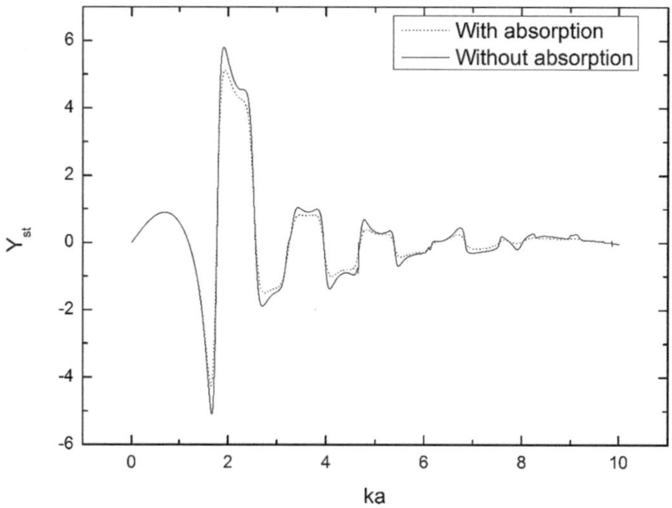

Fig. 4.7 Acoustic radiation force function of a PMMA sphere in water under Gaussian standing wave field

4.2.4 Acoustic Radiation Force of Solid Elastic Spheres in Water Under Gaussian Standing Wave Field

We also studied and discussed the acoustic radiation force acting on solid elastic materials, which can provide a theoretical basis for industrial microstructure equipment, such as the acoustic levitation effect of Gaussian standing waves on solid elastic spheres. Figures 4.8 and 4.9 show the acoustic radiation force function curves of elastic stainless steel spheres and copper spheres under a Gaussian standing wave field, respectively. It can be seen that the $Y_{st} - ka$ curve also shows a series of peaks and troughs according to the resonance frequencies of the elastic vibrations of the two materials. At the same time, the waist width of the Gaussian standing wave has an impact on the peaks and troughs, and this impact decreases as the waist width increases. When the waist width is much larger than the wavelength of the incident wave ($W_0 > 3\lambda$), the acoustic radiation force function of the Gaussian standing wave is consistent with that of the plane standing wave.

4.2.5 Acoustic Radiation Force of Stainless Steel Spheres in Water Under Quasi-Gaussian Standing Wave Field

Figure 4.10 shows the acoustic radiation force function curve of the quasi-Gaussian standing wave field acting on stainless steel spheres. At this time, the waist width of the Gaussian column standing wave is $W_0 = 3\lambda$, and the ratio of the amplitude of

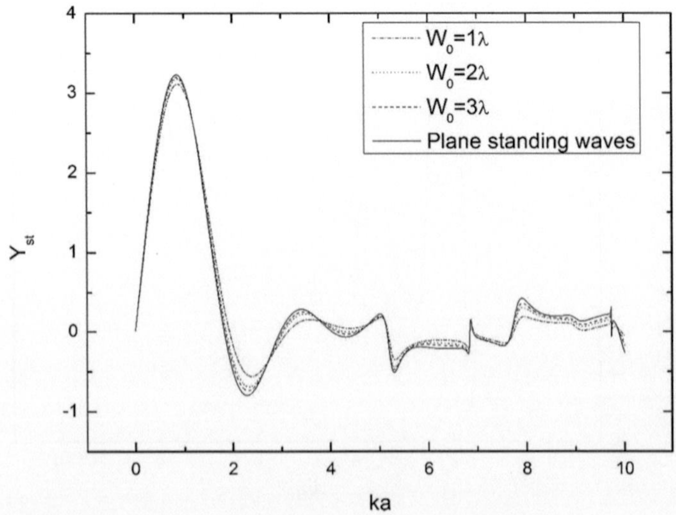

Fig. 4.8 Acoustic radiation force function of a stainless steel sphere in water under Gaussian standing wave field

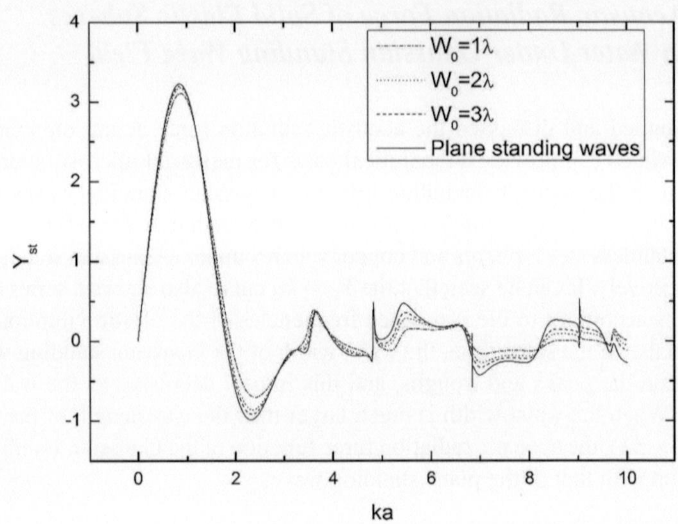

Fig. 4.9 Acoustic radiation force function of a copper sphere in water under Gaussian standing wave field

the Gaussian wave propagating in the negative z direction to that in the positive z direction is $\frac{B}{A}$. It can be seen from the figure that as the ratio $\frac{B}{A}$ decreases, the peak value of the acoustic radiation force curve at $ka = 1$ also decreases, but the subsequent peaks and troughs increase, making the acoustic radiation force function more similar to the Gaussian traveling wave acoustic radiation force function. Moreover, it can be

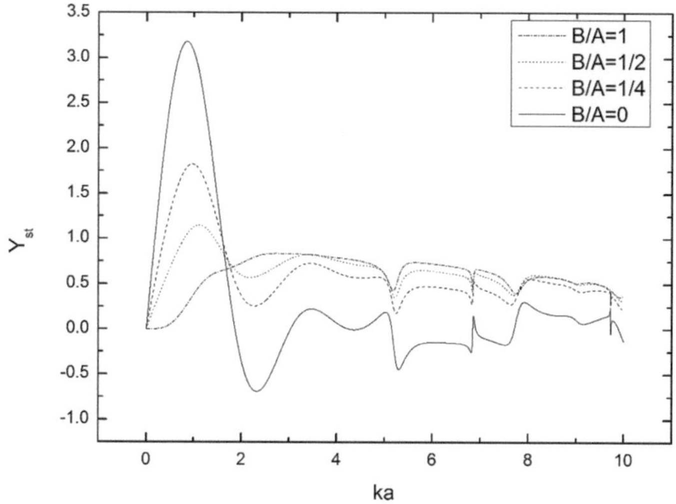

Fig. 4.10 Radiation force function of quasi-Gaussian standing wave field on a stainless steel ball in water

seen that when the amplitude ratio is $B = A$, the acoustic radiation force F_{qst} of the quasi-Gaussian standing wave field completely coincides with the radiation force F_{st} of the Gaussian standing wave field. When the amplitude ratio is $\frac{B}{A} = 0$, the acoustic radiation force F_{qst} of the quasi-Gaussian standing wave field completely coincides with the radiation force F_G of the Gaussian traveling wave field [11]. The simulation results are consistent with the theoretical predictions.

According to formula (4.36), the curve of R_{ps} value is calculated. Since $Y_{st} = Y_p = 0$, R_{ps} is meaningless at ka $= 0$. When $0 < ka \leq 1$, because Y_{st} increases sharply with the increase of ka, the value of R_{ps} is very large. Figure 4.11 depicts the ratio of acoustic radiation force functions experienced by the stainless steel ball in the range of $1 \leq ka \leq 10$ in the Gaussian standing wave field and the Gaussian traveling wave field. As can be seen from Fig. 4.10, the acoustic radiation force function curve of the Gaussian standing wave field at the position of $ka \rightarrow 1$ is much larger than that of the Gaussian traveling wave field, so the value of $ka \rightarrow 1$ is initially quite large near R_{ps}. Near ka $= 1.5$, the acoustic radiation force function curves of the Gaussian standing wave field and the traveling wave field intersect, so in Fig. 4.11, $R_{ps} = 1$ is at the same position. Similar to the discussion on Fig. 4.10, according to the resonance mode of particle material, a series of peaks and troughs appear at the same position. R_{ps} also has a zero point at the position of $Y_{st} = 0$ in Fig. 4.10.

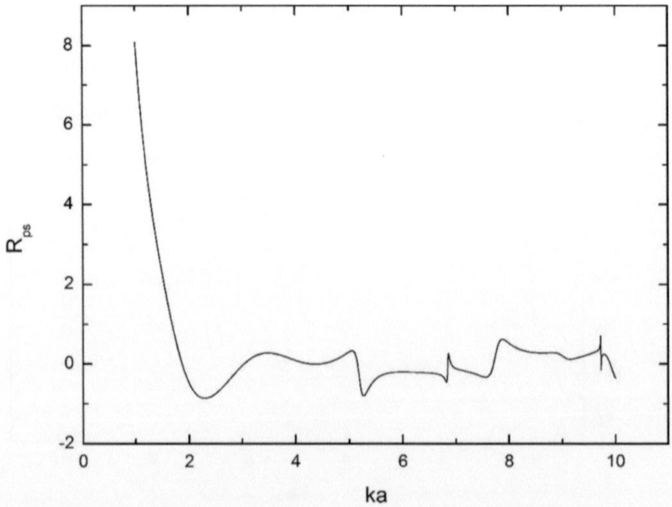

Fig. 4.11 R_{ps} curve for a stainless steel ball for Quasi-Gaussian standing wave

4.2.6 Separation of Different Spherical Particles by Gaussian Standing Wave Field

According to formula (4.39), Fig. 4.12 shows the radiation force function experienced by stainless steel spherical particles and PMMA spherical particles at different positions in the Gaussian standing wave field. At this time, we take ka = 2, and the beam waist width is $W_0 = 3\lambda$.

It can be clearly seen that the acoustic radiation force function of the two particles in the Gaussian standing wave field is sinusoidal, indicating that the particles will be pulled or pushed to the node position of the standing wave field, that is, the zero point of the acoustic radiation force function. Therefore, due to the different resonance modes of the two spherical particles, at ka = 2, the acoustic radiation force function experienced by the PMMA ball is much larger than that of the stainless steel ball. At the same time, the zero point positions of the radiation force functions experienced by different particles are also different. Therefore, this conclusion indicates the possibility of manipulating and separating different spherical cells or particles using a Gaussian standing wave field.

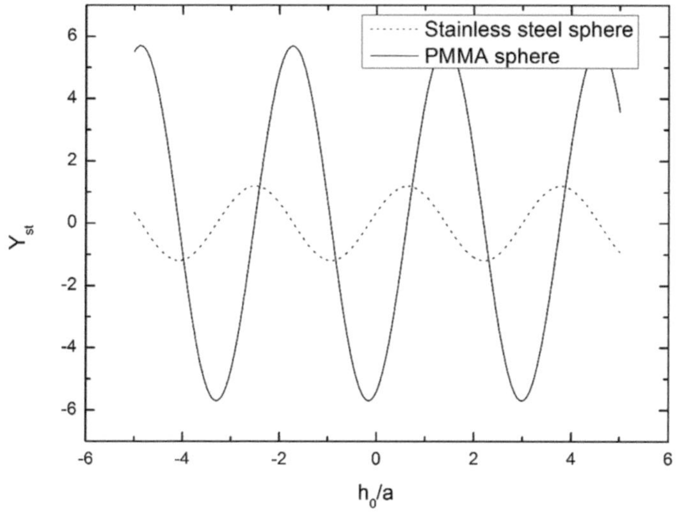

Fig. 4.12 Radiation force function of Gaussian standing wave field on different balls

References

1. Shi JJ, Daniel A, Mao X et al (2009) Acoustical tweezers: patterning cells and micro particles using standing surface acoustic waves (SSAW). Lab Chip 9:2890–2895
2. Shi JJ, Shahrzad Y, Lin S et al (2011) Three-dimensional continuous particle focusing in a microfluidic channel via standing surface acoustic waves (SSAW). Lab Chip 11:2319–2324
3. Tran SBQ, Marmottant P, Thibault P (2012) Fast acoustic tweezers for the two-dimensional manipulation of individual particles in microfluidic channels. Appl Phys Lett 101:114103
4. Meng L, Cai FY (2012) Precise and programmable manipulation of microbubbles by two-dimensional standing surface acoustic waves. Appl Phys Lett 100:173701
5. Ding XY, Huang TJ (2012) On-chip manipulation of single microparticles, cells, and organisms using surface acoustic waves. Proc Natl Acad Sci 109:11105–11109
6. Hasegawa T (1979) Acoustic radiation force on a sphere in a quasistationary wave field-theory. J Acoust Soc Am 65:32–40
7. Mitri FG (2008) Acoustic radiation force on a sphere in standing and quasi-standing zero-order Bessel beam tweezers. Ann Phys 323:1604–1620
8. Mitri FG (2009) Acoustic radiation force of high-order Bessel beam standing wave tweezers on a rigid sphere. Ultrasonics 49:794–798
9. Wu JR (1991) Acoustical tweezers. J Acoust Soc Am 89:2140–2143
10. Zhang XF, Zhang GB (2012) Acoustic radiation force of a Gaussian beam incident on spherical particles in water. Ultrasound Med Biol 38:2007–2017
11. Zhang XF, Song ZG, Chen DM et al (2015) Finite series expansion of a Gaussian beam for the acoustic radiation force calculation of cylindrical particles in water. J Acoust Soc Am 137:1826–1833
12. Wu RR, Cheng KX, Liu XZ et al (2016) Study of axial acoustic radiation force on a sphere in a Gaussian quasi-standing field. Wave Motion 62:63–74
13. Hartmann B, Jarzynski J (1972) Ultrasonic hysteresis absorption in polymers. J Appl Phys 43:4304

Chapter 5
Solving the Acoustic Radiation Force on Multilayer Spherical Particles by Gaussian Plane Waves Based on Acoustic Scattering Method

When sound waves propagating in a fluid medium encounter obstacles, they reflect, refract, and scatter at the surface of the obstacles. Therefore, the interaction between objects in the sound field and the sound waves is the mutual interaction between the object and the incident sound waves, reflected sound waves, refracted sound waves, and scattered sound waves. By calculating the scattering coefficient of the object in the sound field, the acoustic radiation force can be solved according to the theory of sound wave radiation.

In recent years, a large amount of theoretical and experimental research has been conducted on the acoustic radiation force of a single ultrasonic beam on spherical objects [1–5]. Acoustic tweezers based on acoustic radiation force have become a renowned technology for capturing, classifying, and assembling microparticles (guiding and capturing tiny particles) [6–10], especially in cell and molecular biology, where particle manipulation plays a crucial role in studying the biomechanical properties of various cells and molecules. However, in previous literature, these particles are mostly single-layer spherical shell structures. In biomedical engineering and materials science, there are many particles with multilayer structures, such as nanoparticle drugs, cells, etc. In drug delivery applications [11–14], polymer microspheres filled with liquid drugs are used, and the microspheres are effectively manipulated by sound radiation, thereby precisely breaking the shell and used for the affected area. Ordinary cells are typical multilayer tissues composed of cell membranes, cytoplasm, and nuclei, and these layers have different acoustic properties. Therefore, the study of the acoustic radiation force of multilayer spherical shell structures will have more practical application value.

Gaussian beams are widely used to simulate sound fields and light wave fields that converge or diverge in the focal area. According to the source power and propagation distance of the Gaussian sound beam, particles can be captured to the focal area [15, 16]. This chapter proposes a model for calculating the acoustic radiation force on a spherically symmetric multilayer sphere by a Gaussian focusing beam. Then, the radiation force of isotropic double-layer and triple-layer spheres under different layer

X. Liu and H. Zheng, *Acoustic Radiation Force*, Advances in Acoustics,
https://doi.org/10.1007/978-981-95-3152-3_5

material radius ratios was studied. The research results show that the theory proposed in this chapter is more universal for the manipulation of multilayer spherical particles and provides a method for calculating the acoustic radiation force of multilayer spherical particles.

5.1 Spherical Function Expansion of Incident Gaussian Sound Beam

Let the Gaussian sound wave incident on a double-layer or triple-layer sphere immersed in a non-viscous fluid medium, and the wave propagates in the + z direction. The center position of the Gaussian beam is the origin of the coordinates. The center of the sphere is (0,0,0), which is the origin of the spherical coordinate system, as shown in Figs. 5.1 and 5.2, where a, b, and c are the outer, middle, and inner radii of the spherically symmetric double-layer or triple-layer sphere. For simplicity, this chapter assumes that the Gaussian sound beam is weakly focused. In addition, the phase front of the propagation in the Gaussian beam is almost equal to the plane wave near the beam waist. This means that it can be approximately regarded as a sound wave with a Gaussian amplitude distribution. The velocity potential of the incident sound wave can be written as:

$$\phi_i = \phi_0 \exp\left(-\frac{(x^2+y^2)}{w_0^2}\right)\exp(ikz) \tag{5.1}$$

Where ϕ_0 is the amplitude of the velocity potential, W_0 is the width of the Gaussian sound beam, let $x = r\sin\theta\cos\varphi$, $y = r\sin\theta\sin\varphi$, $z = r\cos\theta$, $s = \frac{1}{kW_0}$, $\rho = kr$

In the spherical coordinate system, the velocity potential of the incident wave can be represented by a general characteristic function expansion of the form [17]:

$$\phi_i = \phi_0 e^{-s^2\rho^2\sin^2\theta} e^{i\rho\cos\theta} = \phi_0 \sum_{n=0}^{\infty} Q_n i^n (2n+1) j_n(\rho) P_n(\cos\theta)\exp(i\omega t) \tag{5.2}$$

Here:

$$Q_{2n} = \frac{\Gamma(n+1)}{\Gamma\left(n+\frac{1}{2}\right)} \sum_{j=0}^{n\sum(-4s^2)^j} \frac{\Gamma\left(n+j+\frac{1}{2}\right)}{\Gamma(n-j)!j!} \tag{5.3}$$

$$Q_{2n+1} = \frac{\Gamma(n+1)}{\Gamma\left(n+\frac{3}{2}\right)} \sum_{j=0}^{n\sum(-4s^2)^j} \frac{\Gamma\left(n+j+\frac{3}{2}\right)}{\Gamma(n-j)!j!} \tag{5.4}$$

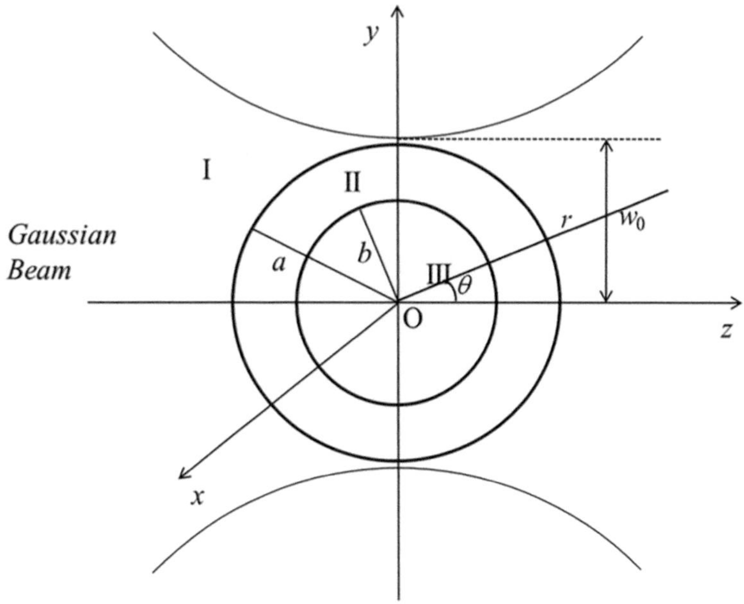

Fig. 5.1 Geometric coordinates of Gaussian sound wave and double-layer spherical model

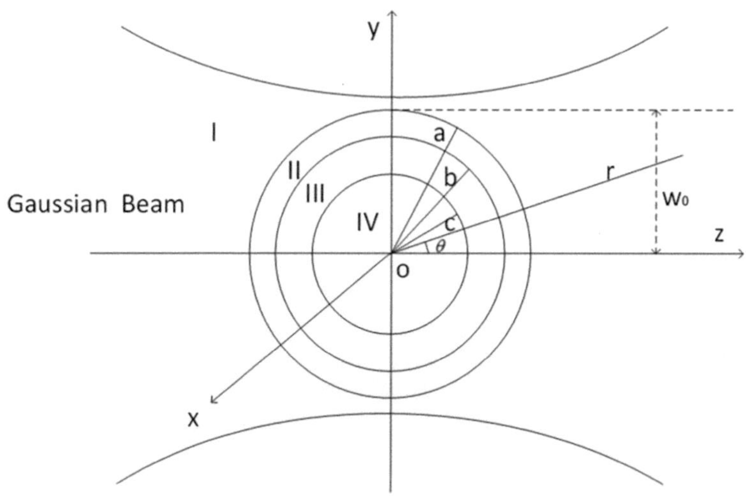

Fig. 5.2 Geometric coordinates of Gaussian sound beam and triple-layer spherical model

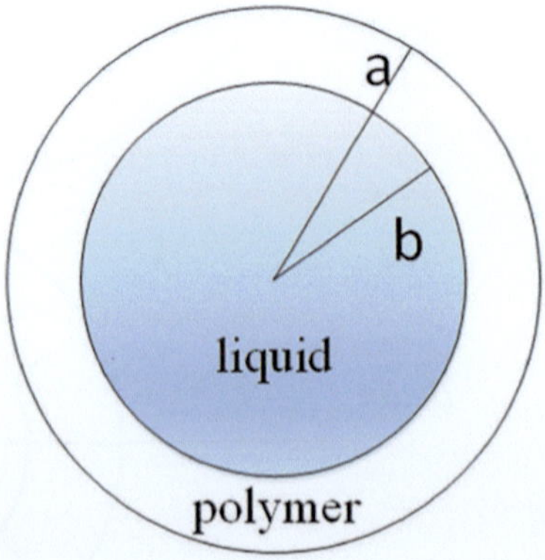

Fig. 5.3 Double-layer solid–liquid model

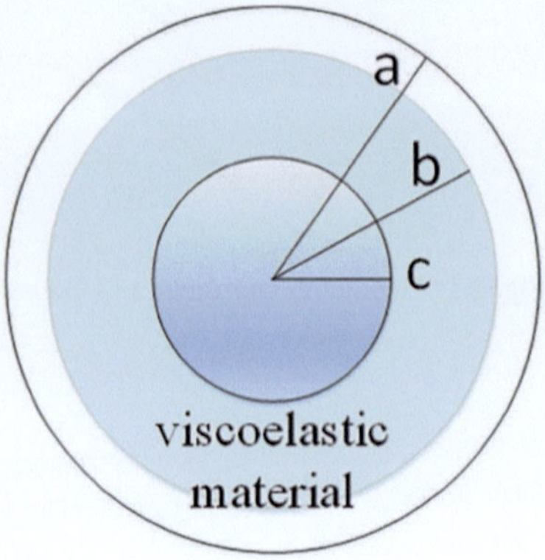

Fig. 5.4 Triple-layer viscoelastic model

where $j_n(\cdot)$ is the nth order spherical Bessel function, $P_n(\cdot)$ is the nth order Legendre function, $\Gamma(\cdot)$ is the Gamma function comes from the equation. From Eqs. (5.3) and (5.4) the beam coefficient Q_n can be given by the dimensionless parameter $1/kW_0$. Different W_0 and λ ratios, different convergent terms.

To verify the effectiveness of the finite series method, we performed calculations for the exact solution and the finite series solution of an incident Gaussian sound beam. The results show that when $W_0 \geq 2\lambda$, the beam calculated by the finite series is consistent with the exact solution [3]

5.2 Scattering Coefficient of Gaussian Wave by Double-Layer Spherical Particles

Considering a Gaussian wave symmetrically incident about the z-axis onto a two-layer sphere suspended in a liquid, we divide the entire region into three areas, as shown in Fig. 5.1. The velocity potential expression of the incident wave in spherical coordinates can be written as:

$$\phi_i = \phi_0 \sum_{n=0}^{\infty} \Lambda_n (-i)^n (2n+1) j_n(k_0 r) P_n(\cos\theta) \tag{5.5}$$

where ϕ_0 represents the amplitude of the velocity potential of the incident wave, j_n is the first kind of spherical Bessel function, k_0 is the wave number of the incident wave in region I, P_n represents the Legendre function.

When the incident wave hits the double-layer sphere shell, reflection occurs, and the velocity potential expression of the reflected wave is:

$$\phi_s = \phi_0 \sum_{n=0}^{\infty} \Lambda_n (2n+1)(-i)^n A_n h_n^{(1)}(k_0 r) P_n(\cos\theta) \tag{5.6}$$

where A_n is the reflection coefficient determined by the boundary conditions.

Therefore, in region I, the total velocity potential expression of the sound wave is:

$$\Phi_1 = \phi_i + \phi_s = \phi_0 \sum_{n=0}^{\infty} \Lambda_n (2n+1)(-i)^n \left(j_n(k_0 r) + A_n h_n^{(1)}(k_0 r) \right) P_n(\cos\theta) \tag{5.7}$$

The incident wave from region I refracts to region II and undergoes mode conversion into two forms: transverse and longitudinal waves. Therefore, in region II, the total velocity potential expression of the longitudinal sound wave is:

$$\Phi_2 = \phi_0 \sum_{n=0}^{\infty} \Lambda_n (2n+1)(-i)^n (B_n j_n(k_1 r) + C_n n_n(k_1 r)) P_n(\cos\theta) \tag{5.8}$$

where n_n is the Neumann function, k_1 is the wave number of the longitudinal wave in region II, B_n, C_n is the unknown coefficient determined by the boundary conditions.

The total velocity potential expression of the transverse sound wave is:

$$\Psi_2 = \phi_0 \sum_{n=0}^{\infty} \Lambda_n (2n+1)(-i)^n (D_n j_n(k_2 r) + E_n n_n(k_2 r)) \frac{dP_n(\cos\theta)}{d\theta} \tag{5.9}$$

where k_2 is the wave number of the transverse wave in region II, D_n, E_n is the unknown coefficient determined by the boundary conditions.

In region III, the velocity potential expression of the longitudinal wave is:

$$\Phi_3 = \phi_0 \sum_{n=0}^{\infty} \Lambda_n (2n+1)(-i)^n (F_n j_n(k_3 r)) P_n(\cos\theta) \tag{5.10}$$

where k_3 is the wave number of the longitudinal wave in region III, F_n is the unknown coefficient determined by the boundary conditions.

$$\Psi_3 = \phi_0 \sum_{n=0}^{\infty} \Lambda_n (2n+1)(-i)^n (G_n j_n(k_4 r)) \frac{dP_n(\cos\theta)}{d\theta} \tag{5.11}$$

where k_4 is the wave number of the longitudinal wave in region III, G_n is the unknown coefficient determined by the boundary conditions.

Next, we calculate the scattering coefficient A_n using the boundary conditions:

At the outer boundary of the double-layer sphere (i.e., between region I and region II, at $r = a$), the normal displacement and normal stress should be continuous, and since region I is an ideal liquid, the tangential stress should be zero.

$$\begin{aligned} &\text{The normal displacement is continuous:} U_r^{(1)} = U_r^{(2)} \\ &\text{The normal stress is continuous:} \sigma_r^{(1)} = \sigma_r^{(2)} \\ &\text{The tangential stress is 0: } \sigma_\theta^{(2)} = 0 \end{aligned} \tag{5.12}$$

where U represents displacement, σ represents stress, the superscript represents the region, r represents the normal direction, θ represents the tangential direction.

At the inner boundary of the double-layer sphere (i.e., between region II and region III, at $r = b$), the normal displacement, tangential displacement, normal stress, and tangential stress should all be continuous.

$$\begin{aligned} &\text{The normal displacement is continuous : } U_r^{(2)} = U_r^{(3)} \\ &\text{The tangential displacement is continuous : } U_\theta^{(2)} = U_\theta^{(3)} \\ &\text{The normal stress is continuous:} \sigma_r^{(2)} = \sigma_r^{(3)} \\ &\text{The tangential stress is continuous:} \sigma_\theta^{(2)} = \sigma_\theta^{(3)} \end{aligned} \tag{5.13}$$

The relationship between displacement and velocity potential is:

$$\begin{aligned} U_r &= \frac{\partial \Phi}{\partial r} + \frac{1}{r\sin\theta} \frac{\partial \Psi}{\partial \theta} \\ U_\theta &= \frac{1}{r}\frac{\partial \Phi}{\partial \theta} - \frac{1}{r}\frac{\partial (r\Psi)}{\partial r} \end{aligned} \tag{5.14}$$

The relationship between stress and velocity potential is:

$$\begin{aligned} \sigma_r &= 2\mu \frac{\partial U_r}{\partial r} + \lambda(\nabla \cdot U) \\ \sigma_\theta &= \mu\left(\frac{1}{r}\frac{\partial U_r}{\partial \theta} + \frac{\partial U_\theta}{\partial r} - \frac{U_\theta}{r}\right) \end{aligned} \tag{5.15}$$

where μ, λ represents the Lame coefficients.

Substituting Eqs. (5.7)–(5.11) and Eqs. (5.14)–(5.15) into the boundary conditions Eqs. (5.12)–(5.13) results in 7 linear equations. The solutions can be obtained as follows:

$$A_n = \frac{\begin{vmatrix} \chi_1 & \chi_{12} & \chi_{13} & \chi_{14} & \chi_{15} & 0 & 0 \\ \chi_2 & \chi_{22} & \chi_{23} & \chi_{24} & \chi_{25} & 0 & 0 \\ 0 & \chi_{32} & \chi_{33} & \chi_{34} & \chi_{35} & 0 & 0 \\ 0 & \chi_{42} & \chi_{43} & \chi_{44} & \chi_{45} & \chi_{46} & \chi_{47} \\ 0 & \chi_{52} & \chi_{53} & \chi_{54} & \chi_{55} & \chi_{56} & \chi_{57} \\ 0 & \chi_{62} & \chi_{63} & \chi_{64} & \chi_{65} & \chi_{66} & \chi_{67} \\ 0 & \chi_{72} & \chi_{73} & \chi_{74} & \chi_{75} & \chi_{76} & \chi_{77} \end{vmatrix}}{\begin{vmatrix} \chi_{11} & \chi_{12} & \chi_{13} & \chi_{14} & \chi_{15} & 0 & 0 \\ \chi_{21} & \chi_{22} & \chi_{23} & \chi_{24} & \chi_{25} & 0 & 0 \\ 0 & \chi_{32} & \chi_{33} & \chi_{34} & \chi_{35} & 0 & 0 \\ 0 & \chi_{42} & \chi_{43} & \chi_{44} & \chi_{45} & \chi_{46} & \chi_{47} \\ 0 & \chi_{52} & \chi_{53} & \chi_{54} & \chi_{55} & \chi_{56} & \chi_{57} \\ 0 & \chi_{62} & \chi_{63} & \chi_{64} & \chi_{65} & \chi_{66} & \chi_{67} \\ 0 & \chi_{72} & \chi_{73} & \chi_{74} & \chi_{75} & \chi_{76} & \chi_{77} \end{vmatrix}} \tag{5.16}$$

where,

$$\chi_{11} = \frac{\rho_1}{\rho_2}\tilde{y}_{22}^2 h_n^{(1)}(y_1)$$

$$\chi_{12} = \left(2n(n+1) - \tilde{y}_{22}^2\right) j_n(\tilde{y}_{21}) - 4\tilde{y}_{21} j_n'(\tilde{y}_{21})$$

$$\chi_{13} = \left(2n(n+1) - \tilde{y}_{22}^2\right) n_n(\tilde{y}_{21}) - 4\tilde{y}_{21} n_n'(\tilde{y}_{21})$$

$$\chi_{14} = 2n(n+1)\left(\tilde{y}_{22} j_n'(\tilde{y}_{22}) - j_n(\tilde{y}_{22})\right)$$

$$\chi_{15} = 2n(n+1)\left(\tilde{y}_{22} n_n'(\tilde{y}_{22}) - n_n(\tilde{y}_{22})\right)$$

$$\chi_{21} = -y_1 h_n^{(1)'}(y_1)$$

$$\chi_{22} = \widetilde{y_{21}} j_n'(\widetilde{y_{21}})$$

$$\chi_{23} = \widetilde{y_{21}} n_n'(\widetilde{y_{21}})$$

$$\chi_{24} = n(n+1) j_n(\widetilde{y_{22}})$$

$$\chi_{25} = n(n+1) n_n(\widetilde{y_{22}})$$

$$\chi_{32} = 2\left(j_n(\widetilde{y_{21}}) - \widetilde{y_{21}} j_n'(\widetilde{y_{21}})\right)$$

$$\chi_{33} = 2\left(n_n(\widetilde{y_{21}}) - \widetilde{y_{21}} n_n'(\widetilde{y_{21}})\right)$$

$$\chi_{34} = 2\widetilde{y_{22}} j_n'(\widetilde{y_{22}}) + \left(\widetilde{y_{22}^2} - 2n(n+1) + 2\right) j_n(\widetilde{y_{22}})$$

$$\chi_{35} = 2\widetilde{y_{22}} n_n'(\widetilde{y_{22}}) + \left(\widetilde{y_{22}^2} - 2n(n+1) + 2\right) n_n(\widetilde{y_{22}})$$

$$\chi_{42} = \widetilde{x_{21}} j_n'(\widetilde{x_{21}})$$

$$\chi_{43} = \widetilde{x_{21}} n_n'(\widetilde{x_{21}})$$

$$\chi_{44} = n(n+1) j_n(\widetilde{x_{22}})$$

$$\chi_{45} = n(n+1) n_n(\widetilde{x_{22}})$$

$$\chi_{46} = -x_{31} j_n'(x_{31})$$

$$\chi_{47} = -n(n+1) j_n(x_{32})$$

$$\chi_{52} = -j_n(\widetilde{x_{21}})$$

$$\chi_{53} = -n_n(\widetilde{x_{21}})$$

$$\chi_{54} = -\tilde{x}_{22} j_n'(\tilde{x}_{22}) - j_n(\tilde{x}_{22})$$

$$\chi_{55} = -\tilde{x}_{22} n_n'(\tilde{x}_{22}) - n_n(\tilde{x}_{22})$$

$$\chi_{56} = j_n(x_{31})$$

$$\chi_{57} = x_{32}j_n'(x_{32}) + j_n(x_{32})$$

$$\chi_{62} = \Omega\big((2n(n+1) - \tilde{x}_{22}^2)j_n(\tilde{x}_{21}) - 4\tilde{x}_{21}j_n'(\tilde{x}_{21})\big)$$

$$\chi_{63} = \Omega\big((2n(n+1) - \tilde{x}_{22}^2)n_n(\tilde{x}_{21}) - 4\tilde{x}_{21}n_n'(\tilde{x}_{21})\big)$$

$$\chi_{64} = 2n(n+1)\Omega\big(\tilde{x}_{22}j_n'(\tilde{x}_{22}) - j_n(\tilde{x}_{22})\big)$$

$$\chi_{65} = 2n(n+1)\Omega\big(\tilde{x}_{22}n_n'(\tilde{x}_{22}) - n_n(\tilde{x}_{22})\big)$$

$$\chi_{66} = 4x_{31}j_n'(x_{31}) - \big(2n(n+1) - x_{32}^2\big)j_n(x_{31})$$

$$\chi_{67} = 2n(n+1)\big(j_n(x_{32}) - x_{32}j_n'(x_{32})\big)$$

$$\chi_{72} = 2\big(j_n(\tilde{x}_{21}) - \tilde{x}_{21}j_n'(\tilde{x}_{21})\big)$$

$$\chi_{73} = 2\big(n_n(\tilde{x}_{21}) - \tilde{x}_{21}n_n'(\tilde{x}_{21})\big)$$

$$\chi_{74} = 2\tilde{x}_{22}j_n'(\tilde{x}_{22}) + \big(\tilde{x}_{22}^2 - 2n(n+1) + 2\big)j_n(\tilde{x}_{22})$$

$$\chi_{75} = 2\tilde{x}_{22}n_n'(\tilde{x}_{22}) + \big(\tilde{x}_{22}^2 - 2n(n+1) + 2\big)n_n(\tilde{x}_{22})$$

$$\chi_{76} = \frac{2}{\Omega}\big(x_{31}j_n'(x_{31}) - j_n(x_{31})\big)$$

$$\chi_{77} = -\frac{1}{\Omega}\big(2x_{32}j_n'(x_{32}) + \big(x_{32}^2 - 2n(n+1) + 2\big) + j_n(x_{32})\big)$$

$$\chi_1 = -\frac{\rho_1}{\rho_2}\tilde{y}_{22}^2 j_n(y_1)$$

$$\chi_2 = y_1 j_n'(y_1)$$

In the above equations, ρ_1, ρ_2 and ρ_3 represent the densities of the media in regions I, II, and III, respectively. $e = \frac{a}{b}$ represents the ratio of the outer radius a to the inner radius b of the double-layer sphere. $x = k_1 b$, $\tilde{x}_{21} = x\frac{c_1}{c_{21}}(1 + i\gamma_{21})$, $\tilde{x}_{22} = x\frac{c_1}{c_{22}}(1 + i\gamma_{22})$, $x_{31} = x\frac{c_1}{c_{31}}$, $x_{32} = x\frac{c_1}{c_{32}}$, c_1 represent the speed of sound in the ideal fluid in region I, c_{21}, c_{22} represent the speeds of longitudinal and transverse waves in the medium in region II, c_{31}, c_{32} represent the speeds of longitudinal and transverse waves in the medium in region III. γ_{21}, γ_{22} represent the absorption coefficients of the medium in region II for longitudinal and transverse waves. $y_1 = xe, \tilde{y}_{21} = \tilde{x}_{21}e, \tilde{y}_{22} = \tilde{x}_{22}e, y_{31} = x_{31}e, y_{32} = x_{32}e, \Omega = \frac{\rho_2}{\rho_3}\left(\frac{c_{22}}{c_{32}}\right)^2$.

5.3 Scattering Coefficient of Three-Layer Spherical Particles for Gaussian Waves

A Gaussian sound beam with a wavelength of λ_1 is scattered by a sphere in an infinite medium. Assuming the incident wave is axisymmetric, the external fluid has a density of ρ_1 and a speed of sound of c_1. In the spherical coordinate system, the velocity potential of the incident wave can be represented by the general eigenfunction expansion form shown in Eq. (5.1).

The scattered velocity potential should be represented as:

$$\phi_s = \phi_0 \sum_{n=0}^{\infty} Q_n(2n+1)(-i)^n A_n h_n^{(1)}(k_1 r) P_n(\cos\theta)\exp(i\omega t) \tag{5.17}$$

where, $h_n^{(1)}(\cdot)$ is the first kind Hankel function, A_n The scattering coefficient is determined by the boundary conditions, hence, the total scalar velocity potential outside the sphere can be written as:

$$\phi_1 = \phi_i + \phi_s = \phi_0 \sum_{n=0}^{\infty} Q_n(2n+1)(-i)^n \left(j_n(k_1 r) + A_n h_n^{(1)}(k_1 r)\right) P_n(\cos\theta)\exp(i\omega t) \tag{5.18}$$

The particle vector displacement in the shell of the elastic material U_i (i = 2, 3, and 4) is also assumed to be [1, 5].

$$U_i = -\nabla \Phi_i + \nabla \times \Psi_i \tag{5.19}$$

where the scalar potential Φ_i and vector potential Ψ_i can be expanded using the unknown coefficients B_n, C_n, D_n, E_n, F_n, G_n, H_n, I_n, K_n, L_n, as shown below:

$$\Phi_2 = \phi_0 \sum_{n=0}^{\infty} Q_n(2n+1)(-i)^n (B_n j_n(k_{21} r) + C_n n_n(k_{21} r)) P_n(\cos\theta)\exp(i\omega t) \tag{5.20}$$

$$\Psi_2 = \phi_0 \sum_{n=0}^{\infty} Q_n(2n+1)(-i)^n (D_n j_n(k_{22} r) + E_n n_n(k_{22} r)) \frac{dP_n(\cos\theta)}{d\theta}\exp(i\omega t) \tag{5.21}$$

$$\Phi_3 = \phi_0 \sum_{n=0}^{\infty} Q_n(2n+1)(-i)^n (F_n j_n(k_{31}r) + G_n n_n(k_{31}r)) P_n(\cos\theta) \exp(i\omega t) \quad (5.22)$$

$$\Psi_3 = \phi_0 \sum_{n=0}^{\infty} Q_n(2n+1)(-i)^n (H_n j_n(k_{32}r) + I_n n_n(k_{32}r)) \frac{dP_n(\cos\theta)}{d\theta} \exp(i\omega t) \quad (5.23)$$

$$\Phi_4 = \phi_0 \sum_{n=0}^{\infty} Q_n(2n+1)(-i)^n K_n j_n(k_{41}r) P_n(\cos\theta) \exp(i\omega t) \quad (5.24)$$

$$\Psi_4 = \phi_0 \sum_{n=0}^{\infty} Q_n(2n+1)(-i)^n L_n j_n(k_{42}r) \frac{dP_n(\cos\theta)}{d\theta} \exp(i\omega t) \quad (5.25)$$

$n_n(\cdot)$ is the second kind of spherical Bessel function, $k_{21}(k_{31}, k_{41})$ and $k_{22}(k_{32}, k_{42})$ are the wave numbers of longitudinal and shear waves in the shell medium (medium 2, 3, and 4), B_n, C_n, D_n, E_n, F_n, G_n, H_n, I_n, K_n, L_n are determined by the boundary conditions.

The boundary conditions derive 11 sets of linear equations. After complex and direct calculations, the scattering coefficients are determined as follows:

$$A_n = \frac{\begin{vmatrix} b_1 & q_{12} & q_{13} & q_{14} & q_{15} & 0 & 0 & 0 & 0 & 0 & 0 \\ b_2 & q_{22} & q_{23} & q_{24} & q_{25} & 0 & 0 & 0 & 0 & 0 & 0 \\ 0 & q_{32} & q_{33} & q_{34} & q_{35} & 0 & 0 & 0 & 0 & 0 & 0 \\ 0 & q_{42} & q_{43} & q_{44} & q_{45} & q_{46} & q_{47} & q_{48} & q_{49} & 0 & 0 \\ 0 & q_{52} & q_{53} & q_{54} & q_{55} & q_{56} & q_{57} & q_{58} & q_{59} & 0 & 0 \\ 0 & q_{62} & q_{63} & q_{64} & q_{65} & q_{66} & q_{67} & q_{68} & q_{69} & 0 & 0 \\ 0 & q_{72} & q_{73} & q_{74} & q_{75} & q_{76} & q_{77} & q_{78} & q_{79} & 0 & 0 \\ 0 & 0 & 0 & 0 & 0 & q_{86} & q_{87} & q_{88} & q_{89} & q_{810} & q_{811} \\ 0 & 0 & 0 & 0 & 0 & q_{96} & q_{97} & q_{98} & q_{99} & q_{910} & q_{911} \\ 0 & 0 & 0 & 0 & 0 & q_{106} & q_{107} & q_{108} & q_{109} & q_{1010} & q_{1011} \\ 0 & 0 & 0 & 0 & 0 & q_{116} & q_{117} & q_{118} & q_{119} & q_{1110} & q_{1111} \end{vmatrix}}{\begin{vmatrix} q_{11} & q_{12} & q_{13} & q_{14} & q_{15} & 0 & 0 & 0 & 0 & 0 & 0 \\ q_{21} & q_{22} & q_{23} & q_{24} & q_{25} & 0 & 0 & 0 & 0 & 0 & 0 \\ 0 & q_{32} & q_{33} & q_{34} & q_{35} & 0 & 0 & 0 & 0 & 0 & 0 \\ 0 & q_{42} & q_{43} & q_{44} & q_{45} & q_{46} & q_{47} & q_{48} & q_{49} & 0 & 0 \\ 0 & q_{52} & q_{53} & q_{54} & q_{55} & q_{56} & q_{57} & q_{58} & q_{59} & 0 & 0 \\ 0 & q_{62} & q_{63} & q_{64} & q_{65} & q_{66} & q_{67} & q_{68} & q_{69} & 0 & 0 \\ 0 & q_{72} & q_{73} & q_{74} & q_{75} & q_{76} & q_{77} & q_{78} & q_{79} & 0 & 0 \\ 0 & 0 & 0 & 0 & 0 & q_{86} & q_{87} & q_{88} & q_{89} & q_{810} & q_{811} \\ 0 & 0 & 0 & 0 & 0 & q_{96} & q_{97} & q_{98} & q_{99} & q_{910} & q_{911} \\ 0 & 0 & 0 & 0 & 0 & q_{106} & q_{107} & q_{108} & q_{109} & q_{1010} & q_{1011} \\ 0 & 0 & 0 & 0 & 0 & q_{116} & q_{117} & q_{118} & q_{119} & q_{1110} & q_{1111} \end{vmatrix}} \quad (5.26)$$

where,

$$b_1 = x_1 j_n'(x_1)$$

$$b_2 = -\frac{\rho_1}{\rho_2} x_{22}^2 j_n(x_1)$$

$$q_{11} = -x_1 h_n^{(1)'}(x_1)$$

$$q_{12} = x_{21} j_n'(x_{21})$$

$$q_{13} = x_{21} n_n'(x_{21})$$

$$q_{14} = -n(n+1) j_n(x_{22})$$

$$q_{15} = -n(n+1) n_n(x_{22})$$

$$q_{21} = \frac{\rho_1}{\rho_2} x_{22}^2 h_n^{(1)}(x_1)$$

$$q_{22} = 2x_{21}^2 j_n''(x_{21}) - \left(x_{22}^2 - 2x_{21}^2\right) j_n(x_{21})$$

$$q_{23} = 2x_{21}^2 n_n''(x_{21}) - \left(x_{22}^2 - 2x_{21}^2\right) n_n(x_{21})$$

$$q_{24} = 2n(n+1)\left(j_n(x_{22}) - x_{22} j_n'(x_{22})\right)$$

$$q_{25} = 2n(n+1)\left(n_n(x_{22}) - x_{22} n_n'(x_{22})\right)$$

$$q_{32} = 2\left(x_{21} j_n'(x_{21}) - j_n(x_{21})\right)$$

$$q_{33} = 2\left(x_{21} n_n'(x_{21}) - n_n(x_{21})\right)$$

$$q_{34} = \left(2 - n^2 - n\right) j_n(x_{22}) - x_{22}^2 j_n''(x_{22})$$

$$q_{35} = \left(2 - n^2 - n\right) n_n(x_{22}) - x_{22}^2 n_n''(x_{22})$$

$$q_{42} = y_{21} j_n'(y_{21})$$

$$q_{43} = y_{21}n_n'(y_{21})$$

$$q_{44} = -n(n+1)j_n(y_{22})$$

$$q_{45} = -n(n+1)n_n(y_{22})$$

$$q_{46} = -y_{31}j_n'(y_{31})$$

$$q_{47} = -y_{31}n_n'(y_{31})$$

$$q_{48} = n(n+1)j_n(y_{32})$$

$$q_{49} = n(n+1)n_n(y_{32})$$

$$q_{52} = j_n(y_{21})$$

$$q_{53} = n_n(y_{21})$$

$$q_{54} = -y_{22}j_n'(y_{22}) - j_n(y_{22})$$

$$q_{55} = -y_{22}n_n'(y_{22}) - n_n(y_{22})$$

$$q_{56} = -j_n(y_{31})$$

$$q_{57} = n_n(y_{31})$$

$$q_{58} = y_{32}j_n'(y_{32}) + j_n(y_{32})$$

$$q_{59} = y_{32}n_n'(y_{32}) + n_n(y_{32})$$

$$q_{62} = 2\mu_2 y_{21}^2 j_n''(y_{21}) - \lambda_2 y_{21}^2 j_n(y_{21})$$

$$q_{63} = 2\mu_2 y_{21}^2 n_n''(y_{21}) - \lambda_2 y_{21}^2 n_n(y_{21})$$

$$q_{64} = 2\mu_2 n(n+1)\left(j_n(y_{22}) - y_{22}j_n'(y_{22})\right)$$

$$q_{65} = 2\mu_2 n(n+1)\left(n_n(y_{22}) - y_{22} n_n'(y_{22})\right)$$

$$q_{66} = -2\mu_3 y_{31}^2 j_n''(y_{31}) + \lambda_3 y_{31}^2 j_n(y_{31})$$

$$q_{67} = -2\mu_3 y_{31}^2 n_n''(y_{31}) + \lambda_3 y_{31}^2 n_n(y_{31})$$

$$q_{68} = 2\mu_3 n(n+1)\left(y_{32} j_n'(y_{32}) - j_n(y_{32})\right)$$

$$q_{69} = 2\mu_3 n(n+1)\left(y_{32} n_n'(y_{32}) - n_n(y_{32})\right)$$

$$q_{72} = 2\mu_2\left(y_{21} j_n'(y_{21}) - j_n(y_{21})\right)$$

$$q_{73} = 2\mu_2\left(y_{21} n_n'(y_{21}) - n_n(y_{21})\right)$$

$$q_{74} = \left(2 - n^2 - n\right)\mu_2 j_n(y_{22}) - \mu_2 y_{22}^2 j_n''(y_{22})$$

$$q_{75} = \left(2 - n^2 - n\right)\mu_2 n_n(y_{22}) - \mu_2 y_{22}^2 n_n''(y_{22})$$

$$q_{76} = 2\mu_3\left(j_n(y_{31}) - y_{31} j_n'(y_{31})\right)$$

$$q_{77} = 2\mu_3\left(n_n(y_{31}) - y_{31} n_n'(y_{31})\right)$$

$$q_{78} = \left(n^2 + n - 2\right)\mu_3 j_n(y_{32}) + \mu_3 y_{32}^2 j_n''(y_{32})$$

$$q_{79} = \left(n^2 + n - 2\right)\mu_3 n_n(y_{32}) + \mu_3 y_{32}^2 n_n''(y_{32})$$

$$q_{86} = z_{31} j_n'(z_{31})$$

$$q_{87} = z_{31} n_n'(z_{31})$$

$$q_{88} = -n(n+1) j_n(z_{32})$$

$$q_{89} = -n(n+1) n_n(z_{32})$$

$$q_{810} = -z_{41} j_n'(z_{41})$$

$$q_{811} = n(n+1)j_n(z_{42})$$

$$q_{96} = j_n(z_{31})$$

$$q_{97} = n_n(z_{31})$$

$$q_{98} = -z_{32}j_n'(z_{32}) - j_n(z_{32})$$

$$q_{99} = -z_{32}n_n'(z_{32}) - n_n(z_{32})$$

$$q_{910} = -j_n(z_{41})$$

$$q_{911} = z_{42}j_n'(z_{42}) + j_n(z_{42})$$

$$q_{106} = 2\mu_3 z_{31}^2 j_n''(z_{31}) - \lambda_3 z_{31}^2 j_n(z_{31})$$

$$q_{107} = 2\mu_3 z_{31}^2 n_n''(z_{31}) - \lambda_3 z_{31}^2 n_n(z_{31})$$

$$q_{108} = 2\mu_3 n(n+1)\big(j_n(z_{32}) - z_{32}j_n'(z_{32})\big)$$

$$q_{109} = 2\mu_3 n(n+1)\big(n_n(z_{32}) - z_{32}n_n'(z_{32})\big)$$

$$q_{1010} = -2\mu_4 z_{41}^2 j_n''(z_{41}) + \lambda_4 z_{41}^2 j_n(z_{41})$$

$$q_{1011} = 2\mu_4 n(n+1)\big(z_{42}j_n'(z_{42}) - j_n(z_{42})\big)$$

$$q_{116} = 2\mu_3\big(z_{31}j_n'(z_{31}) - j_n(z_{31})\big)$$

$$q_{117} = 2\mu_3\big(z_{31}n_n'(z_{31}) - n_n(z_{31})\big)$$

$$q_{118} = \big(2 - n^2 - n\big)\mu_3 j_n(z_{32}) - \mu_3 z_{32}^2 j_n''(z_{32})$$

$$q_{119} = \big(2 - n^2 - n\big)\mu_3 n_n(z_{32}) - \mu_3 z_{32}^2 n_n''(z_{32})$$

$$q_{1110} = 2\mu_4\big(j_n(z_{41}) - z_{41}j_n'(z_{41})\big)$$

$$q_{1111} = \left(n^2 + n - 2\right)\mu_4 j_n(z_{42}) + \mu_4 z_{42}^2 j_n''(z_{42})$$

In the above formula, $\rho_1, \rho_2, \rho_3, \rho_4$ represent the densities of the media in regions I, II, III, and IV, respectively. $e = \frac{b}{a}, f = \frac{c}{a}$ represent the ratios of the inner diameters b, c to the outer diameter a of the three-layer sphere. $x_1 = k_1 a$, $x_{21} = x_1 \frac{c_1(1+i\gamma_{21})}{c_{21}}$, $x_{22} = x_1 \frac{c_1(1+i\gamma_{22})}{c_{22}}$, $x_{31} = x_1 \frac{c_1(1+i\gamma_{31})}{c_{31}}$, $x_{32} = x_1 \frac{c_1(1+i\gamma_{32})}{c_{32}}$, $x_{41} = x_1 \frac{c_1}{c_{41}}$, $x_{42} = x_1 \frac{c_1}{c_{42}}$, $y_{21} = x_{21}e$, $y_{22} = x_{22}e$, $y_{31} = x_{31}e$, $y_{32} = x_{32}e$, $z_{31} = x_{31}f$, $z_{32} = x_{32}f$, $z_{41} = x_{41}f$, $z_{42} = x_{42}f$, c_1 represents the speed of sound in the ideal liquid in region I, c_{21}, c_{22} represent the speeds of longitudinal and transverse waves in the medium in region II, c_{31}, c_{32} represent the speeds of longitudinal and transverse waves in the medium in region III, c_{41}, c_{42} represent the speeds of longitudinal and transverse waves in the medium in region IV. γ_{21}, γ_{22} represent the absorption coefficients of the medium in region II for longitudinal and transverse waves. γ_{31}, γ_{32} represent the absorption coefficients of the medium in region III for longitudinal and transverse waves. μ_i, λ_i is the Lamé coefficient for the corresponding region i.

5.4 Acoustic Radiation Force on a Multilayer Sphere

Yosioka and Kawasima [18] calculated the acoustic radiation force caused by sound wave scattering.

$$\langle \boldsymbol{F} \rangle = -\iint\limits_{S_0} \langle P - P_0 \rangle \boldsymbol{n} dS = -\langle \iint \rho(v_n \boldsymbol{n} + v_t \boldsymbol{t}) v_n dS \rangle$$
$$+ \left\langle \iint \frac{1}{2} \rho_1 v^2 \boldsymbol{n} dS - \right\rangle \left\langle \iint \frac{1}{2} \frac{\rho_1}{c_1^2} \left(\frac{\partial \phi_1}{\partial t} \right)^2 \boldsymbol{n} dS \right\rangle \tag{5.27}$$

where S $_0$ is the cross-section surrounding the scattering body, v_n and v_t are the normal and tangential components of the boundary particle velocity, respectively, where the symbol $\langle \cdot \rangle$ represents the time average.

After calculation, it can be expressed as:

$$\langle F \rangle = \pi a^2 E Y_p \tag{5.28}$$

and

$$Y_p = -\frac{4}{(ka)^2} \sum_{n=0}^{\infty} (n+1) \left\{ \begin{array}{l} Re\left[Q_n Q_{n+1}^*\right](\alpha_n + \alpha_{n+1} + 2\alpha_n \alpha_{n+1} + 2\beta_n \beta_{n+1}) \\ +Im\left[Q_n Q_{n+1}^*\right]\left[\beta_{n+1}(1 + 2\alpha_n) - \beta_n(1 + 2\alpha_n)\right] \end{array} \right\} \tag{5.29}$$

In the formula, Yp is a dimensionless factor called the radiation force function [4], * represents the complex conjugate, α_n and β_n are the real and imaginary parts of the scattering coefficient defined by the double-layer sphere (5.16) or the triple-layer sphere (5.26).

5.5 Simulation

The acoustic radiation force function Y_P of the solid sphere given by formula (5.29) was calculated, where the physical constants are as shown in Table 5.1. This chapter mainly discusses the double-layer and triple-layer microsphere models. In the simulation, $a = 5*10^{-6}$ m (the outer radius of the sphere), and the detailed analysis of the acoustic radiation force acting on the microsphere is as follows (Figs. 5.3 and 5.4):

To verify the validity of the above theory, we take a single-layer sphere as an example. When $e = f = 1$ ($a = b = c$), and $z_0 = 0$, the triple-layer model becomes a single-layer model located at the origin. Figure 5.5 shows the Yp-ka curve of a stainless steel sphere in water. The numerical results are consistent with the results of single-layer spheres in the papers by Takahi-Hasegawa [4] and Zhang [5].

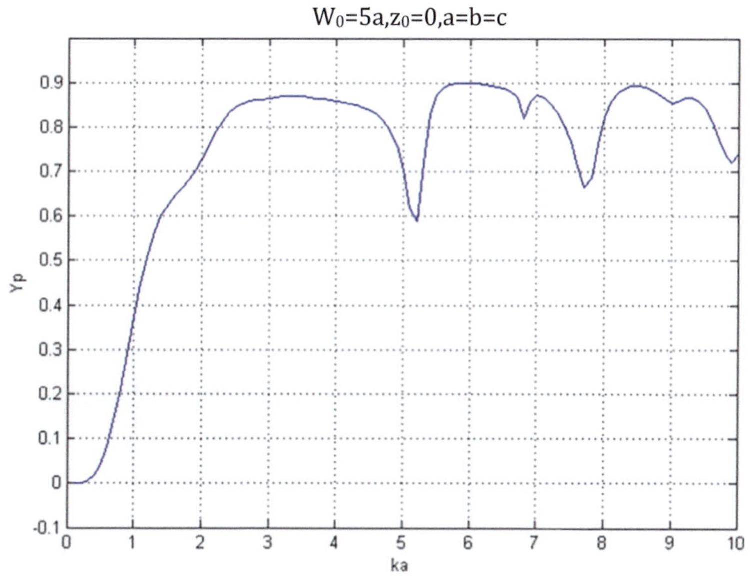

Fig. 5.5 Relationship between the acoustic radiation force function of a stainless steel ball in water and ka

5.5.1 Double-Layer Solid–liquid Model

The acoustic radiation force acting on a liquid drug encapsulated by a spherical polymer layer can also be used for drug delivery applications, where the shell is precisely broken and used in the affected area. Figures 5.6, 5.7, 5.8 and 5.9 show the axial radiation force diagram in the low-density polymethyl methacrylate double-layer model (b = c) filled with water.

To study the effect of beam center distance deviation on the axial radiation force, we provide the Yp~ka curve for different z_0 values. In Fig. 5.6, e = 0.99, $W_0 = 4a$. The results show that when ka < 1, the radiation force decreases with the increase of z_0 value, and when 3 < ka < 10, the radiation force has a series of significant peaks, which do not change with the change of z_0 value.

Figures 5.7 and 5.8 show the relationship between the radiation force and z_0 under different b/a and ka = 0.2 in the Gaussian beam width. According to the simulation results, as the value of z_0 increases, the radiation force tends to zero. When the particle deviates from the beam center on the positive half-axis of the z-axis, the radiation force is positive, and when the particle deviates from the beam center on the left axis of the z-axis, the radiation force is negative. Therefore, the backward force experienced by the polymer-coated sphere is on the left side of the origin, that is, the potential trap is located in the converging part of the Gaussian focusing beam, not the diverging part [22].

Figure 5.9 shows the Yp-ka curve calculated from e from 0.99 to 0.80, where $z_0 = 0$, $W_0 = 5a$. It is found that in the case of shell structure with 3 < ka < 10, the

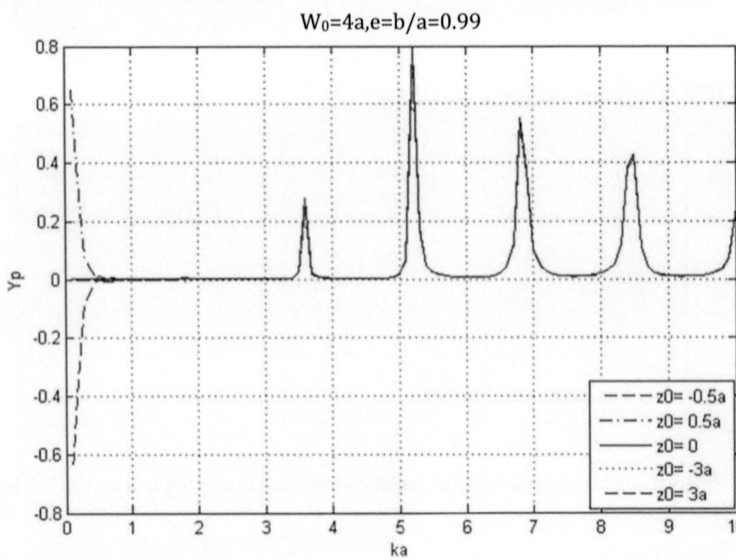

Fig. 5.6 Relationship between the acoustic radiation force function and ka of a double-layer polymer-coated sphere in water at different distances

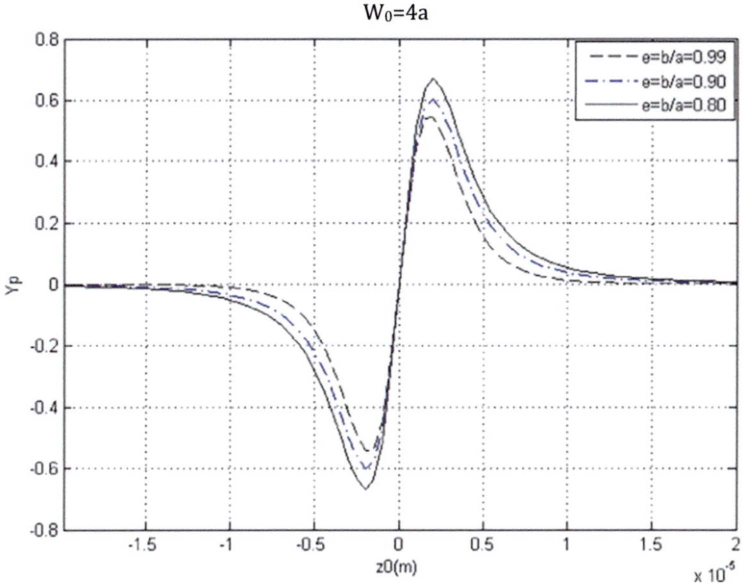

Fig. 5.7 Relationship between the acoustic radiation force function and z_0 of a double-layer polymer-coated sphere in water under different b/a

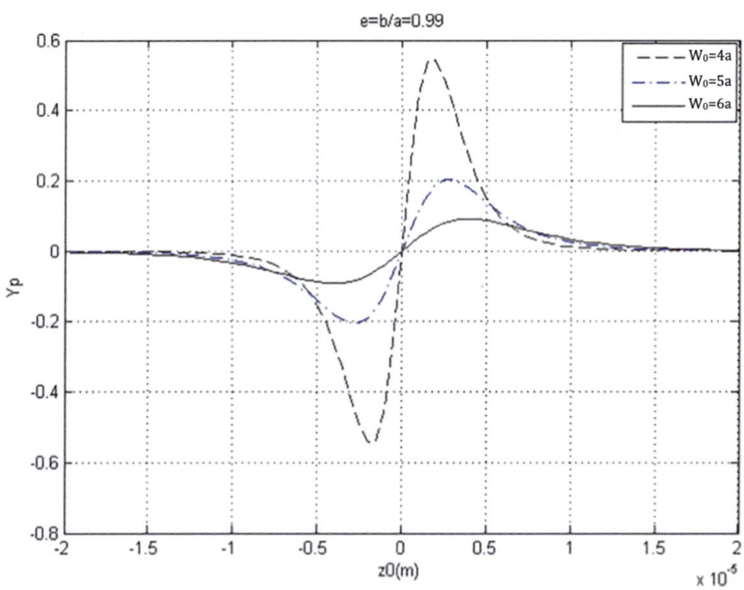

Fig. 5.8 Relationship between the acoustic radiation force function and z_0 of a double-layer polymer-coated sphere in water at different distances

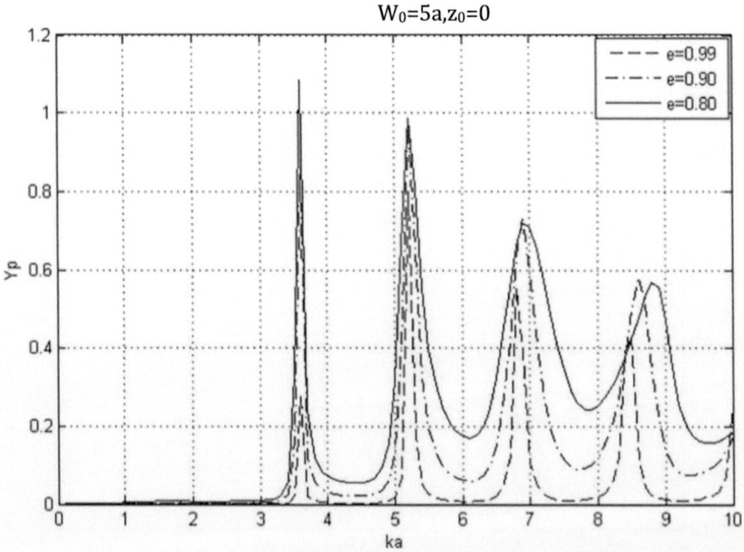

Fig. 5.9 Relationship between the acoustic radiation force function and ka of a double-layer polymer-coated sphere in water under different b/a when z0 = 0

Yp~ka curve has a series of prominent peaks and valleys. The peaks and valleys on the Yp~ka curve are caused by the resonant vibration of the shell, and their positions and amplitudes change with the change of b/a value, and the peak value decreases with the increase of ka value.

5.5.2 Three-Layer Viscoelastic Solid Sphere Model

Figures 5.10, 5.11, 5.12, 5.13, 5.14 and 5.15 show the simulation results of the three-layer viscoelastic solid sphere model. Here, polymethyl methacrylate, polystyrene, and phenolic polymer are selected as the materials for the three-layer structure for simulation.

In Fig. 5.10, the Yp-ka curve under different z_0 values was calculated, where $e = 0.90$, $f = 0.60$, $W_0 = 5a$. It is found that when $ka < 0.5$, the radiation force fluctuates significantly and decreases with the increase of z_0 value. When the sphere is on the positive half-axis of z, the value of the radiation force is positive, otherwise it is negative. The acoustic radiation force function has a series of significant peak and valley values at $0.5 < ka < 10$, which do not change with z_0 and tend to a finite value.

Figures 5.11, 5.12 and 5.13 show the change of radiation force with b/a, c/a and Gaussian beam width W_0 at $ka = 0.2$. Similar to Figs. 5.7 and 5.8, according to the simulation results, as the value of z_0 increases, the radiation force will approach zero. When the sphere deviates from the origin on the positive half-axis of z, the radiation

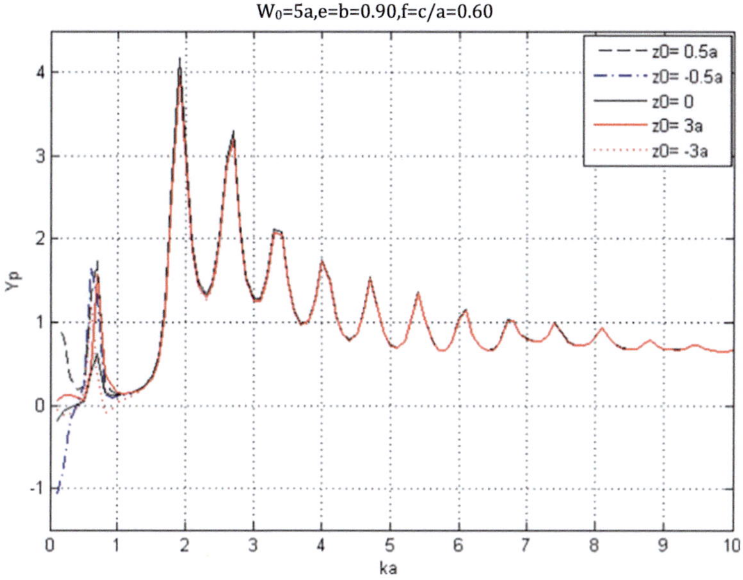

Fig. 5.10 Relationship between the acoustic radiation force function and ka of a three-layer viscoelastic sphere in water at different distances

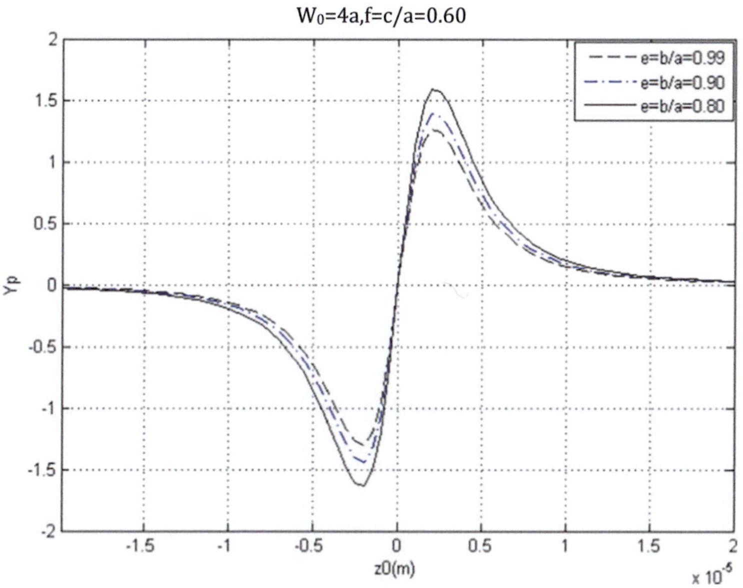

Fig. 5.11 Relationship between the acoustic radiation force function and z0 of a three-layer viscoelastic sphere in water under different b/a

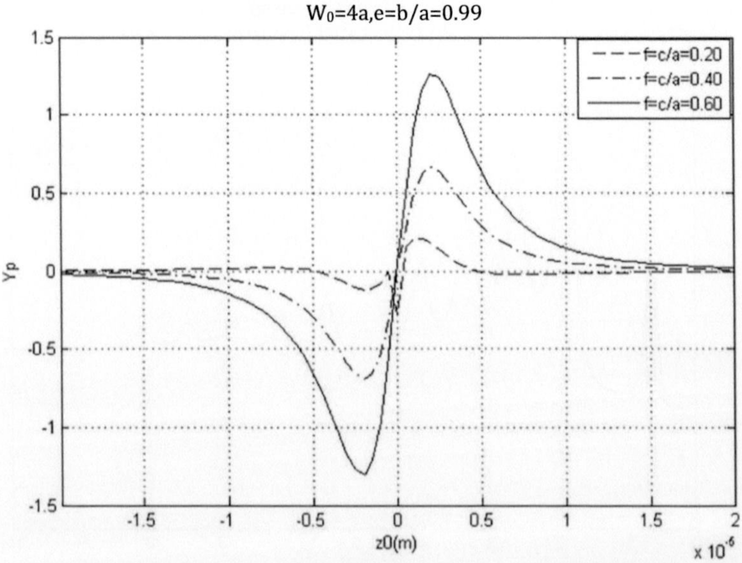

Fig. 5.12 Variation of the acoustic radiation force function of different c/a underwater three-layer viscoelastic spheres with z_0

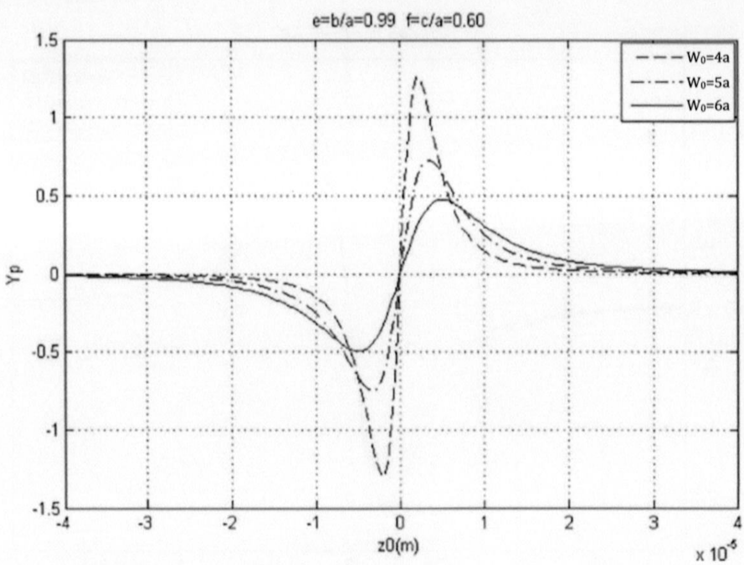

Fig. 5.13 Relationship between the acoustic radiation force function of a three-layer viscoelastic sphere underwater at different W_0 and z_0

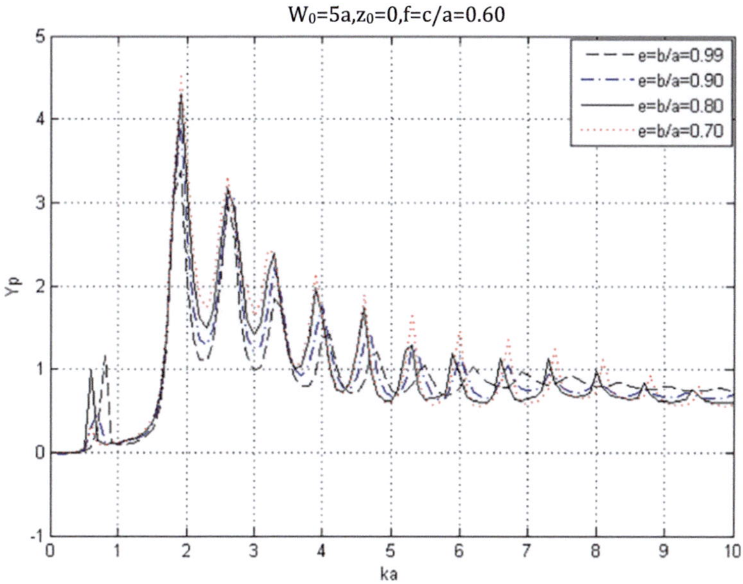

Fig. 5.14 Relationship between the acoustic radiation force function and ka of a three-layer viscoelastic sphere underwater at different b/a when $z_0 = 0$

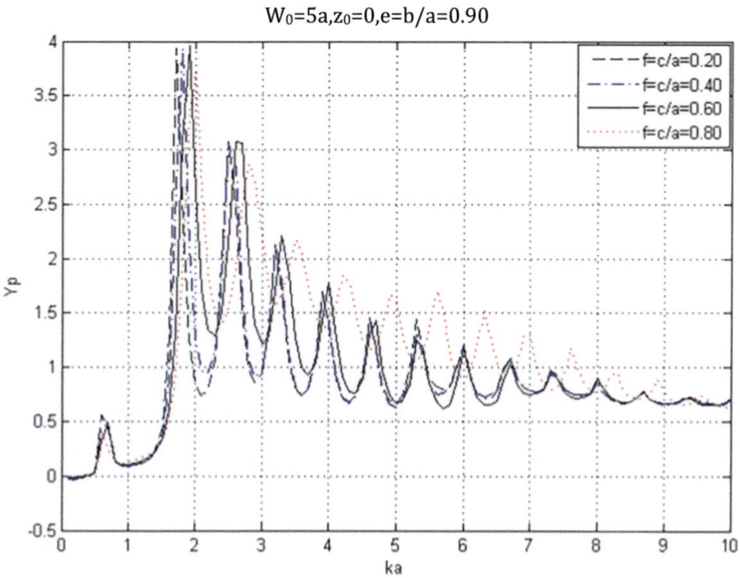

Fig. 5.15 Relationship between the acoustic radiation force function and ka of a three-layer viscoelastic sphere underwater at $z_0 = 0$, with different c/a

Table 5.1 Physical constants of materials used in sample calculations [5, 19–21]

Material	Density (× 10^3 kg/m^3)	Longitudinal wave speed (m/s)	Transverse wave speed (m/s)	Normalized longitudinal wave attenuation coefficient γ_{21}	Normalized shear wave attenuation coefficient γ_{22}
Water	1	1500	–	–	–
Stainless steel	7.90	5240	2978	–	–
Polystyrene	1.05	2340	1150	–	–
Polymethyl methacrylate	1.19	2690	1340	0.0035	0.0053
Phenolic polymer	1.22	2840	1320	0.0119	0.0255

force is positive. The radiation force is negative when it deviates from the origin on the negative half-axis of z, indicating that the backward force experienced by the three-layer viscoelastic sphere is also on the left side of the origin.

Figure 5.14 shows the Yp~ka curve calculated in e with a step length of 0.10 from 0.99 to 0.70, where f = 0.60, $z_0 = 0$ and $W_0 = 5a$. It is easy to find that the curve also has a series of significant peaks and valleys. The size of the resonance peak decreases with the increase of ka value. In addition, it can be observed that the resonance peaks and valleys shift with the change of e, indicating that the thickness of the sphere also affects the resonance frequency of the sphere. It is mainly observed that the damping of the resonance peak decreases when the thickness of the outer material decreases.

In Fig. 5.15, the Yp~ka curve is calculated from f = 0.80 to 0.20, where e = 0.90, $z_0 = 0$, $W_0 = 5a$. Similar to Fig. 5.14, some similar conclusions can be drawn by analyzing the curve. It is not difficult to find through the above analysis that the radiation force function changes significantly with e and f, that is, the thickness of each layer affects the radiation force function.

This chapter is based on the finite series method, using spherical harmonic function to expand Gaussian wave, and obtains the beam factor of Gaussian beam. Then the acoustic radiation force function of multi-layer spherical particles in fluid is represented by Gaussian sound field, and numerical analysis is carried out for double-layer and three-layer spherical shells, focusing on the influence of the ratio of the outer radius a to the inner radius b and c on the acoustic radiation force. We also discussed the influence of the width of Gaussian wave on the radiation force function. The results show that when the microsphere is located on the left z-axis, the acoustic radiation force is negative, which means that it is possible to manipulate multi-layer spheres with known mechanical and acoustic characteristics using Gaussian sound waves, which provides motivation for further research on cell biomechanics and drug delivery. Of course, if greater progress can be made in obtaining the structure of cells and drug carrier microspheres, such as the acoustic characteristics of cell components, the application of this theory will yield some more valuable and meaningful results.

References

1. Hasegawa T, Hino Y, Annou A et al (1993) Acoustic radiation pressure acting on spherical and cylindrical shells. J Acoust Soc Am 93:154–161
2. Wu R, Du GH (1990) Acoustic radiation force on a small compressible sphere in a focused beam. J Acoust Soc Am 87:997–1003
3. Zhang XF, Song ZG, Chen DM (2015) Finite series expansion of a Gaussian beam for the acoustic radiation force calculation of cylindrical particles in water. J Acoust Soc Am 137:1826–1833.
4. Zhang XF, Zhang GB (2012) Acoustic radiation force of a Gaussian beam incident on spherical particles in water. Ultrasound Med Biol 38:2007–2017
5. Hasegawa T, Yosioka K (1969) Acoustic-radiation force on a solid elastic sphere. J Acoust Soc Am 46:1139–1143
6. Wu RR, Cheng KX, Liu XZ (2016) Study of axial acoustic radiation force on a sphere in a Gaussian quasi-standing field. Wave Motion 63–74
7. Shi JJ, Daniel A (2009) Acoustical tweezers: patterning cells and micro particles using standing surface acoustic waves (SSAW). Lab Chip 9:2861–3024
8. Lee JW, Lee CY, Shung KK (2010) Calibration of sound forces in acoustic traps. IEEE Trans Ultrason Ferroelectr Freq Control 57:2305–2310
9. Lee JW (2011) Targeted cell immobilization by ultrasound microbeam. Biotechnol Bioeng 108:1643–1650
10. Li S (2013) An On-Chip, multichannel droplet sorter using standing surface acoustic waves. Anal Chem 85:5468–5474
11. Brannon-Peppas L (1997) Polymers in controlled drug delivery. Med Plastics Biomater Magaz 97:34–45
12. Hu Y, Qin S, Jiang Q (2004) Characteristics of acoustic scattering rom a double-layered micro shell for encapsulated drug delivery. IEEE Trans Ultrason Ferroelectr Freq Control 51:809–821
13. Jeffers R (1995) Ultrasound activation of anti-cancer drugs. J Acoust Soc Am 98:2380
14. Munshi N, Rapoport N, Pitt WG (1997) Ultrasonic activation of drug delivery from Pluronic P-105 micelles. Cancer Lett 118:13–19
15. Wu J (1991) Acoustical tweezers. J Acoust Soc Am 89:2140–2143
16. Lee J, Teh SY, Lee A et al (2009) Single beam acoustic trapping. Appl Phys Lett 7:073701–073703
17. Hasegawa T (1979) Acoustic radiation force on a sphere in a quasistationary wave field-theory. J Acoust Soc Am 65:32–40
18. Yosioka K, Kawasima Y (1955) Acoustic radiation pressure on a compressible sphere. Acustica 5:167–173
19. Norris AN (2017) An inequality for longitudinal and transverse wave attenuation coefficients. J Acoust Soc Am 14:475–479
20. Hartmann B, Jarzynski J (1972) Ultrasonic hysteresis absorption in polymers. J Appl Phys 11:4304–4312
21. Hartmann B (1975) Ultrasonic properties of phenolic and poly(phenylquinoxaline) polymers. J Appl Polym Sci 12:3241–3255
22. Thompson RB, Lopes EF (1984) The effects of focusing and refraction on Gaussian ultrasonic beams. J Nondestr Eval 4:107–123

Chapter 6
Acoustic Radiation Force of Other Sound Sources

Scholars often use plane waves, focused waves, and Gaussian waves. In these acoustic fields, the acoustic radiation forces of various types of particles in all directions have been thoroughly studied theoretically. The study of acoustic tweezers requires new types of sound waves. This chapter investigates the acoustic radiation force of hollow focused sound sources, annular piston transducers, Bessel-Gaussian waves, and Airy waves on cylindrical particles.

6.1 Acoustic Radiation Force of Hollow Focused Sound Sources

6.1.1 Theoretical Calculation

As shown in Fig. 6.1, assume there is a hollow focused sound source in space, where the point o in space is the geometric focus of the focused sound source. The geometric radius of the focused sound source is set to R_0, the minimum cone angle is θ_1, and the maximum cone angle is θ_2. When the axis of the focused sound source is set as the z-axis, the center of the particle placed on the axis is represented as o_2, and the distance to oo_2 is represented as z. Assume that any point on the surface of the particle with o_2 as the center is P, the radius of the spherical particle is r_0, and the distance from this point to any point Q on the focused sound source is R, and the distance from point Q to the center of the sphere is r. Assuming that any point on the sound source has the same vibration speed v_0 and the same vibration frequency ω, then in the spherical coordinate system (r, θ, γ), according to the Rayleigh-Sommerfeld diffraction integral formula and the Fresnel-Kirchhoff integral approximation principle [1–4], the velocity potential generated by the focused sound source on the point P on the sphere surface is:

X. Liu and H. Zheng, *Acoustic Radiation Force*, Advances in Acoustics,
https://doi.org/10.1007/978-981-95-3152-3_6

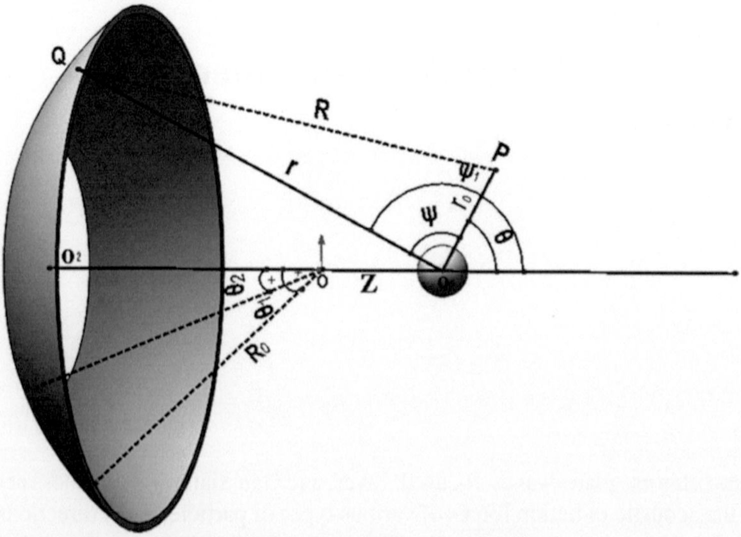

Fig. 6.1 Schematic diagram

$$\phi_i = \frac{1}{2\pi} \iint_{S_0} \frac{v_0 e^{i(kR-\omega t)}}{R} dS \tag{6.1}$$

Where S_0 represents the integral area of the focused sound source, k represents the wave number of the sound wave. To better understand the calculation process, assume that plane Poo_2 is plane A, then ψ_1 represents the angle between the line connecting the point mapped from point Q on the focused sound source to plane A and the center of the sphere, and the z-axis, γ represents the azimuth angle of the vibrating point with the reference plane. According to the geometric relationship:

$$\begin{aligned} R^2 &= r_0^2 + r^2 - 2rr_0\cos\psi, \\ \cos\psi &= \sin\psi_1\sin\theta\cos(\gamma_1 - \gamma) + \cos\psi_1\cos\theta \end{aligned} \tag{6.2}$$

The surface differential element dS at the non-geometric focus $z \neq R_0$ can be decomposed as follows

$$dS = \frac{rR_0}{z} drd\gamma \tag{6.3}$$

To simplify the calculation process, discard the calculation of $e^{-j\omega t}$, according to the decomposition formula of the spherical harmonic function of the point sound source $\frac{e^{ikR}}{R} = ik\sum_{n=0}^{\infty}(2n+1)h_n^{(1)}(kr)j_n(kr_0)P_n(\cos$ where $j_n(\cdot)$ is the nth order spherical Bessel function, $P_n(\cdot)$ is the n order Legendre function. From the integral formula (6.1), the velocity potential generated by the focused sound source on the point P on the sphere surface can be expressed as:

$$\phi_i = \frac{ikR_0 v_0}{2\pi} \sum_{n=0}^{\infty} (2n+1) j_n(kr_0) \int_{R_2}^{R_1} h_n^{(1)}(kr) r dr \int_0^{2\pi} P_n(\cos\psi) d\gamma \tag{6.4}$$

The area integral is decomposed into two parts, the angle integral $\int_0^{2\pi} P_n(\cos\psi)d\gamma$ and the length integral $\int_{R_2}^{R_1} h_n^{(1)}(kr) r dr$. According to the basic properties of the Legendre function [6, 7], the angle integral part $\int_0^{2\pi} P_n(\cos\psi)d\gamma = 2\pi P_n(\cos\psi_1)P_n(\cos\theta)$, angle θ represents the angle between Po_2 in plane A and the z-axis. The upper limit of the length integral part is $R_1^2 = R_0^2 + z^2 - 2R_0 z\cos\theta_1$, and the lower limit is $R_2^2 = R_0^2 + z^2 - 2R_0 z \cos\theta_2$.

Based on geometric relationships, the integral result of Eq. (6.4) can be expressed as:

$$\phi_i = \phi_0 R_0 \sum_{n=0}^{\infty} i^n (2n+1) j_n(kr_0) P_n(\cos\theta) f_n \tag{6.5}$$

And we have $f_n = (-1)^n i^{n+1} \frac{1}{kz} \int_{kR_2}^{kR_1} h_n^{(1)}(x) x P_n(\cos\psi_1) dx$, integral $\Lambda_n = \int_{kR_2}^{kR_1} h_n^{(1)}(x) x P_n(\cos\psi_1) dx$ the result can be obtained through rigorous calculation [8]:

$$\Lambda_n = \begin{cases} -e^{ikR_1} + e^{ikR_2} & n = 0 \\ \frac{1}{A}\left(2 + i\left(\frac{B}{kR_2} - kR_2\right)\right)e^{ikR_2} - \frac{1}{A}\left(2 + i\left(\frac{B}{kR_1} - kR_1\right)\right)e^{ikR_1} & n = 1 \\ -\frac{2(2n-1)}{A}\Lambda_{n-1} - \Lambda_{n-2} - kR_1 h_{n-1}(kR_1)\left(P_n(-\cos\theta_1) - P_{n-2}(-\cos\theta_1)\right) & \\ +kR_2 h_{n-1}(kR_2)\left(P_n(-\cos\theta_2) - P_{n-2}(-\cos\theta_2)\right) & n \geq 2 \end{cases} \tag{6.6}$$

The parameters in Eq. (6.6) $A = 2kz$, $B = k^2R_0^2 - k^2z^2$. The calculation results prove that the velocity potential caused by the focused sound source at point P on the sphere surface is related to multiple factors such as the radius of the focused sound source, the upper and lower limit cone angles, the size and position of the spherical particles. Although Eq. (6.1) could only be applied in the case of weak focusing in the early stage, it was found that the equation is still valid at a distance far from the transducer [9–11]. At the geometric focus $z = R_0$, the previous area integral result can be expressed as [10]:

$$\Lambda_n = \begin{cases} ie^{ikR_0}(\cos\theta_1 - \cos\theta_2) & n = 0 \\ (-1)^n n^{-1}\left[\cos\theta_1 P_n(\cos\theta_1) - \cos\theta_2 P_n(\cos\theta_2) - P_{n+1}(\cos\theta_1) + P_{n+1}(\cos\theta_2)\right] & n \geq 1 \end{cases} \tag{6.7}$$

If the incident velocity potential on the surface of the spherical particle can be expressed as $\phi = \phi_0 \sum_{n=0}^{\infty} i^n (2n+1) j(kr_0) P_n(\cos\theta) \Lambda_n$, then the reflected sound field after reflection from the spherical surface is represented as $\phi =$

$\phi_0 \sum_{n=0}^{\infty} i^n(2n+1)S_n h_n^{(1)}(kr_0)P_n(\cos\theta)\Lambda_n$,, the incident sound field and the reflected sound field together form the combined sound field on the surface of the spherical particle. Under the action of this sound field, if the radius of the sphere is set to a, the radiation force of the focused sound field on the spherical particle is $F = S_c EY$, where $S_c = \pi a^2$ is the cross-sectional area of the sphere, $E = \rho k^2 \phi_0^2 R^2/2$ is the characteristic energy density of the sound radiation force, Y is the dimensionless sound radiation force function, which can be expressed as [12]:

$$Y = \frac{4}{(ka)^2} \sum_{n=0}^{\infty} (n+1)(-1)^n \begin{bmatrix} Re\left(\Lambda_n \Lambda_{n+1}^*\right)(\alpha_n + \alpha_{n+1} + 2\alpha_n\alpha_{n+1} + 2\beta_n\beta_{n+1}) \\ +Im\left(\Lambda_n \Lambda_{n+1}^*\right)(\beta_{n+1} - \beta_n + 2\alpha_n\beta_{n+1} - 2\alpha_{n+1}\beta_n) \end{bmatrix} \tag{6.8}$$

α_n and β_n are the real and imaginary parts of the parameter S_n respectively, from which the sound radiation force function of particles with different properties on the axis of the hollow focusing transducer can be obtained.

6.1.2 Simulation Results and Discussion

In order to study the relationship between the sound radiation force and the distance between the spherical particle and the transducer, and to discuss the influence of the size of the inner and outer radii of the annular focusing transducer on the sound radiation force. We set the spherical particle as a rigid stationary sphere, fixing the wave number of the vibration signal k, then the scattering coefficient $S_n = -\frac{j_n'(ka)}{h_n^{(1)'}(ka)}$ [13]. To simplify the calculation, we set the parameter $kR_0 = 30$, parameter $ka = 0.5$. According to Eqs. (6.6), (6.7) and (6.8), the sound radiation force can be calculated and discussed. As shown in Fig. 6.2, because the focus of the focusing transducer is o, the distance from the center of the sphere on the axis to the geometric focus of the transducer is z, so the distance from the center of the sphere to the transducer is $z+R_0$, here the distance from the center of the sphere to the transducer is represented by the dimensionless $kz + kR_0$, and is used as the horizontal coordinate. The vertical coordinate is the size of the sound radiation force function, where the positive value represents thrust (pushing the particle away from the transducer), and the negative value represents pull (pulling the particle closer to the transducer). In order to study the influence of the hollow focusing transducer on the sound radiation force of the spherical particles on the axis, set the maximum cone angle $\theta_1 = \frac{\pi}{4}$, the minimum cone angle θ_2 starts from zero and increases step by step according to $\frac{\pi}{20}$, then the radiation force received by the small ball on the axis is shown in Fig. 6.2a. When $\theta_2 = 0$ the sound radiation force curve is shown as a solid line, it is clearly visible that the sound radiation force on the axis before the focus point approximates a sinusoidal variation, and several maximum and minimum values appear. It can be seen in the figure that there are sound radiation pull force areas 3.5–5.1, 8.5–11.8, 17.0–18.5 on the axis, the intersection of the sound radiation pull force and thrust is the force balance point of the suspended particles, which can be called the sound suspension point. The sound radiation force reaches its maximum near the focus, then gradually decreases and tends to stabilize. As the inner angle increases, the maximum and minimum values of the sound radiation force move towards the direction closer to the transducer. The sound radiation pull force area moves to the left and the width slightly increases, and the number of sound suspension points decreases. As shown in Fig. 6.2a, when $\theta_2 = \frac{3\pi}{20}$, the acoustic radiation tension region changes to 9.4–15.3. The inner angle of the hollow transducer is set to $\theta_2 = \frac{\pi}{6}$, and the outer angle increases from $\theta_1 = \frac{\pi}{4}$ to $\theta_1 = \frac{\pi}{2}$ in steps of $\frac{\pi}{12}$. The acoustic radiation force is shown in Fig. 6.2b. It can be observed that as the outer angle increases, the maximum and minimum values of the axial acoustic radiation force move away from the transducer, the number of acoustic radiation tension regions increases, and the range of the tension region decreases. For example, when $\theta_1 = \frac{\pi}{4}$, the acoustic radiation tension region is 7–10.3, and when $\theta_1 = \frac{\pi}{2}$, the acoustic radiation tension region is 11.2–14.2, 16.1–21.4, 24.1–29.2. When $\theta_1 = \frac{\pi}{2}$, it can be found that the direction of the acoustic radiation force before and after the geometric focus of the transducer is opposite and symmetrical to this point, so this point is a relatively perfect suspension point, which can guide the design of acoustic tweezers. Figure 6.2a, b show that the increase in the inner angle and the outer angle cause opposite changes. Set the difference between the inner and outer angles to a constant value $\theta_1 - \theta_2 = \frac{\pi}{6}$, the change curve of the radiation force on the axis is shown in Fig. 6.2c. It can be clearly seen that as the inner and outer angles increase simultaneously, the number of acoustic radiation tension regions (i.e., suspension points) hardly changes, the acoustic radiation force curve as a whole moves away from the transducer and shows a widening trend, and this change result causes the acoustic radiation tension region to widen. For example, when $\theta_2 = \frac{\pi}{12}, \theta_1 = \frac{\pi}{4}$, the acoustic radiation tension region is 8.3–9.1, 11.4–12.8. When $\theta_2 = \frac{\pi}{3}, \theta_1 = \frac{\pi}{2}$, the acoustic radiation tension region is 8.5–12.3, 17.8–29.4. The alternating long and short curves in Fig. 6.2c indicate that when $\theta_2 = \frac{\pi}{3}, \theta_1 = \frac{\pi}{2}$, there is a relatively perfect suspension point near the geometric focus.

In order to study the radiation force of the hollow transducer on the axial spherical particles, we calculated the radiation force received by the acoustic radiation force of different radius particles under the condition of fixed inner and outer angles $\theta_2 = \frac{\pi}{20}, \theta_1 = \frac{\pi}{4}$, and the focus radius parameter of the hollow transducer $kR_0 = 30$, as shown in Fig. 6.3. The X-axis represents the distance from the particle center to the transducer, the y-axis represents the particle radius size, and the z-axis represents the size of the acoustic radiation force function. As can be seen from the figure, the acoustic radiation tension region in the area before the geometric focus does not change significantly with the change of the sphere radius, for example, there are clearly three acoustic radiation tension regions in front of the transducer focus regardless of the sphere radius size. In addition, the amplitude of the acoustic radiation force function will increase approximately linearly with the increase of the particle radius. When the small ball is near the focus of the transducer, the radiation force received by the small radius spherical particles ($ka \leq 0.3$) is significantly smaller than that of the large radius spherical particles ($ka \geq 0.3$). This indicates that when

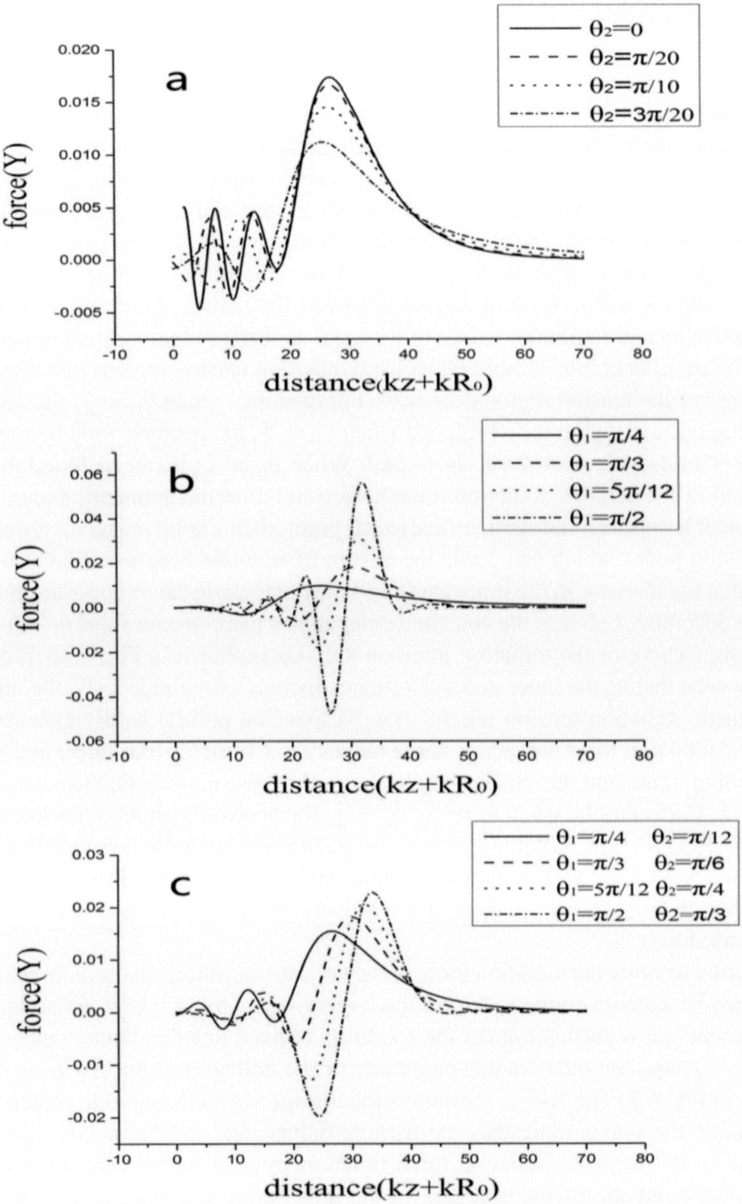

Fig. 6.2 Acoustic radiation force curve **a** outer angle $\theta_1 = \frac{\pi}{4}$, inner angle θ_2 increase sequentially **b** inner angle $\theta_2 = \frac{\pi}{6}$, outer angle θ_1 increase sequentially **c** difference between inner and outer angles $\theta_1 - \theta_2 = \frac{\pi}{6}$, inner and outer angles increase simultaneously

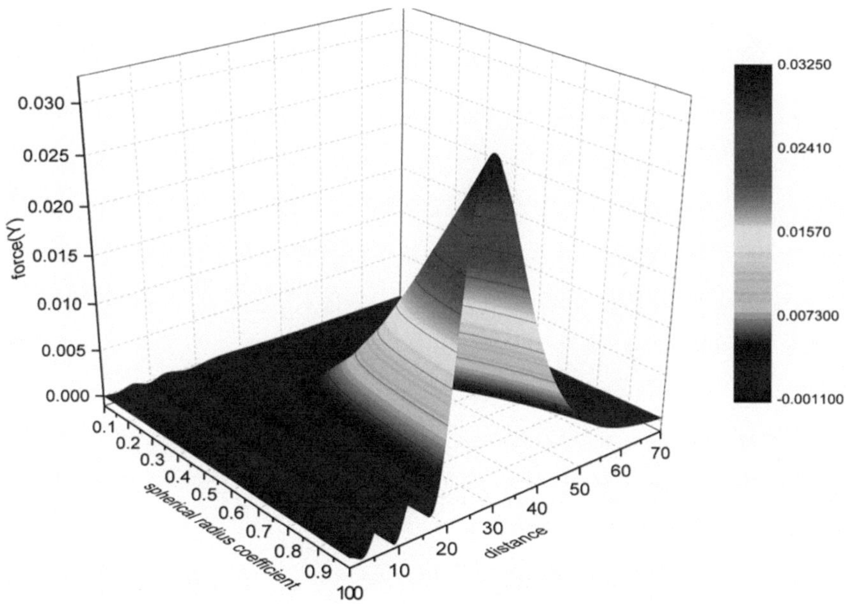

Fig. 6.3 Axial acoustic radiation force for a hollow focusing transducer when $\theta_2 = \frac{\pi}{20}, \theta_1 = \frac{\pi}{4}$ on a rigid spherical particle

the spherical particles are not too small, the acoustic radiation force near the focus is the largest and is most likely to be used as the position of the acoustic suspension point.

6.2 Acoustic Radiation Force Caused by Annular Transducer

6.2.1 Theoretical Part

Similar to the focusing transducer, as shown in Fig. 6.4, it is assumed that the annular transducer is immersed in water, and its inner and outer radii are respectively R_2 and R_1. Under the condition that the thickness of the transducer is negligible, a spherical particle with a radius of r is placed on its axis. In the three-dimensional coordinate system, in order to better display the geometric relationship in the figure, any point P on the surface of the sphere is not marked on the surface of the sphere. Then the distance symbols in the figure can be represented as shown in the figure. The distance from any point P on the surface of the sphere to any point on the surface of the transducer is set to R, the distance from the center of the sphere to the same point of the transducer is r_0, and the distance from the point on the surface of the

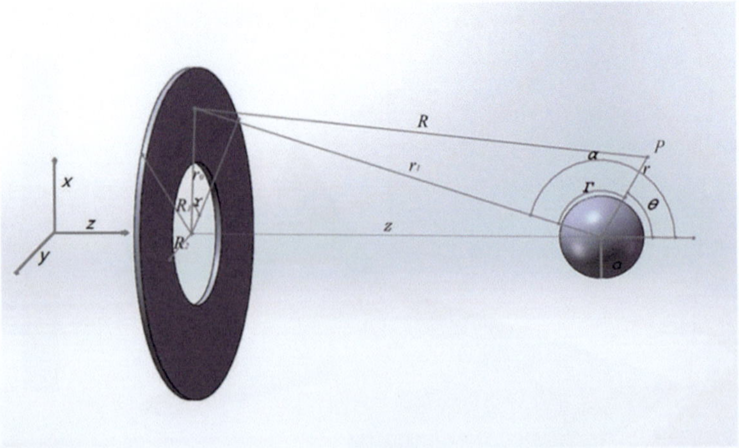

Fig. 6.4 Schematic diagram of the annular transducer

transducer to the center of the ring is represented as r_0. In addition to the distance symbols shown in the figure, there are corresponding angles. To clearly illustrate the geometric meanings of these angles, the x–z plane in the diagram is set as the reference plane. In a three-dimensional coordinate system, the angle between any point on the transducer in the x–z plane and the line connecting it to the center of the sphere, and the z-axis is represented as α which represents the angle between the line connecting the point on the transducer surface within the reference plane and the center of the sphere, and the z-axis, β represents the angle between the line connecting P within the reference plane and the center of the sphere, and the z-axis, γ represents the azimuthal angle of the vibrating point on the transducer with respect to the reference plane.

Ideally, all vibrating points on the annular transducer vibrate at the same initial phase and the same vibration speed v_0 [14–16], so the combined velocity potential ϕ_i generated by the annular transducer at any point P on the sphere surface can be represented in the following integral form [17]

$$\phi_i = \frac{1}{2\pi} \iint\limits_{s} \frac{v_0 e^{i(\omega t - kR)}}{R} ds \tag{6.9}$$

$k = \frac{\omega}{c}$ is the wave number of the sound wave, where ω is the angular frequency of the sound wave generated by all points on the annular transducer, ds is the differential unit of the annular transducer surface. In the spherical coordinate system (r, θ, γ), the relationship between the angles and lengths in Fig. 6.4 can be represented as

$$R^2 = r_1^2 + r^2 - 2rr_1 \cos\Gamma \tag{6.10}$$

$$\cos\Gamma = \cos\alpha\cos\theta + \sin\alpha\sin\theta\cos\gamma \tag{6.11}$$

By substituting Eq. (6.10) into (6.11), according to the spherical coordinate expansion principle of point sound source, the integral in Eq. (6.9) can be represented in the form of spherical coordinates

$$\phi_i = \frac{-ikv_0}{2\pi}\sum_{n=0}^{\infty}(2n+1)j_n(kr)\iint_s h_n^{(2)}(kr_1)P_n(\cos\Gamma)dS \tag{6.12}$$

where P_n, j_n and $h_n^{(2)}$ are the n order Legendre function, spherical Bessel function, second kind spherical Hankel function and Legendre function, and the area integral $dS = r_1 dr1d\gamma$. By Eq. (6.11) and the integral property of the Legendre function, integrating the angle γ from 0 to 2π, we have [18, 19]

$$\int_0^{2\pi} P_n(\cos\Gamma)d\gamma = 2\pi P_n(cos\alpha)P_n(cos\theta) = 2\pi(-1)^n P_n(\cos(\pi-\alpha))P_n(cos\theta) \tag{6.13}$$

As can be seen from Fig. 6.4, $\cos(\pi-\alpha) = \frac{z}{r_1}$. Assuming $kr_1 = x$, according to Eq. (6.13), the area integral in Eq. (6.12) can be represented as

$$\iint_s h_n^{(2)}(kr_1)P_n(\cos\Gamma)dS = \frac{2\pi}{k^2}(-1)^n P_n(cos\theta)\int_{kr_a}^{kr_b} xh_n^{(2)}(x)P_n\left(\frac{kz}{x}\right)dx \tag{6.14}$$

where the upper limit of the integral is $r_b = \sqrt{R_1^2+z^2}$, and the lower limit is $r_a = \sqrt{R_2^2+z^2}$. According to the formula for calculating sound radiation force, the final integral result in Eq. (6.14) can be represented as

$$\phi_i = \frac{v_0}{k}\sum_{n=0}^{\infty}(2n+1)(-i)^n j_n(kr)P_n(\cos\theta)\Lambda_n \tag{6.15}$$

where $\Lambda_n = (-i)^{n+1}f_n$, $f_n = \int_{kr_a}^{kr_b} xh_n^{(2)}(x)P_n\left(\frac{kz}{x}\right)dx$, and the integral result of f_n is [20, 21]

$$f_n = \begin{cases} e^{(-ikr_a)} - e^{(-ikr_b)} & n=0 \\ kz_0(h_0^{(2)}(kr_a) - h_0^{(2)}(kr_b)) & n=1 \\ -f_{n-2} + kr_a h_{n-1}^{(2)}(kr_a)\left(P_n\left(\frac{kz}{kr_a}\right) - P_{n-2}\left(\frac{kz}{kr_a}\right)\right) & \\ -kr_b h_{n-1}^{(2)}(kr_b)\left(P_n\left(\frac{kz}{kr_b}\right) - P_{n-2}\left(\frac{kz}{kr_b}\right)\right) & n \geq 2 \end{cases} \tag{6.16}$$

If the incident combined velocity potential can be represented in the form of Eq. (6.15), then the corresponding spherical scattering velocity potential ϕ_s can be represented as:

$$\phi_s = \frac{v_0}{k}\sum_{n=0}^{\infty}(2n+1)(-i)^n S_n h_n^{(2)}(kr)P_n(cos\theta)\Lambda_n \tag{6.17}$$

where the material characteristics of the spherical particles determine the scattering coefficient $S_n = \alpha_n + i\beta|_n$. Here, to make the formula clear and concise, use ϕ_0 to replace $\frac{v_0}{k}$. Then the combined sound field on the surface of the spherical particle is represented as:

$$\phi = \phi_0\sum_{n=0}^{\infty}(2n+1)(-i)^n A_n P_n(\cos\theta)\Lambda_n \tag{6.18}$$

where $A_n = (1+\alpha_n)j_n(kr) + \beta_n n_n(kr) + i(\beta_n j_n(kr) - \alpha_n n_n(kr))$. Assuming that the radius of the spherical particle on the axis is a fixed value a, we obtain the acoustic radiation force of the annular piston transducer on spherical particles $F = S_c EY$, where $S_c = \pi a^2$ is the cross-sectional area of the sphere, $E = \rho k^2\phi_0^2/2$ is the characteristic energy density of the acoustic radiation force, Y is the dimensionless acoustic radiation force function, which can be expressed as [12]

$$Y = -\frac{4}{(ka)^2}\sum_{n=0}^{\infty}(n+1)\begin{bmatrix}\Re(\Lambda_n\Lambda_{n+1}^*)(\alpha_n + \alpha_{n+1} + 2\alpha_n\alpha_{n+1} + 2\beta_n\beta_{n+1}) \\ +Im(\Lambda_n\Lambda_{n+1}^*)(\beta_{n+1} - \beta_n + 2\alpha_n\beta_{n+1} - 2\alpha_{n+1}\beta_n)\end{bmatrix} \tag{6.19}$$

6.2.2 Simulation Analysis and Discussion

This section simulates the acoustic radiation force on spherical particles on the axis of the annular transducer. Since the aluminum sphere is relatively hard compared to water, the results of the elastic sphere made of metal material are very similar to those of the rigid immovable sphere, so we mainly calculate and discuss the acoustic radiation force received by the rigid immovable sphere and the liquid sphere. The focus is on studying the relationship between the tensile force area on the axis of the annular transducer and the inner and outer radii of the transducer and the radius of the sphere, thereby providing a theoretical basis for the design and manufacture of acoustic tweezers. Compared to the acoustic radiation thrust, we are more concerned about the range and position of the acoustic radiation tensile force area and the suspension point, so the calculation results of this part only retain the part where the radiation force function is less than zero.

6.2.2.1 Acoustic Radiation Force of the Transducer on the Axial Rigid Sphere

The sum of the vibration speeds on the surface of the rigid sphere is zero, and its scattering coefficient can be expressed as $S_n = -\frac{j_n'(ka)}{h_n^{(1)'}(ka)}$. According to this scattering coefficient and the dimensionless acoustic radiation force function formula (6.19), Fig. 6.5 shows the calculated results of the acoustic radiation force received by the rigid immovable sphere on the axis with the radius coefficient kR_1 increasing by 10 steps. Where x the abscissa represents kz, y the ordinate represents kr_0, z the z-axis represents the dimensionless acoustic radiation force function Y.

Figure 6.5 is a distribution diagram of the radiation tensile force area of different radius piston transducers on the axial rigid immovable sphere. Theoretically speaking, the near field of the circular piston transducer is defined as $z \le \frac{R_1^2}{\lambda}$, which can be expressed in another form as $kz \le \frac{(kR_1)^2}{k\lambda}$, and because of the identity $k\lambda = 2\pi$, its near field is $kz \le \frac{(kR_1)^2}{2\pi}$. The outer radius variable increases by 10 steps, and its corresponding near field can be calculated separately as (a) $kz \le 63.66$,

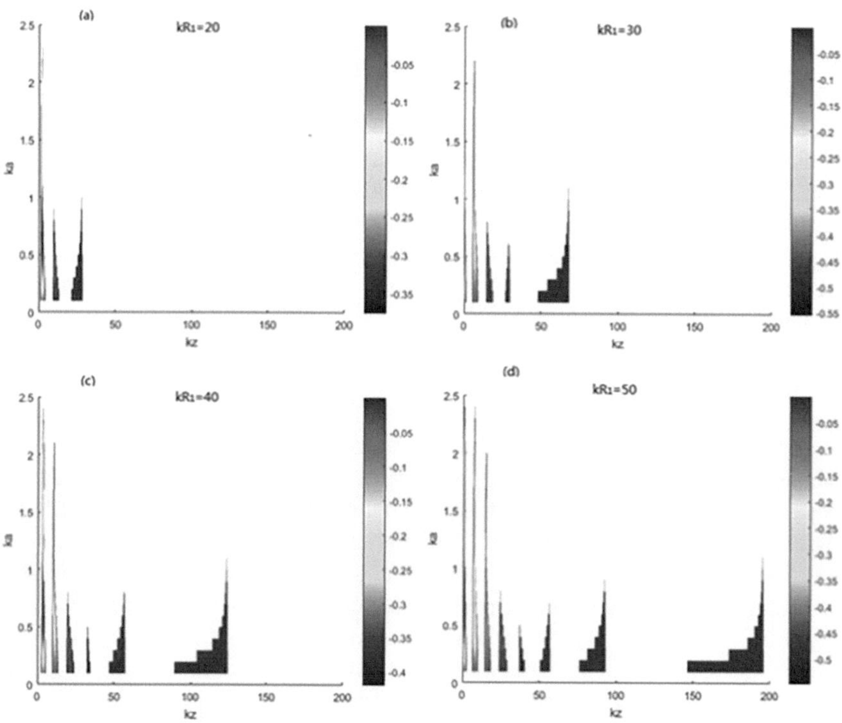

Fig. 6.5 The distribution diagram of acoustic radiation tensile force received by rigid sphere in water for a piston transducer kR_1 increasing by 10 in sequence,

(b) $kz \leq 143.24$, (c) $kz \leq 254.65$, (d) $kz \leq 389.89$. Since all parts greater than 0 in Fig. 6.5 are invalid values, the remaining areas are all negative acoustic radiation forces, that is, the acoustic radiation tensile force area. It can be seen that under different outer radius conditions, the acoustic radiation tensile force area is all within the near field area of the piston transducer. For the rigid immovable sphere, the distribution of the acoustic radiation force has the following two characteristics: First, when the sphere radius is larger, the number and range of the acoustic radiation tensile force area on the axis are reduced. For example, when $R_1 = \frac{25\lambda}{\pi}$ $(kR_1 = 50)$, the sphere radius $a = \lambda/\pi(ka = 2)$, its acoustic tensile force area kz ranges from 0.2–1.4, 7–8, and 15–15.4. But when the sphere radius is $a = \lambda/4\pi$ $(ka = 0.5)$, the range of its acoustic radiation tensile force area is 0.2–2.4, 6.6–9.6, 14.4–17, 24.4–25.7, 55.6–56.5, 90.5–93, and 192.2–195.6, etc. Second, the acoustic radiation force on the axis changes approximately sinusoidally along the axis, and the closer to the transducer, the greater the amplitude of the maximum and minimum values of the acoustic radiation force. For example, when $R_1 = \frac{25\lambda}{\pi}$ and $a = \lambda/4\pi$ As shown in Fig. 6.6, the distribution of sound radiation force decreases with the increase of distance from the transducer until it reaches a state of equilibrium. In addition, we focus on the relationship between the sound radiation force region and the outer radius. As can be seen from the four comparisons of the sound radiation force region in Fig. 6.5, with the increase of the outer radius of the piston transducer, the sound radiation force region on the axis generally expands outward, and new force regions will be generated near the transducer, and the number of sound radiation force regions will gradually increase. As the force region expands outward, the width of the sound radiation force region slightly increases for particles with a small radius.

In discussing the sound radiation force on particles along the axis of a hollow focusing transducer, it can be seen that the inner and outer angles jointly control the position and width of the sound radiation force region. The inner radius of the ring transducer also affects the sound radiation force region. Figure 6.7 shows the distribution of the radiation force region on the axial rigid sphere of the ring transducer under the condition that the outer radius is fixed at $R_1 = \frac{25\lambda}{\pi}$ $(kR_1 = 50)$, and the inner radius parameter kR_2 increases step by step from zero to 30. The four comparison figures show that the relationship between the sound force region of the ring transducer and the inner radius has the following two characteristics: First, when the inner radius is very small $(kR_2 \leq 10)$, the change in the inner radius will not significantly cause changes in the sound radiation force region on the axis of the ring transducer. However, when the inner radius is larger, the change in the inner radius significantly causes changes in the sound radiation force region on the axis. For example, compared with the inner radius $R_2 = 0$ $(kR_2 = 0)$ and $R_2 = 5\lambda/\pi$ $(kR_2 = 10)$, the change in the force region caused by the ring transducer is not very obvious. Compared with the inner radius $R_2 = 10\lambda/\pi$ $(kR_2 = 20)$ and $R_2 = 15\lambda/\pi$ $(kR_2 = 30)$, the change in the force region caused by the ring transducer is very obvious. Second, the sound radiation force region on the axis of the ring transducer gradually moves towards the direction close to the transducer as the inner radius increases. The force region close to the transducer will "disappear" one by one as the inner radius increases. The sound radiation force region farther from it will gradually

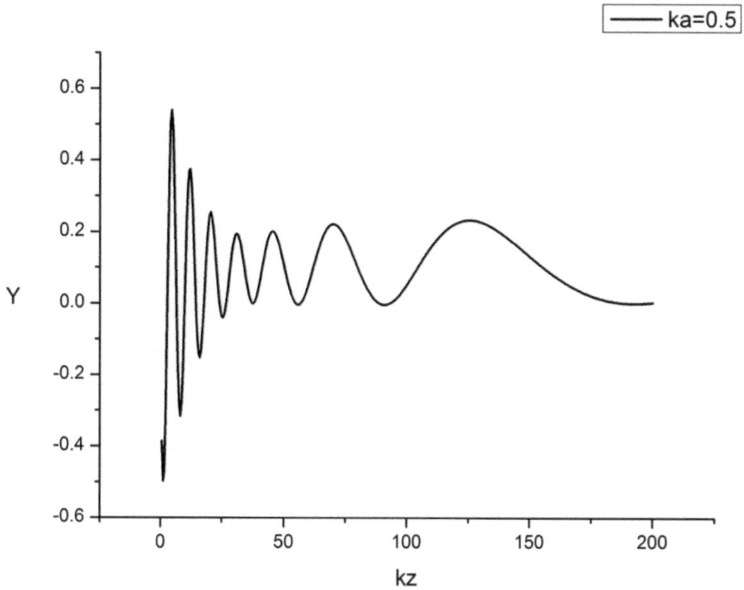

Fig. 6.6 The distribution of acoustic radiation force on a rigid sphere with radius $a = \lambda/4\pi$ ($ka = 0.5$) in water for a piston transducer with outer radius $R_1 = \frac{25\lambda}{\pi}$ ($kR_1 = 50$)

approach the transducer and the range of each force region will correspondingly increase, and the radiation force will be slightly larger than before. For example, Fig. 6.8 is the distribution of sound radiation force on a rigid sphere with radius $a = \lambda/4\pi$ ($ka = 0.5$) in water, generated by a ring piston transducer with outer radius $R_1 = \frac{25\lambda}{\pi}$ ($kR_1 = 50$) and inner radius $R_2 = \frac{10\lambda}{\pi}$ ($kR_2 = 20$). Compared with Fig. 6.6, the distribution of the maximum and minimum values of the sound radiation force function curve is more sparse, and the amplitude is relatively smaller. In addition, for larger radius spherical particles, their sound radiation force region is limited. As the inner radius gradually increases, the sound radiation force on large particles weakens and disappears.

6.2.2.2 Sound Radiation Force of Transducer on Axial Liquid Sphere

Because only longitudinal waves exist inside the liquid sphere, the calculation of this boundary condition should follow the following conditions. First, the stress on the surface of the liquid sphere should be continuous. Second, the displacement on the surface of the liquid sphere should be continuous. Its scattering coefficient can be expressed as

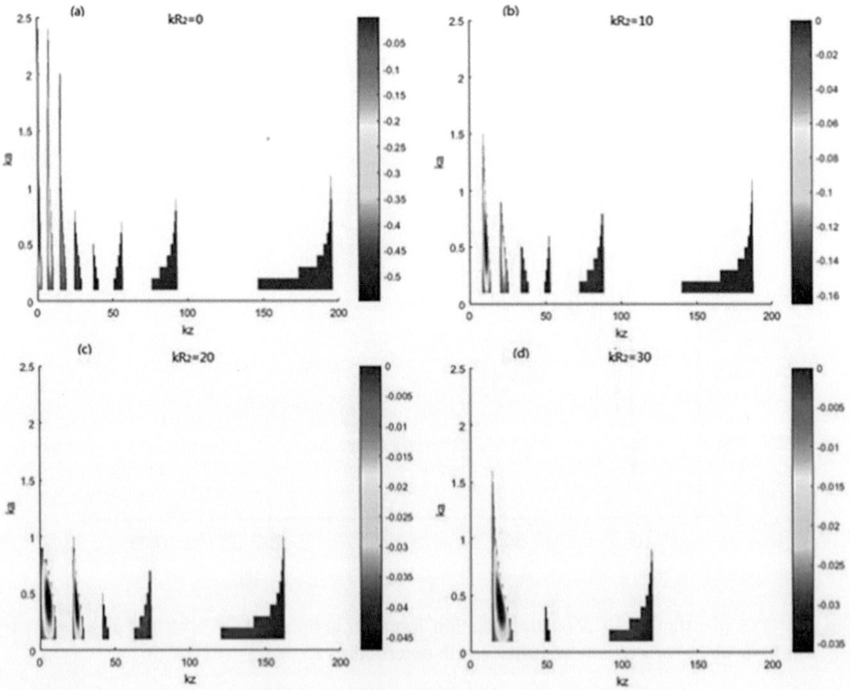

Fig. 6.7 The distribution of acoustic radiation force on a rigid sphere in water for a ring transducer with fixed kR_1 at 50, and kR_2 increasing by 10 each time

$$S_n = -\frac{\begin{vmatrix} \gamma j_n(ka) & j_n(k_1 a) \\ j_n'(ka) & j_n'(k_1 a) \end{vmatrix}}{\begin{vmatrix} -\gamma h_n^{(1)}(ka) & j_n(k_1 a) \\ -h_n^{(1)'}(ka) & j_n'(k_1 a) \end{vmatrix}} \tag{6.20}$$

where $\gamma = \rho c/\rho_1 c_1$, ρ_1 and c_1 represent the density and sound speed of the liquid inside the sphere, and k_1 This represents the wave number inside the sphere. Suppose a liquid sphere submerged in water is a benzene sphere, the density and speed of sound of water can be found to be 1000 kg/m^3 and 1500 m/s respectively, and the density and speed of sound of benzene are 880 kg/m^3 and 1298 m/s respectively. According to Eq. (6.19), the acoustic radiation force on a benzene sphere in water from a piston transducer with different outer radii is calculated, as shown in Fig. 6.9. The radiation force experienced by the benzene sphere is slightly larger than that of a rigid immovable sphere under the same conditions. In addition, the acoustic radiation tension zone of the transducer on the benzene sphere in water has different characteristics compared to the tension zone of the rigid sphere: First, for a large-radius benzene sphere, the sphere particles will still be affected by the acoustic radiation tension force when they are far from the transducer, but they may not be

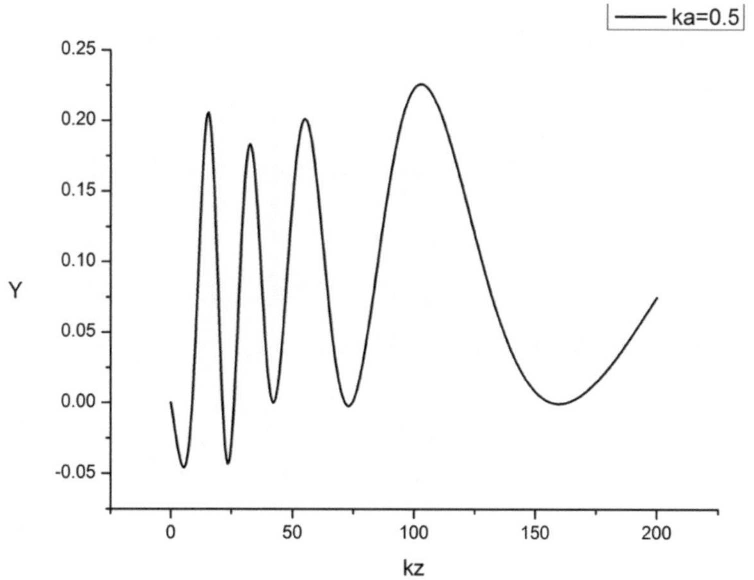

Fig. 6.8 The distribution of acoustic radiation force on a rigid sphere with radius $a = \lambda/4\pi$ ($ka = 0.5$) in water for a ring piston transducer with outer radius $R_1 = \frac{25\lambda}{\pi}$ ($kR_1 = 50$) and inner radius $R_2 = \frac{10\lambda}{\pi}$ ($kR_2 = 20$)

affected by the radiation tension force when they are close to the transducer, which is exactly the opposite of the tension zone characteristics of the rigid immovable sphere. For example, when $a = 5\lambda/4$ ($ka = 2.5$), $R_1 = \frac{25\lambda}{\pi}$ ($kR_1 = 50$), the acoustic radiation tension zones kz from near to far are 20.0–23.2, 32.2–36.4, 50.6–56.2, 86.4–92.4, 189.4–195, etc. When the radius is $a = \lambda/4$ ($ka = 1$), $R_1 = \frac{25\lambda}{\pi}$ ($kR_1 = 50$), the acoustic radiation tension zones kz from near to far are 3.6–6.4, 11.0–14.2, 20.2–24, 51.8–56.8, 88.0–93.0, 190.8–195.6, etc. Second, compared to the rigid immovable sphere, the distribution of the acoustic radiation tension zone of the benzene sphere is more regular, and the change in the radius of the sphere particles will not significantly affect the position and size of the tension zone, ensuring the universality of sphere particles of different radii. In addition, the comparison of the four figures in Fig. 6.9 shows that, similar to the rigid immovable sphere, an increase in the outer diameter will cause the acoustic radiation tension zone to move away from the transducer, and a new sound tension zone will be generated in the area close to the transducer. To compare the differences in the radiation force curves experienced by the benzene sphere and the rigid sphere, we have drawn the acoustic radiation force diagram of the piston transducer on the axial radius $R_1 = \frac{25\lambda}{\pi}$ ($kR_1 = 50$) benzene sphere, as shown in Fig. 6.10. Compared to the rigid sphere, the distribution curve of the acoustic radiation force function of the benzene sphere is symmetrical about the horizontal axis in the near field, and as the distance from the sphere to the transducer increases, its maximum and minimum values continue to strictly decrease. The acoustic radiation

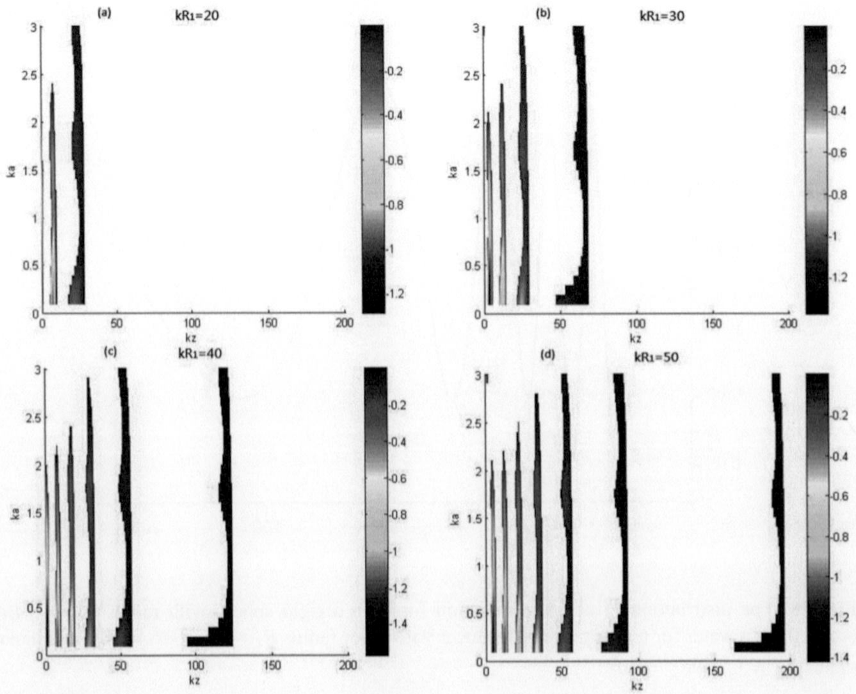

Fig. 6.9 The distribution of the acoustic radiation tension force on the benzene sphere in water as the piston transducer kR_1 increases by 10 each time

force of the piston transducer is relatively dense when it is close to the transducer, and as the kz value increases, the acoustic radiation force curve gradually loosens.

In order to observe the relationship between the inner radius and the radiation tension zone of the benzene sphere, we set the outer radius to $R_1 = \frac{25\lambda}{\pi}$ ($kR_1 = 50$), and the inner radius R_2 increases from 0 to $R_2 = \frac{15\lambda}{\pi}$ ($kR_2 = 30$). The acoustic radiation tension zone of the benzene sphere on the central axis is shown in Fig. 6.11. The overall trend of change is similar to that of the rigid sphere, which is not repeated here. Similarly, Fig. 6.12 shows the distribution of the acoustic radiation force on the benzene sphere with a radius of $a = \lambda/4\pi$ (ka = 0.5) in water from the annular piston transducer with an outer radius of $R_1 = \frac{25\lambda}{\pi}$ ($kR_1 = 50$) and an inner radius of $R_2 = \frac{10\lambda}{\pi}$ ($kR_2 = 20$). Compared to Fig. 6.10, its acoustic radiation force curve becomes sparser, and the maximum and minimum values decrease. This indicates that as the inner radius increases, the number of acoustic radiation force tension zones decreases and the width increases.

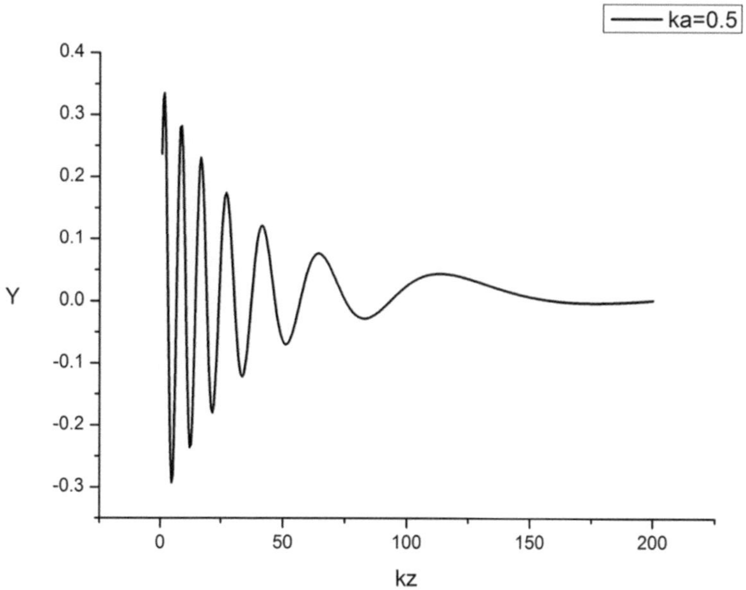

Fig. 6.10 The distribution of the acoustic radiation tension force on the benzene sphere with a radius of $a = \lambda/4\pi$ ($ka = 0.5$) in water for the piston transducer with an outer radius of $R_1 = 25\lambda/\pi$ ($kR_1 = 50$)

6.3 History of Research on Bessel Wave Acoustic Radiation Force

Non-diffractive waves, such as Bessel waves, Airy waves, X-waves, etc., maintain their energy localization distribution during propagation, and have characteristics such as energy concentration and long propagation distance. These non-diffractive waves can be widely used in the fields of ultrasound imaging and communication. The acquisition of Bessel waves has been successfully achieved in optical experiments. In 2003, J.C. Gutiérrez-Vega and others successfully obtained Bessel-Gaussian waves using a spherical coupling prism [22]. Subsequently, A. Hakola and others directly used diode-pumped high-power Nd-Yag lasers and plane mirrors and diffraction mirrors to achieve low-order Bessel-Gaussian wave light [23]. In recent years, the research on the acoustic radiation force of Bessel waves on spherical particles has been flourishing. Mitri and Wang proposed to obtain the Bessel-Gaussian wave of sound waves using a finite aperture method [24, 25]. Since 2006, Marston first studied the low-order Bessel wave's acoustic radiation force on axial particles and discovered the existence of acoustic traction. The torque of Bessel waves and the acoustic radiation force of high-order Bessel waves have also been further studied [26–32].

When the distribution of Bessel waves is affected by Gaussian components, the Bessel-Gaussian wave has the non-diffractive characteristics of Gaussian waves and the energy focusing characteristics of Gaussian waves. The acoustic radiation force

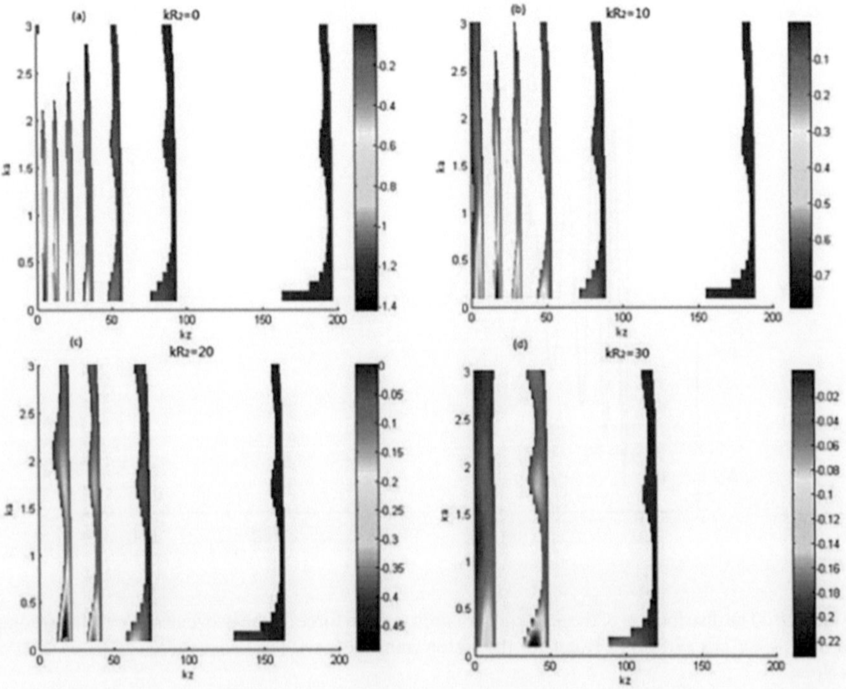

Fig. 6.11 The distribution of the acoustic radiation tension force on the benzene sphere submerged in water as the annular transducer is fixed at $kR_1 = 50$, and kR_2 increases by 10 each time

will show phenomena different from general Bessel waves. Similar to Gaussian waves, the beam waist radius varies at the focus and non-focus points. Its complete form is expressed as

$$G(R, z) = \frac{\exp\left(-\left(\frac{R}{w}\right)^2\right)\exp[ikz + ik(R^2/2(rwf))]}{(1 + iz/b)} \tag{6.21}$$

where $w = W_0[1 + (z/b)^2]^{1/2}$, $rwf = z + b^2/z$ and $b = k\left(W_0{}^2\right)/2$. Here we only discuss the Bessel-Gaussian wave at the focus point, that is, the acoustic radiation force of the quasi-Bessel-Gaussian wave on the small ball on the axis. Then the condition $z/b << 1$ needs to be satisfied. This section uses the method of series expansion to obtain the beam factor of the quasi-Bessel-Gaussian wave and calculate the acoustic radiation force on the spherical particles on the axis.

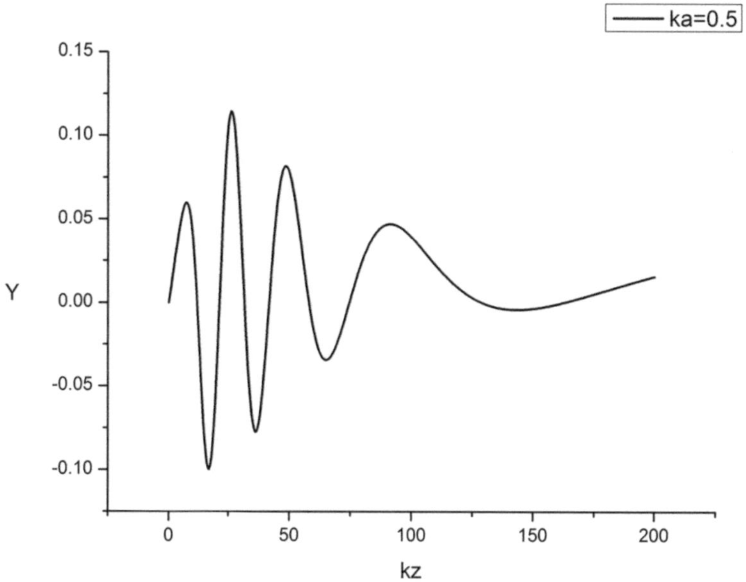

Fig. 6.12 The distribution of acoustic radiation force on a phenyl sphere with radius $a = \lambda/4\pi(ka = 0.5)$ in water for a ring-shaped piston transducer with outer radius $R_1 = \frac{25\lambda}{\pi}(kR_1 = 50)$ and inner radius $R_2 = \frac{10\lambda}{\pi}(kR_2 = 20)$

6.3.1 Theoretical Calculation of Acoustic Radiation Force of Zero-Order Quasi-Bessel-Gaussian Wave

The schematic diagram of the zero-order quasi-Bessel-Gaussian incident wave and the axial particle is shown in Fig. 6.13. The mathematical expression of the zero-order quasi-Bessel-Gaussian wave is

$$\phi_i = \phi_0 J_0(k_r r_\perp) e^{-(k_r r_\perp / k W_0)^2} e^{i(k_z z - \omega t)} \tag{6.22}$$

where ϕ_0 represents the amplitude of the velocity potential, $k_z = k\cos\beta$ and $k_r = k\sin\beta$ represent the axial and radial wave numbers respectively. $z = r\cos\theta$ and $r_\perp = r\sin\theta$ represent the axial and radial lengths respectively, the angle β represents the semi-cone angle of the Bessel wave. ω is the vibration angular frequency. W_0 is the beam waist radius of the Gaussian component at the focus, J_0 is the 0th order cylindrical Bessel function, representing the Bessel component in the Bessel-Gaussian wave. When the beam waist radius W_0 is infinitely large, the original waveform degenerates into a 0th order Bessel wave. From Eq. (6.21), the quasi-Gaussian wave needs to satisfy the condition $z/b << 1$, where $b = k(W_0^2)/2$. Therefore, Eq. (6.22) needs to finally satisfy the condition $2kz \ll (kW_0)^2$.

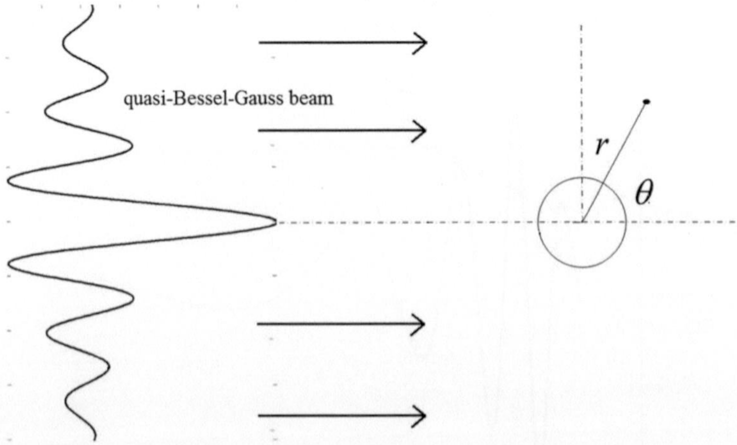

Fig. 6.13 Geometric schematic diagram of zero-order quasi-Bessel-Gaussian incident wave and axial particle

This section uses the method of series expansion to obtain the beam factor of the quasi-Bessel-Gaussian wave and calculate the acoustic radiation force on the spherical particles on the axis. According to the strict spherical decomposition formula of the zero-order Bessel wave

$$\begin{aligned}\phi_i =&\phi_0 J_0(kr\sin\theta\sin\beta)e^{ikr\cos\theta\cos\beta}\\=&\phi_0\sum_{n=0}^{\infty}(2n+1)i^n j_n(kr)P_n(\cos\theta)P_n(\cos\beta)\end{aligned}\tag{6.23}$$

Assume the spherical decomposition formula of the quasi-Bessel-Gaussian beam

$$\begin{aligned}\phi_i &= \phi_0 e^{-i\omega t}J_0(kr\sin\theta\sin\beta)e^{-\left(\frac{kr\sin\theta\sin\beta}{kW_0}\right)^2}e^{ikr\cos\theta\cos\beta}\\&= \phi_0 e^{-i\omega t}\sum_{n=0}^{\infty}(2n+1)i^n j_n(kr)P_n(\cos\theta)C_n P_n(\cos\beta)\end{aligned}\tag{6.24}$$

where C_n represents the beam factor of this waveform, which is related to the waist radius of the Gaussian component. Whether in electromagnetic fields or acoustic fields, the finite series expansion method has been used to calculate the beam factor of Gaussian waves [33–35]. Here we use this method to calculate the beam factor of the quasi-Bessel-Gaussian wave. For convenience of calculation, we define $s = \frac{1}{kW_0}$, and we can obtain the beam factor C_n.

It is known that $J_0(x)e^{-(sx)^2}$ Taylor series can be expressed as

$$J_0(x)e^{-(sx)^2} = \sum_{n=0}^{\infty}(-1)^n\frac{x^{2n}}{2^{2n}n!n!}\sum_{m=0}^{\infty}\left[\frac{(-1)^m s^{2m}x^{2m}}{m!}\right]$$

$$= \sum_{n=0}^{\infty}\sum_{m=0}^{n} \frac{(-1)^n s^{(2n-2m)} x^{2n}}{\left[2^{2m} m! m! (n-m)!\right]} \tag{6.25}$$

Then, according to the special function section in Watson's book [36], suppose there is

$$x^{\frac{1}{2}} g(x) = \sum_{n=0}^{\infty} c_n J_{n+\frac{1}{2}}(x) \tag{6.26}$$

$$j_n(x) = \sqrt{\frac{\pi}{2x}} J_{n+\frac{1}{2}}(x) \tag{6.27}$$

There is also the Maclaurin series of $g(x)$:

$$g(x) = \sum_{n=0}^{\infty} b_n x^n \tag{6.28}$$

Then we can obtain the beam factor c_n

$$c_n = \left(n + \frac{1}{2}\right) \sum_{m=0}^{\le n/2} 2^{\left(n-2m+\frac{1}{2}\right)} \frac{\Gamma\left(\frac{1}{2+n-m}\right)}{m!} b_{n-2m} \tag{6.29}$$

We set in Eq. (6.24) $\theta = \frac{\pi}{2}$, $\beta = \frac{\pi}{2}$, use x to replace kr to simplify the calculation process, then the equation can be transformed into

$$J_0(x) e^{-(sx)^2} = \sum_{n=0}^{\infty} (2n+1) i^n j_n(x) P_n(0) C_n P_n(0) \tag{6.30}$$

According to Eq. (6.25), the left side of Eq. (6.30)

$$J_0(x) e^{-(sx)^2} = \sum_{l=0}^{\infty} B_{2l} x^{2l} \tag{6.31}$$

According to Eq. (6.25), we can obtain the value of B_n under even conditions

$$B_{2l} = \sum_{j=0}^{l} \frac{(-1)^l s^{2l-2j}}{\left(j!^2 4^j (l-j)!\right)} \tag{6.32}$$

Multiply both sides of Eq. (6.30) by $x^{\frac{1}{2}}$

$$x^{\frac{1}{2}} J_0(x) e^{-(sx)^2} = \sum_{n=0}^{\infty} (2n+1) i^n \sqrt{\frac{\pi}{2}} J_{n+\frac{1}{2}}(x) P_n(0) C_n P_n(0) \tag{6.33}$$

Finally, we obtain the beam factor under even conditions C_n. When n is an even number, we use $2l$. to replace n, then the beam factor can be expressed as

$$C_{2l} = (-1)^l \sum_{j=0}^{l} \frac{2^{(2l-2j)} \Gamma\left(\frac{1}{2} + 2l - j\right) B_{2l-2j}}{\left(\sqrt{\pi} P_{2l}^2(0) j!\right)} \tag{6.34}$$

Similarly, when n is an odd number, differentiate both sides of Eq. (6.24) with respect to θ and β in turn. Then set $\theta = \frac{\pi}{2}$, $\beta = \frac{\pi}{2}$. Using the same method, we obtain the beam factor when n is an odd number.

$$C_{2l+1} = (-1)^l \sum_{j=0}^{l} \frac{2^{(2l-2j+1)} \Gamma\left(\frac{3}{2} + 2l - j\right) B_{2l-2j+1}}{\left(\sqrt{\pi} P_{2l+1}^2(0) j!\right)} \tag{6.35}$$

where Γ is the Gamma function, $P_{2l}(0)$ is the $2l$ order Legendre function, $P'_{2l+1}(0)$ is the derivative of the $2l + 1$ order Legendre function.

To verify the correctness of this method, we calculate the waveform comparison between the series expansion method and the theoretical value under different waist radius conditions, as shown in Figs. 6.14, 6.15 and 6.16. The results show that the two results are almost identical, proving the feasibility of this method.

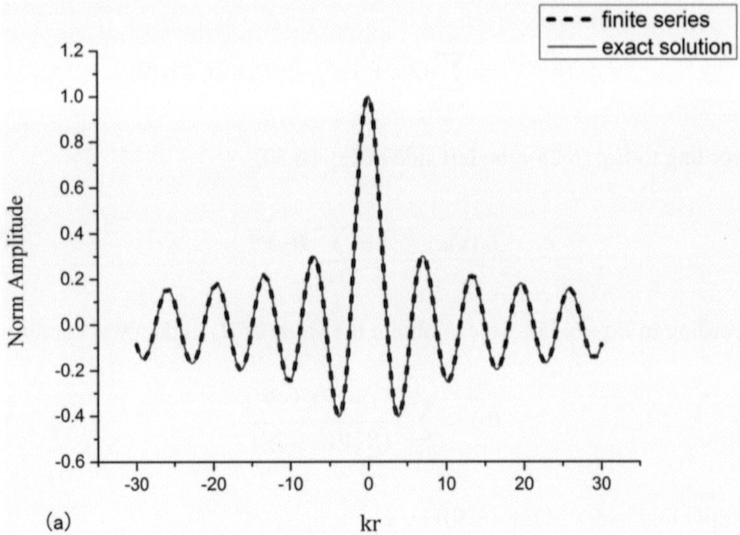

Fig. 6.14 Comparison of waveform between series expansion method and theoretical value when $W_0 = 10\lambda$

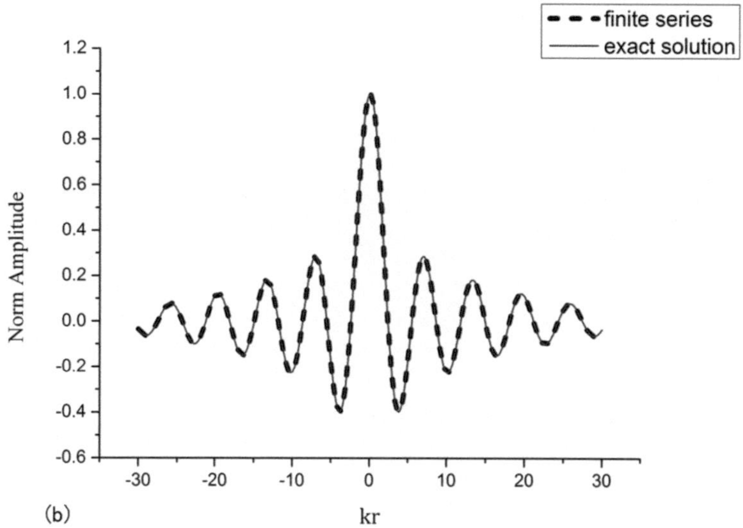

Fig. 6.15 Comparison of waveform between series expansion method and theoretical value when $W_0 = 5\lambda$

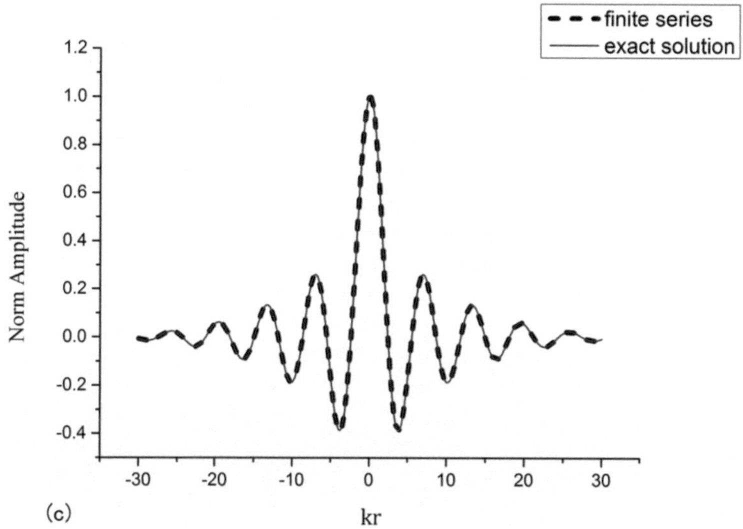

Fig. 6.16 Comparison of waveform between series expansion method and theoretical value when $W_0 = 3\lambda$

The acoustic radiation force function of the zero-order Bessel-Gaussian wave can be obtained:

$$Y = \frac{4}{(ka)^2}\sum_{n=0}^{\infty}(n+1)(-1)^n \left[\begin{array}{l} Re\left(\Lambda_n\Lambda_{n+1}^*\right)(\alpha_n+\alpha_{n+1}+2\alpha_n\alpha_{n+1}+2\beta_n\beta_{n+1}) \\ +Im\left(\Lambda_n\Lambda_{n+1}^*\right)(\beta_{n+1}-\beta_n+2\alpha_n\beta_{n+1}-2\alpha_{n+1}\beta_n) \end{array}\right] P_n(\cos\beta)P_{n+1}(\cos\beta) \tag{6.36}$$

6.3.2 Simulation and Discussion

In this section, we separately calculate the simulation of the quasi-zero-order Bessel-Gaussian beam with waist radius divided into $W_0 = 3\lambda$, $W_0 = 5\lambda$ and $W_0 = 10\lambda$ are used to analyze the acoustic radiation force on spherical particles of different properties on the axis. It should be noted that for three different waist radii, the condition $2kr_0 \ll (kW_0)^2$ is satisfied.

Using the boundary conditions and scattering method to calculate the general formula for the acoustic radiation force of a sphere, we calculated the acoustic radiation force curve of a rigid sphere on the axis under the quasi-zero-order Bessel-Gaussian traveling wave of different waist radii, as shown in Fig. 6.17. With a semi-cone angle of $\beta = \frac{\pi}{6}$, the acoustic radiation force curve of a rigid sphere on the axis under the quasi-zero-order Bessel-Gaussian traveling wave with different waist radii changes with the sphere radius, as shown in Fig. 6.17a. It can be observed that the radiation force gradually increases with the radius, reaching a maximum at $kr_0 = 2$ and then rapidly decreasing. After $kr_0 \geq 5$, the influence of the waist radius on the radiation force begins to manifest, for example, when $W_0 = 3\lambda$, as shown in Fig. 6.17b, there are noticeable fluctuations within the radius range $7 \leq kr_0 \leq 10$. When the semi-cone angle is $\beta = \frac{\pi}{3}$ it is also found that the radiation force gradually increases with the radius, reaching a maximum at $kr_0 = 2$ and then slowly decreasing. Similarly, after $kr_0 \geq 5$ the influence of the waist radius on the radiation force begins to manifest, and as the waist radius gradually decreases, the acoustic radiation force received by the large radius spherical particles is greater than that of the small waist radius. This change may be due to the fact that when kr_0 is very small compared to kW_0 the energy reduction at the edge of the Bessel wave caused by the waist radius has little effect on the acoustic radiation force. When kr_0 increases to not far from kW_0, the decrease in edge energy may cause a significant change in radiation force.

Unlike rigid spheres, liquid spheres are widely used in biomedicine. For example, manipulating blood cells in plasma and achieving directional movement of blood cells has significant practical significance. The acoustic radiation force curve of a liquid sphere on the axis under the quasi-zero-order Bessel-Gaussian traveling wave with

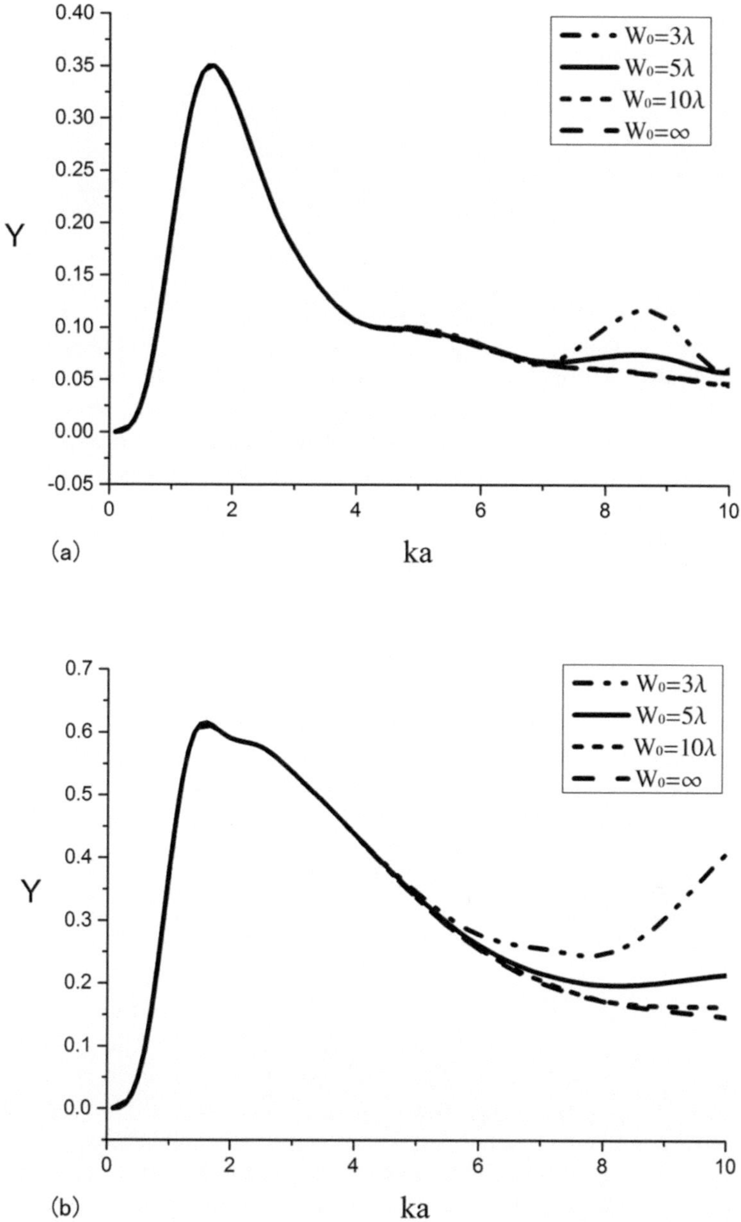

Fig. 6.17 The acoustic radiation force of a rigid sphere on the axis under the quasi-Bessel-Gaussian wave with different waist radii ($W_0 = 3\lambda$, $W_0 = 5\lambda$, $W_0 = 10\lambda$, $W_0 = \infty$) under the condition of a semi-cone angle $\beta = \frac{\pi}{6}$ (**a**) and $\beta = \frac{\pi}{3}$ (**b**)

different waist radii is shown in Fig. 6.18. Assuming that the liquid sphere immersed in water is a benzene sphere, the density and sound speed of water are 1000 kg/m^3 and 1500 m/s respectively, and the density and sound speed of benzene are 880 kg/m^3 and 1298 m/s respectively. Compared to rigid spheres, due to the vibration characteristics of the liquid sphere itself [37], the acoustic radiation force curve of the benzene sphere has several irregular protrusions and depressions. Comparing the two figures in Fig. 6.18, it can be seen that the positions of the irregular protrusions and depressions in the acoustic radiation force curve do not change much when the semi-cone angles β are inconsistent. Figure 6.18b shows that there is a reverse pulling force at $ka = 2.55$, which was first discovered by Marston [30]. When the waist radius decreases, the change in the acoustic radiation force curve with the sphere radius is similar to that of a rigid sphere, that is, after $kr_0 \geq 5$, the influence of the waist radius on the radiation force begins to manifest, and as the waist radius gradually decreases, the acoustic radiation force received by the large radius spherical particles is greater than that of the small waist radius. The study results of the acoustic radiation pulling force of the zero-order quasi-Bessel-Gaussian wave on the axial liquid sphere particles are beneficial for the realization of single sound source acoustic tweezers in the future.

6.4 Acoustic Radiation Force of Airy-Gaussian Waves

Airy-Gaussian waves are a generalized form of Airy waves, possessing similar characteristics: the shape of the sound intensity distribution remains approximately constant during sound propagation (non-diffractive propagation), they can self-repair when encountering obstacles (self-healing), and the sound waves propagate along a curve (self-accelerating). Airy-Gaussian waves were introduced by Bandres and Gutierrez-Vega [38] who proposed that both infinite energy [39] and finite energy [40] Airy sound waves are special cases of Airy-Gaussian waves. In optics, Airy-Gaussian waves have been extensively studied [41], such as the propagation characteristics of a single Airy-Gaussian wave in different media [42–44], and the interaction of two Airy-Gaussian waves in Kerr media [45] and nonlocal nonlinear media [46]. However, in acoustics, although Mitri studied the acoustic radiation force of Airy waves on cylindrical particles in water [47] or elastic media [48] in 2015, research on the acoustic radiation force of Airy-Gaussian waves is still lacking.

This section will introduce the application of Airy-Gaussian waves in acoustics using the scattering method, provide the expression of the acoustic radiation force on cylindrical particles, and demonstrate its acoustic tweezer effect on cylindrical particles.

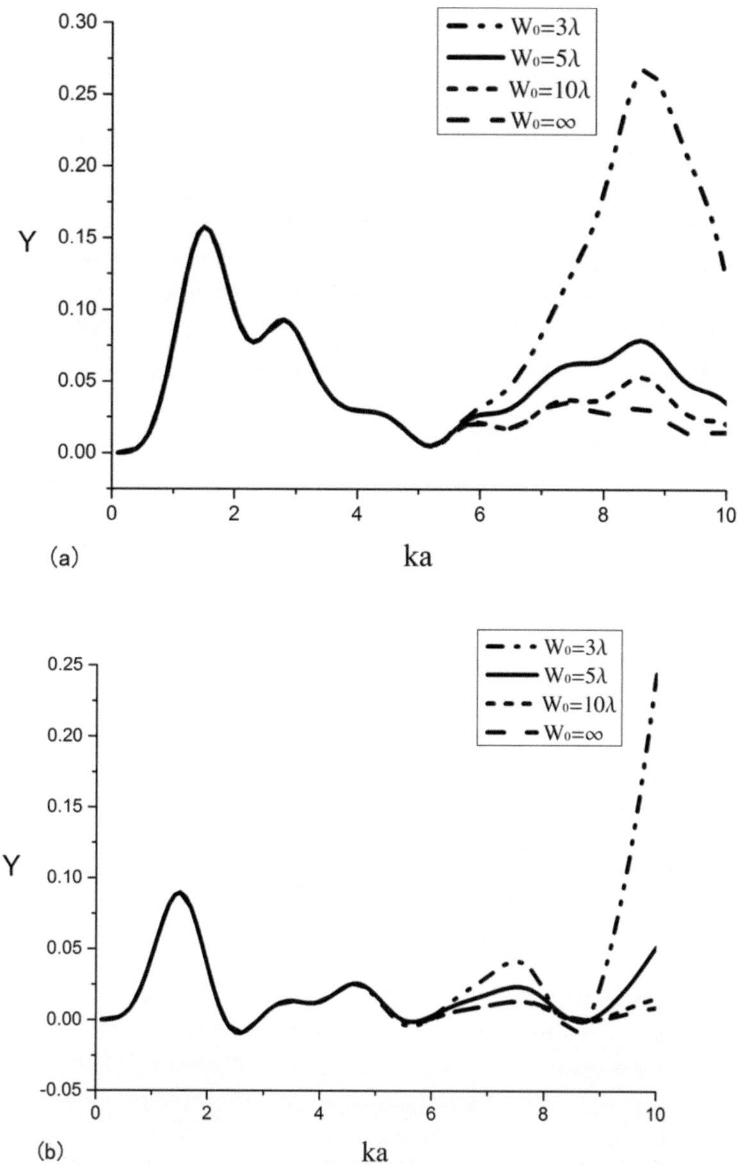

Fig. 6.18 The acoustic radiation force of a benzene sphere on the axis under the quasi-Bessel-Gaussian wave with different waist radii ($W_0 = 3\lambda$, $W_0 = 5\lambda$, $W_0 = 10\lambda$, $W_0 = \infty$) under the condition of a semi-cone angle $\beta = \frac{\pi}{6}$ (**a**) and $\beta = \frac{\pi}{3}$ (**b**)

6.4.1 Model and Calculation

As shown in Fig. 6.19, ideally, it is assumed that two Airy-Gaussian waves in water medium are directly facing the Z-axis of the cylindrical particle, propagating in the x-axis direction. The cylindrical microparticle is located in the Cartesian coordinate system $x_0 - o - y_0$, where a and (x_c, y_c) are the radius and center coordinates of the microparticle, respectively. Figure 6.19a shows the incident velocity potential of two Airy-Gaussian sound beams, where the phase difference is $Q = 0$. When the sound wave encounters the particle, scattering and refraction occur. For convenience, all formulas here omit the time factor $\left(e^{-iwt}\right)$, and the velocity potential of the two Airy-Gaussian waves at the sound field $x = 0$ is:

$$\Phi_i^{(0)}(x=0,y) = A_1 Ai\left(\frac{y-B}{y_0}\right)e^{\alpha\left(\frac{y-B}{y_0}\right)}e^{-\frac{(y-B)^2}{w_0^2}} + A_2 e^{iQ} Ai\left(\frac{-y-B}{y_0}\right)e^{\alpha\left(\frac{-y-B}{y_0}\right)}e^{-\frac{(-y-B)^2}{w_0^2}} \tag{6.37}$$

where, A_1, A_2 are the amplitudes of the initial velocity potential of the two Airy-Gaussian waves, $Ai(\cdot)$ is the Airy function, y_0 is the control factor of the Airy function, α is the decay coefficient, Q is the phase difference between the two Airy-Gaussian waves, B is the distance between the two Airy-Gaussian waves, and w_0 is the Gaussian factor beam width.

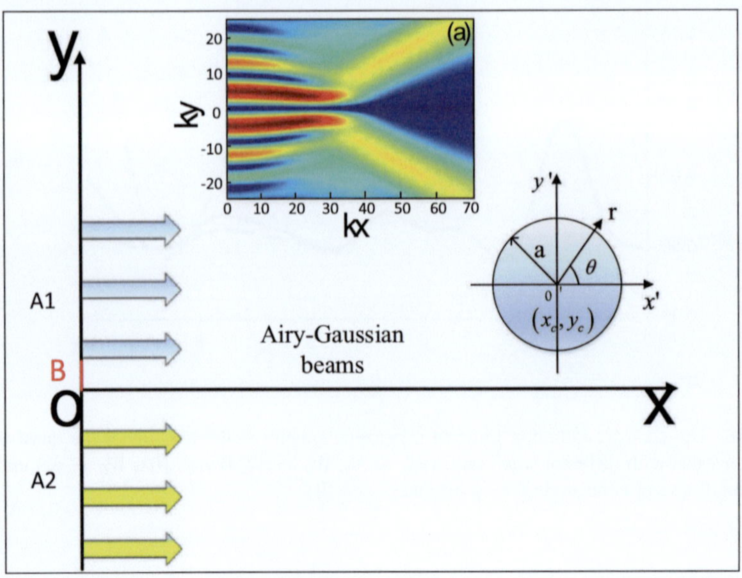

Fig. 6.19 Schematic diagram of the interaction between two Airy-Gaussian sound beams and a cylindrical particle

Based on the angular spectrum analysis method [49], the incident velocity potential of the sound field in the Cartesian coordinate system can be expressed as:

$$\Phi_i = \int_{-\infty}^{\infty} g(p, q) e^{ik(px+qy)} dq \tag{6.38}$$

where, k is the wave number, β is the propagation direction of the incident wave and $p = \cos\beta$, $q = \sin\beta$, $g(p, q)$ is the angular spectrum function, which is determined by the initial value of the incident velocity potential. The angular spectrum function can be obtained through the inverse Fourier transform:

$$g(p, q) = \frac{\phi_0}{\lambda} \int_{-\infty}^{\infty} \Phi_i^{(0)}(x = 0, y) e^{-ik(px+qy)} dy \tag{6.39}$$

where, λ is the wavelength of the sound wave. By substituting Eq. (6.37) into Eq. (6.39) and applying the Airy transformation formula [50], the angular spectrum function formula is obtained as follows:

$$\begin{aligned} g(p, q) = {} & \frac{ky_0\sqrt{\pi}}{2\pi\chi} e^{\frac{1}{96\chi^6}} \left[A_1 e^{\left(\frac{\alpha - ikqy_0}{2\chi}\right)^2 + \frac{\alpha - ikqy_0}{8\chi^4} - ikqB} Ai\left(\frac{\alpha - ikqy_0}{2\chi^2} + \frac{1}{16\chi^4} \right) \right. \\ & \left. - A_2 e^{iQ} e^{\left(\frac{\alpha + ikqy_0}{2\chi}\right)^2 + \frac{\alpha + ikqy_0}{8\chi^4} + ikqB} Ai\left(\frac{\alpha + ikqy_0}{2\chi^2} + \frac{1}{16\chi^4} \right) \right] \end{aligned} \tag{6.40}$$

where, $\chi = \frac{y_0}{w_0}$ and ky_0 is a dimensionless factor.

By substituting Eq. (6.40) into Eq. (6.38), the incident velocity potential in the sound field is:

$$\begin{aligned} \Phi_i(x, y) = {} & \frac{ky_0\sqrt{\pi}}{2\pi\chi} e^{\frac{1}{96\chi^6}} \left[A_1 \int_{-\infty}^{\infty} e^{\left(\frac{\alpha - ikqy_0}{2\chi}\right)^2 + \frac{\alpha - ikqy_0}{8\chi^4} - ikqB} Ai\left(\frac{\alpha - ikqy_0}{2\chi^2} + \frac{1}{16\chi^4} \right) e^{ik(px+qy)} dq \right. \\ & \left. - A_2 e^{iQ} \int_{-\infty}^{\infty} e^{\left(\frac{\alpha + ikqy_0}{2\chi}\right)^2 + \frac{\alpha + ikqy_0}{8\chi^4} + ikqB} Ai\left(\frac{\alpha + ikqy_0}{2\chi^2} + \frac{1}{16\chi^4} \right) e^{ik(px+qy)} dq \right] \end{aligned} \tag{6.41}$$

Using the Jacobian expression [51], Eq. (6.43) can be rewritten in the cylindrical coordinate system as:

$$\Phi_i(r, \theta) = \sum_{n=-\infty}^{\infty} J_n(kr) e^{in\theta} \Lambda_n \tag{6.42}$$

where, $J_n(\cdot)$ is the n-th order Bessel function, Λ_n is the beam factor:

$$\Lambda_n = \frac{ky_0}{2\pi}\frac{\sqrt{\pi}}{\chi}e^{\frac{1}{96\chi^6}}i^n(a_n - b_n)$$

$$a_n = A_1\int_{-\infty}^{\infty} e^{\left(\frac{\alpha-ikqy_0}{2\chi}\right)^2+\frac{\alpha-ikqy_0}{8\chi^4}-ikqB}e^{\left(x_c\sqrt{1-q^2}+y_cq\right)}Ai\left(\frac{\alpha-ikqy_0}{2\chi^2}+\frac{1}{16\chi^4}\right)e^{-in\sin^{-1}q}dq$$

$$b_n = A_2e^{iQ}\int_{-\infty}^{\infty} e^{\left(\frac{\alpha+ikqy_0}{2\chi}\right)^2+\frac{\alpha+ikqy_0}{8\chi^4}+ikqB}e^{\left(x_c\sqrt{1-q^2}+y_cq\right)}Ai\left(\frac{\alpha+ikqy_0}{2\chi^2}+\frac{1}{16\chi^4}\right)e^{-in\sin^{-1}q}dq$$

Similarly, the scattering velocity potential of sound waves on cylindrical particles can be expressed as:

$$\Phi_s(r,\theta) = \sum_{n=-\infty}^{\infty} s_n H_n^{(1)}(kr)e^{in\theta}\Lambda_n \tag{6.43}$$

where, s_n is the scattering coefficient determined by the boundary conditions of the cylindrical particles, $H_n^{(1)}(\cdot)$ is the n-th order Hankel function of the first kind.

Based on the incident velocity potential and scattering velocity potential in the sound field, the longitudinal (Y_x) and transverse (Y_y) acoustic radiation force functions on cylindrical particles in the field of two Airy-Gaussian waves can be obtained [52]:

$$Y_x = \frac{1}{ka}\Im\left\{\sum_{n=-\infty}^{\infty}\Lambda_n(1+s_n)\left(\Lambda_{n+1}^* s_{n+1}^* - \Lambda_{n-1}^* s_{n-1}^*\right)\right\} \tag{6.44}$$

$$Y_y = -\frac{1}{ka}\Re\left\{\sum_{n=-\infty}^{\infty}\Lambda_n(1+s_n)\left(\Lambda_{n+1}^* s_{n+1}^* + \Lambda_{n-1}^* s_{n-1}^*\right)\right\} \tag{6.45}$$

where, a is the radius of the cylindrical particle, and the superscript $*$ denotes the complex conjugate.

6.4.2 Simulation and Analysis

In this section, we calculate and analyze the acoustic radiation force of two Airy-Gaussian beams on cylindrical particles, and discuss the influence of various beam parameters on the characteristics of the particle's acoustic radiation force. The cylindrical particles in this section are made of transparent synthetic resin with a density of $\rho_1 = 1191\,\text{kg/m}^3$, shear wave speed of $c_1 = 2690\,\text{m/s}$, longitudinal wave speed of $c_2 = 1340\,\text{m/s}$, the resin's absorption coefficient for shear waves is $\gamma_1 = 0.0035$, and for longitudinal waves is $\gamma_2 = 0.0053$ [53], the medium liquid is water with a density of $\rho_0 = 1000\,\text{kg/m}^3$, and sound speed of $c_0 = 1500\,\text{m/s}$. In the following simulations, we uniformly take $ky_0 = 4$, $\alpha = 0.01$, and the particle's radius size $ka = 0.1$.

6.4.2.1 The Influence of Phase Difference on the Acoustic Radiation Force of Particles

Figure 6.20 shows the radiation force field ($Y_\perp = Y_x\vec{e}_x + Y_y\vec{e}_y$) of two Airy-Gaussian beams on cylindrical particles at different phase differences, as well as the motion trajectories of particles under these force fields. The size and direction of the force field are indicated by vector arrows, the circle represents the initial position of the particle, and the triangle represents the position of the particle after a period of motion. Here, we take four phase differences: $Q = 0, \frac{\pi}{2}, \pi, \frac{3\pi}{2}$, and the beam width is $w_0 = \lambda$.

From Fig. 6.20a, c, it can be seen that when the phase difference is $Q = 0$, the acoustic radiation force vector is symmetrically distributed along the x-axis, and in the part near the x-axis, the radiation force vector points to the x-axis, these force vectors can push the particles towards the x-axis; when the phase difference is $Q = \pi$,

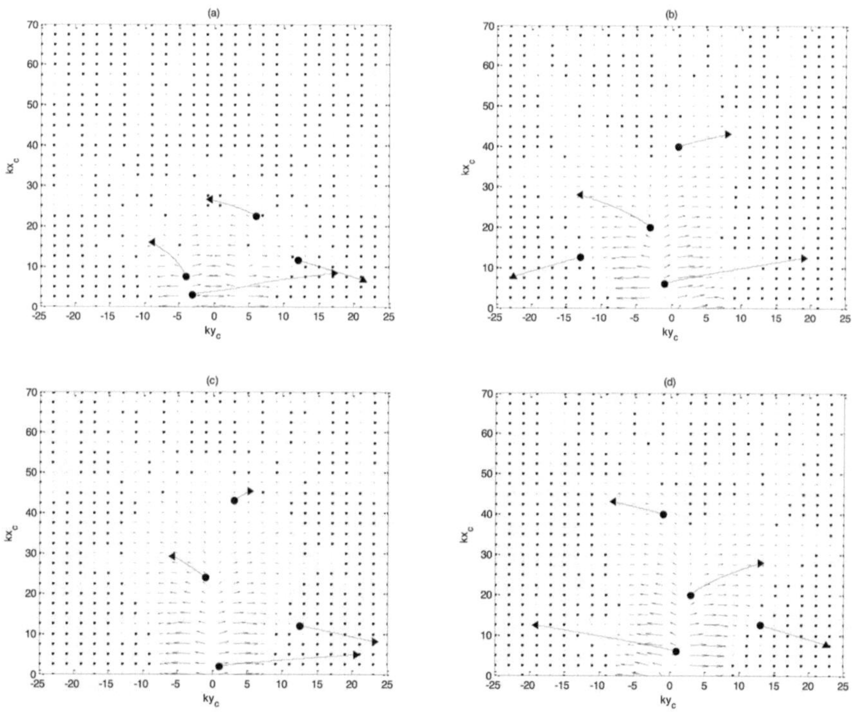

Fig. 6.20 The acoustic radiation force vector field of particles and the particle motion trajectory, where **a** $Q = 0$, **b** $Q = \frac{\pi}{2}$, **c** $Q = \pi$, **d** $Q = \frac{3\pi}{2}$

the radiation force vector is also symmetrically distributed along the x-axis, but unlike the phase difference $Q = 0$, here the radiation force vectors near the x-axis deviate from the x-axis, they will push the particles away from the x-axis; similarly, from Fig. 6.20b, d, it can be seen that when the phase difference is $Q = \frac{\pi}{2}, \frac{3\pi}{2}$, the radiation force vector is not symmetrically distributed along the x-axis, but if the force field at phase difference $Q = \frac{\pi}{2}$ is flipped 180° along the x-axis, it will be found that the force field at this time is consistent with the force field at phase difference $Q = \frac{3\pi}{2}$, and their radiation force vector arrows all deviate from the x-axis, they will push the particles away from the x-axis.

Here, we only consider the effects of radiation force and viscous force on the particle moving in the x-o-y plane. According to Newton's second law, the equation of motion for the particle is as follows:

$$m\mathbf{r}'' = \mathbf{F}_{rad} + \mathbf{F}_{drag} \tag{6.46}$$

where, m is the mass of the particle, $\mathbf{r}''\left(= \frac{d^2\mathbf{r}}{dt^2}\right)$ is the particle acceleration, t(s) is the time, $\mathbf{F}_{rad}(= \mathbf{Y}_{\perp} S_c E_0)$ is the radiation force vector, $\mathbf{F}_{drag}\left(= -32\eta a\mathbf{r}'/3\right)$ is the viscous force, $S_c(= 2al)$ is the cross-sectional area of the cylindrical particle, $E_0\left(= 1/2\rho_0 k^2 |\phi_0|^2\right)$ is the unit energy density, $\eta(= 0.001Pa \cdot s)$ is the dynamic viscosity of the medium, $\mathbf{r}' = d\mathbf{r}/dt$ is the velocity. The trajectory of the particle motion is obtained using Matlab software, implemented by the multi-step Runge–Kutta method. This method can obtain the position and velocity of the particle at the next moment based on the position and velocity of the particle at the previous moment. In the simulation, the initial velocity and initial time of the particle are assumed to be 0, and the beam intensity is assumed to be $3.3 \times 10^8 \mathrm{W/m^2}$.

In Fig. 6.20, the initial position of the particle is set to four in each phase simulation, and the motion of the particle at each position is assumed to consume the same time. As can be seen from Fig. 6.20, under the action of the radiation force, the particle is pushed away from the y-axis along a curved trajectory, or pulled closer to the y-axis. When the particle passes through an area with a larger radiation force, its trajectory is longer than that of other positions. It is worth noting that when the initial position of the particle in the phase difference $Q = \frac{\pi}{2}$ is symmetrical to the initial position of the particle in the phase difference $Q = \frac{3\pi}{2}$ with respect to the x-axis, their particle trajectories are also symmetrical to the x-axis, as shown in Fig. 6.20b, d.

Figure 6.21 shows the changes in the longitudinal and lateral acoustic radiation forces and the changes in the particle motion trajectory when the phase difference changes. Figure 6.21a describes the lateral acoustic radiation force under different phase differences at $kx_c = 10$. As can be seen from the figure, the lateral acoustic radiation force of each phase difference is both positive and negative, which can pull back or push away the particle; while Fig. 6.21b shows the longitudinal acoustic radiation force of different phase differences at $ky_c = 3$, their amplitudes are basically positive, only pushing the particle away. Figure 6.21c, d show the changes in the particle motion trajectory when the beam phase difference changes. In Fig. 6.21c, the particle first moves forward along the positive y-axis under the radiation force of phase difference $Q = \frac{\pi}{2}$ while maintaining forward along the x-axis. When the phase difference changes from $Q = \frac{\pi}{2}$ to $Q = 0$, the particle slows down until it turns to the negative y-axis. When the phase difference changes to $Q = \pi$, the particle turns to the positive y-axis again. Similarly, in Fig. 6.21d, as the beam phase difference changes, the direction of particle motion changes back and forth on the negative y-axis, while the particle continues to move forward along the positive x-axis. By observing the motion trajectory of the particle in Fig. 6.21 and the changes in radiation force with the change of phase difference, we can find that by changing the phase difference, the direction of particle motion on the y-axis can be changed, which means that the lateral acoustic radiation force plays an important role in changing the direction of particle motion.

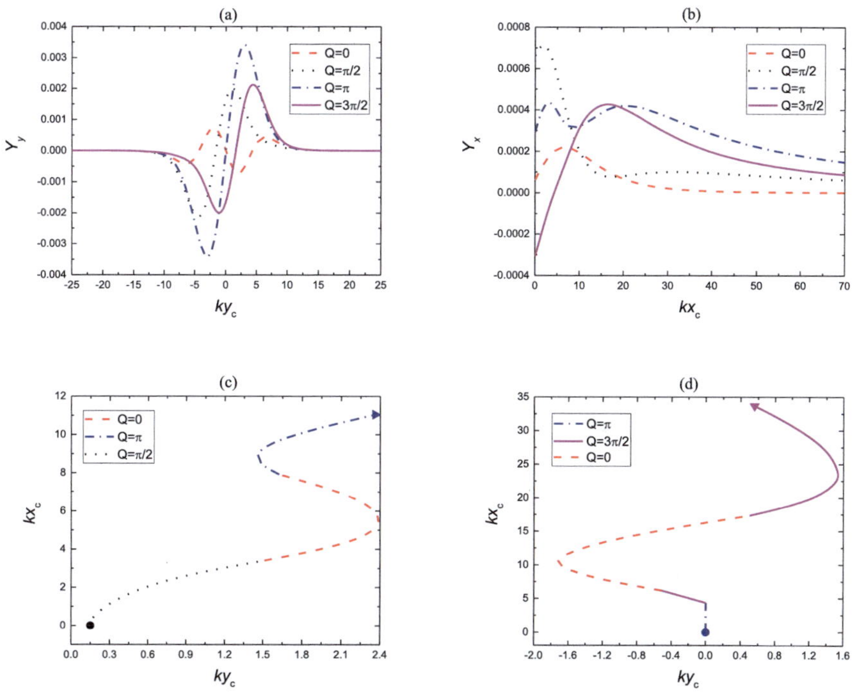

Fig. 6.21 The radiation force function curve and motion trajectory of the particle at different phase differences. **a** The lateral acoustic radiation force at $kx_c = 10$; **b** The longitudinal acoustic radiation force at$ky_c = 3$; **c** and **d** show the motion trajectory of the particle under the change of phase difference

6.4.2.2 The Effect of Beam Spacing on the Acoustic Radiation Force of Particles

The force of sound waves acting on particles generally manifests as a pushing force, but acoustic tweezers require not only a pushing force but also a pulling force. Here, not only can the pulling force be applied to the particle by adjusting the phase difference between the two Gaussian beams, but the sound radiation force that presents the pulling effect can also be obtained by changing the spacing between the two beams. Figure 6.22 shows the change in the acoustic radiation force acting on the particle at $x = 0$ with the change in spacing, where the amplitude of the two beams is set to $A_1/A_2 = 1/3$, and the phase difference is fixed at $Q = 0$.

Looking at the changes in the horizontal and vertical acoustic radiation forces with respect to the spacing in Fig. 6.22a–c, when the spacing varies between zero and negative values, the vertical acoustic radiation force is always positive, appearing in the positive y-axis region, with little change in amplitude. However, the curve of the horizontal acoustic radiation force shifts to the right. That is, initially, only a small portion of the horizontal acoustic radiation force on the right side of the y-axis exhibits a tensile effect. As the spacing decreases, the area exhibiting the tensile effect correspondingly shifts to the right. By controlling the spacing, the horizontal acoustic radiation force in the target area can be made to exhibit a tensile effect. In Fig. 6.22d–f, the change in spacing is opposite to that in Fig. 6.22a–c, with the spacing varying between 0 and 10. It can be observed that the vertical acoustic radiation force is essentially positive and appears in the negative y-axis region. As the spacing increases, the area on the negative y-axis exhibiting a tensile effect correspondingly shifts to the left. This means that by controlling the positive value of the spacing, any area on the negative y-axis can be made to exhibit a tensile effect.

6.4.2.3 The Impact of Amplitude on Acoustic Radiation Force on Particles

The previous section discussed the situation where the target area exhibits a tensile effect by controlling the spacing between the beams. This is achieved by shifting the curve of the horizontal acoustic radiation force to the left or right as the spacing changes. Here, we will study the effect of changes in beam amplitude on the acoustic radiation force. Figure 6.23 shows the horizontal and vertical acoustic radiation forces on a particle at $x = 0$, with a fixed phase difference of 0.

In Figs. 6.23a, b, the beam spacing is -2, and the amplitude ratios of the two beams are $A_1/A_2 = 1/3$ and $A_1/A_2 = 3/1$, respectively. It can be observed that their amplitude ratios are opposite, and the corresponding radiation force calculations are symmetrically distributed along the y-axis. That is, when the beam spacing remains constant, swapping the amplitude ratio can make the horizontal acoustic radiation force in the target area exhibit a tensile effect. In Fig. 6.23c, d, increasing the ratio A_1/A_2, it can be observed that the direction of the horizontal and vertical acoustic radiation forces remains unchanged, and the amplitude increases correspondingly.

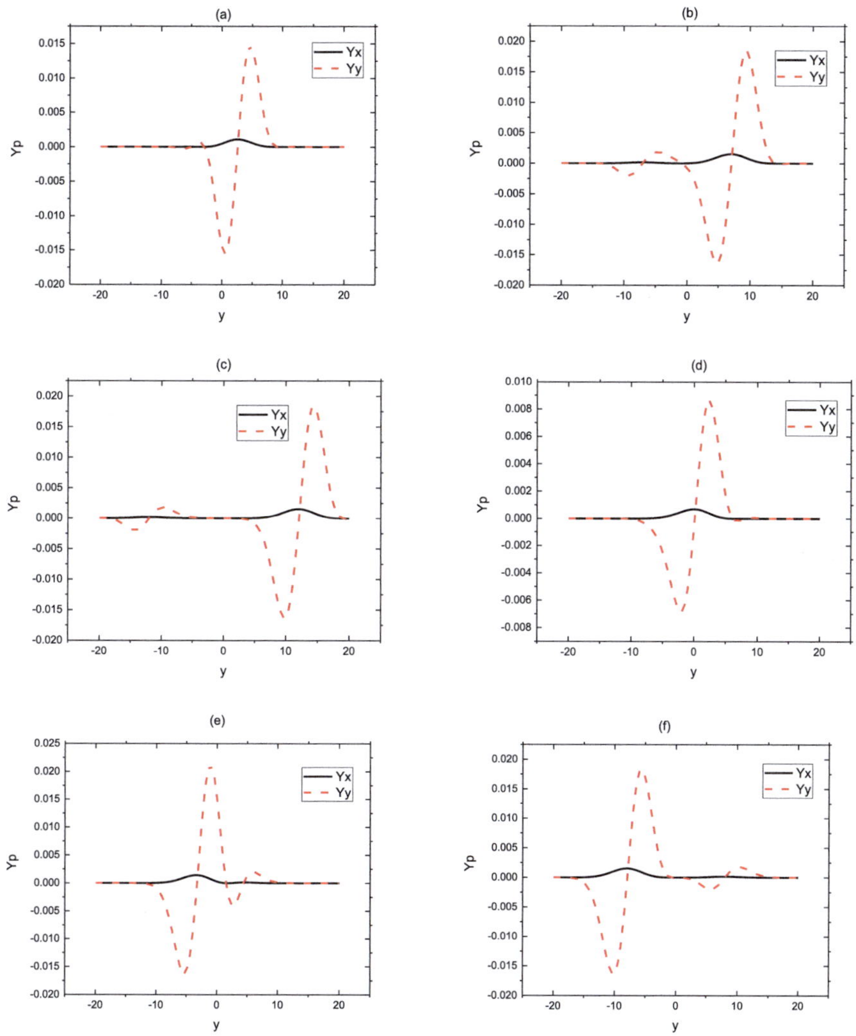

Fig. 6.22 The change in the lateral and longitudinal acoustic radiation force of the microparticle at x = 0 with the change in beam spacing, **a** $kB = 0$, **b** $kB = -5$, **c** $kB = -10$, **d** $kB = 2$, **e** $kB = 5$, **f** $kB = 10$

That is, by changing the amplitude ratio of the two beams, a larger amplitude of acoustic radiation force can be obtained. In the previous section, when discussing the effect of beam spacing on the acoustic radiation force, we fixed the amplitude ratio of the two beams at $A_1/A_2 = 1/3$. If the ratio is changed to $A_1/A_2 = 3/1$, and the beam spacing is negative, as shown in Figs. 6.23e, f, it can be observed that the smaller the spacing, the more the curve of the horizontal acoustic radiation force shifts to the left, and the area exhibiting a tensile effect also shifts to the

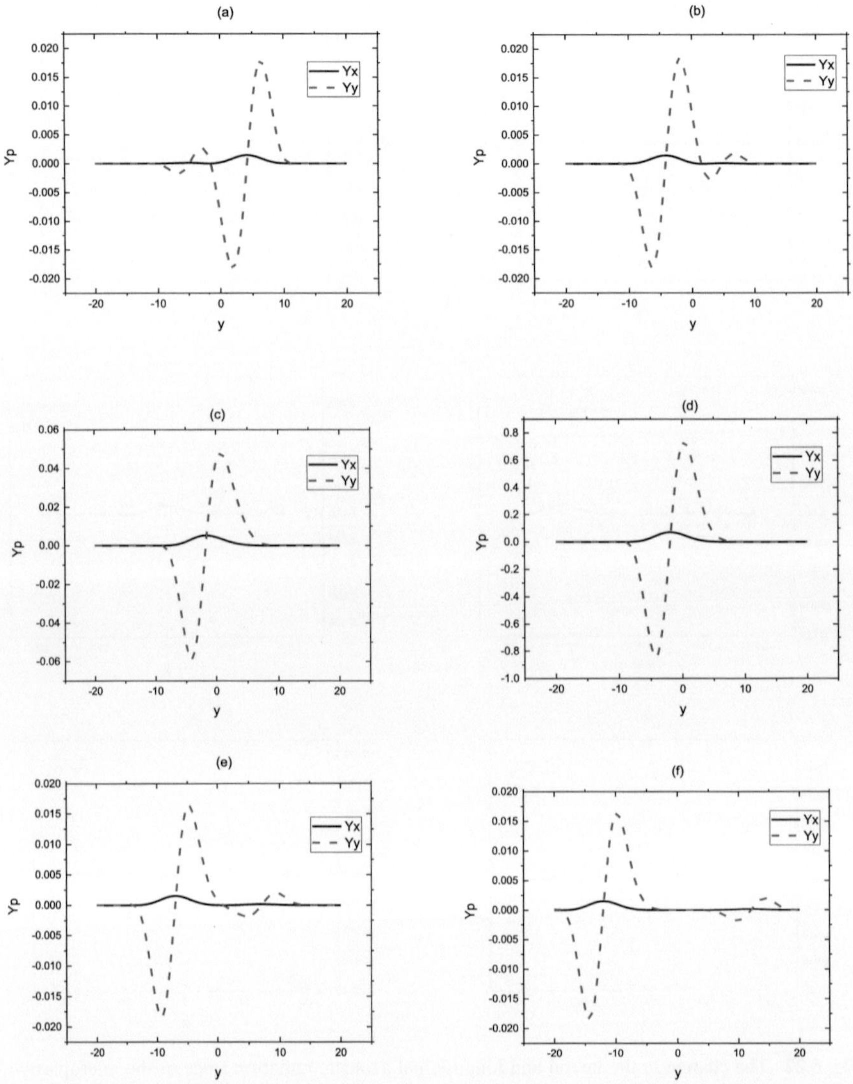

Fig. 6.23 Relationship between the horizontal and vertical acoustic radiation forces on a particle and the amplitude ratio of two sound waves. **a** $A_1/A_2 = 1/3$, $kB = -2$ **b** $A_1/A_2 = 3/1$, $kB = -2$ **c** $A_1/A_2 = 5/1$, $kB = 0$ **d** $A_1/A_2 = 10/1$, $kB = 0$ **e** $A_1/A_2 = 3/1$, $kB = -5$ **f** $A_1/A_2 = 3/1$, $kB = -10$

left. This is contrary to what was shown in Fig. 6.22a–c in the previous section. By comparing these calculation results, it can be found that the change in amplitude ratio not only affects the amplitude of the acoustic radiation force, but can also be used in conjunction with beam spacing to control the direction of the horizontal acoustic radiation force in the target area, i.e., whether a tensile or pushing force appears.

6.4.2.4 The Impact of Beam Width on Acoustic Radiation Force on Particles

We have already discussed the effects of phase difference, spacing, and amplitude ratio between two beams. Next, we will discuss the impact of the Gaussian factor in the beam, i.e., the effect of beam width on the acoustic radiation force. Here, we take the phase difference $Q = 0$, beam spacing $kB = 0$, and amplitude ratio $\frac{A_1}{A_2} = 1$.

The background of Fig. 6.24a is the standard normalized acoustic intensity of two Airy Gaussian incident waves at beam width $w_0 = \lambda(\frac{I}{I_0})$, where $I_0 = \frac{\rho_0 w^2 |\phi_0|^2}{(2c_0)}$, and the black arrows on the background image represent the acoustic radiation force vector. Similarly, Fig. 6.24c shows the standard normalized acoustic intensity and acoustic radiation force vector at beam width $w_0 = 4\lambda$. When discussing the impact of phase difference on particle motion, it was pointed out that the transverse acoustic radiation force plays a major role. Here, the changes in the transverse acoustic radiation force are shown separately, as in Fig. 6.24b, d, where the beam widths are $w_0 = \lambda$ and $w_0 = 4\lambda$ respectively.

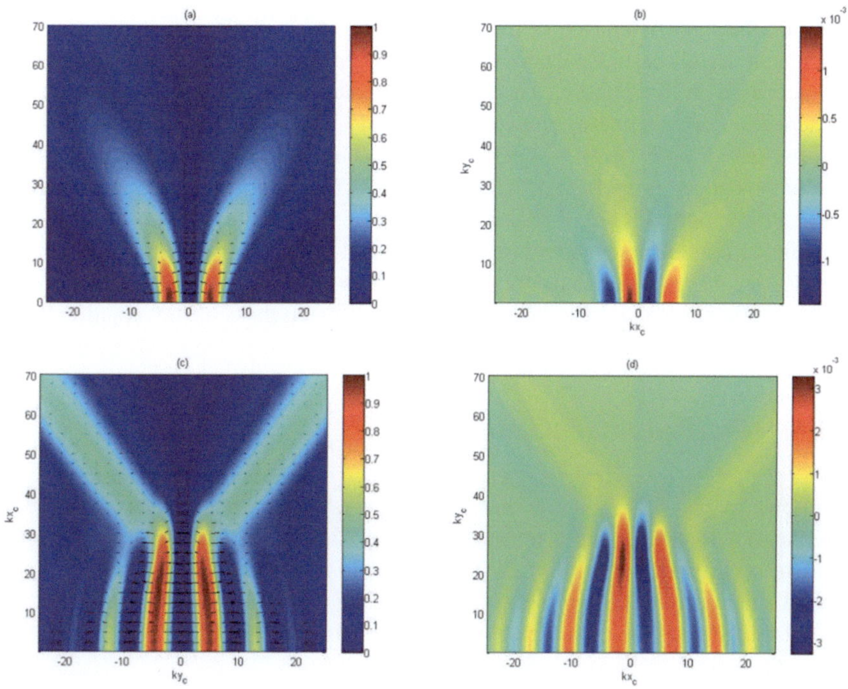

Fig. 6.24 The backgrounds of **a** and **c** represent the standard normalized acoustic intensity of two Airy Gaussian beams at different beam widths, where the black arrows indicate the acoustic radiation force vector; **b** and **d** represent the transverse radiation force function at different beam widths. **a–b** $w_0 = \lambda$, **c–d** $w_0 = 4\lambda$

By comparing Fig. 6.24a, c, it can be found that when the beam width decreases, the sidelobes of the standard normalized acoustic intensity of the incident wave decrease due to the truncation effect of the Gaussian factor. Similarly, by comparing the acoustic radiation force vectors, it can be found that as the beam width increases, the distribution of the acoustic radiation force vector becomes more complex, similar to the distribution of acoustic intensity, and the amplitude of the transverse acoustic radiation force will increase. This is because the acoustic radiation force vector is closely related to the distribution of acoustic intensity.

6.5 Interaction of Airy Waves with a Pair of Cylindrical Microparticles

In acoustic experiments, multiple scattering of sound waves between scatterers and boundaries [54–58] is common. When studying multiple scattering, secondary scattering is generally considered, while tertiary and higher order scattering is ignored due to the small energy of the sound waves. If there are multiple target scatterers in the medium, the suspension [59] or manipulation [60] of multiple particles needs to be considered. In this case, the theory of acoustic tweezers after the external sound field interacts with multiple particles needs to be studied. In 2000, Roumeliotis discussed the acoustic scattering between two particles [61], and in 2017, Mitri introduced the acoustic scattering between two particles into the field of acoustic tweezers [62–64]. Due to the special properties of Airy waves [39, 65], the results of their interaction with multiple particles are of research value.

This section discusses the multiple scattering problem by examining the interaction of Airy waves with a pair of cylindrical microparticles, focusing on the effects of changes in particle radius size and particle spacing on the acoustic radiation force.

6.5.1 Model and Calculation

As shown in Fig. 6.25, assume that there are a pair of particles 1 and 2 on the x-o-y plane, with radii a and b respectively. The Airy wave in the water medium is incident on the two particles at an incident angle β perpendicular to the Z-axis of the cylindrical particles, propagating in the x-axis direction. For convenience, all formulas here omit the time factor $\left(e^{-iwt}\right)$. The initial velocity potential of the Airy wave at x = 0 is:

$$\Phi_i^{(0)}(x=0,y) = \phi_0 Ai\left(\frac{y}{y_0}\right)e^{\alpha\left(\frac{y}{y_0}\right)} \tag{6.47}$$

where, ϕ_0 is the amplitude, Ai(.) is the first kind of Airy function, y_0 is the control factor of the Airy function, α is the decay coefficient.

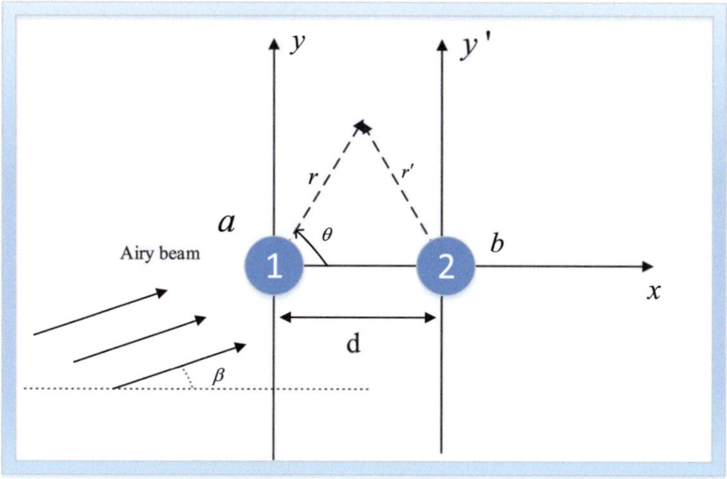

Fig. 6.25 Principle diagram of the interaction of an Airy sound wave with a pair of cylindrical particles at any incident angle, where the radii of particles 1 and 2 are a and b respectively

According to the angular spectrum analysis method [66], the Airy transformation formula [67] and the Jacobian formula [68], in the cylindrical coordinate system, the velocity potential of the incident wave can be expressed as:

$$\Phi_{inc}(r,\theta) = \phi_0 \sum_{n=-\infty}^{n=\infty} \Lambda_n J_n(kr) e^{in\theta} \tag{6.48}$$

where, $J_n(\cdot)$ is the first kind of Bessel function, $k = \frac{w}{c}$ is the wave number, c is the speed of sound in water, Λ_n is the beam factor, $\Lambda_n = i^n \frac{ky_0}{2\pi} \int_{-\infty}^{\infty} e^{(\alpha - iky_0 q)^3/3} e^{-in\sin^{-1} q} dq$

Similarly, in the cylindrical coordinate system (r, θ) and (r', θ') The scattering velocity potentials of particles 1 and 2 can be obtained separately:

$$\Phi_{sca,1}(r,\theta) = \phi_0 \sum_{n=-\infty}^{n=\infty} C_{n,1} H_n^{(1)}(kr) e^{in\theta} \tag{6.49}$$

$$\Phi_{sca,2}(r',\theta') = \phi_0 \sum_{n=-\infty}^{n=\infty} C_{n,2} H_n^{(1)}(kr\prime) e^{in\theta'} \tag{6.50}$$

where, $C_{n,1}$ and $C_{n,2}$ are unknown scattering parameters, $H_n^{(1)}(\cdot)$ is the Hankel function of the first kind of order n. Equation (6.49) not only includes the scattering wave of the initial incident wave acting on particle 1, but also the secondary reflection wave, which is the scattering wave of the wave reflected from particle 2 acting on particle 1. Similarly, Eq. (6.50) includes not only the direct scattering wave of the initial incident wave acting on particle 2, but also the scattering wave of the reflection wave from particle 1 acting on particle 2.

Since the solution of the particle scattering coefficient is determined according to the boundary conditions of the particle, the velocity potential of the transmission wave inside the particle is required:

$$\Phi_{int,1}(r,\theta) = \phi_0 \sum_{n=-\infty}^{n=\infty} C_{n,1}^{int_n k_1 r) e^{in\theta}} \tag{6.51}$$

$$\Phi_{int,2}\left(r',\theta'\right) = \phi_0 \sum_{n=-\infty}^{n=\infty} C_{n,2}^{int_n k_2 r\prime e^{in\theta'}} \tag{6.52}$$

where, k_1, k_2 are the longitudinal wave numbers inside particles 1 and 2, $C_{n,1}^{int_{n,2}^{int}}$ are the unknown parameters of the transmission wave inside particles 1 and 2, determined by the boundary conditions.

Next, we use the boundary conditions to solve the scattering coefficients $C_{n,1}$ and $C_{n,2}$. Since $C_{n,1}$ and $C_{n,2}$ are related to the secondary scattering waves of both particles, the boundary conditions of both particles need to be considered when calculating the scattering coefficients. Since the reference frames of the two particles are different, when operating in one reference frame, the formula in the other reference frame needs to be transformed into the same reference frame. Therefore, using the addition theorem [69], the Hankel functions in different cylindrical coordinate systems can be transformed into each other as follows:

$$H_n^{(1)}(kr\prime)e^{in\theta'} = \begin{cases} \sum\limits_{m=-\infty}^{\infty} J_{n-m}(kd)H_m^{(1)}(kr)e^{im\theta}, r > d \\ \sum\limits_{m=-\infty}^{\infty} J_m(kr)H_{n-m}^{(1)}(kd)e^{im\theta}, r < d \end{cases} \tag{6.53}$$

$$H_n^{(1)}(kr)e^{in\theta} = \begin{cases} \sum\limits_{m=-\infty}^{\infty} J_{n-m}(kd)H_m^{(1)}(kr\prime)e^{im\theta'}, r' > d \\ \sum\limits_{m=-\infty}^{\infty} J_m(kr\prime)H_{n-m}^{(1)}(kd)e^{im\theta'}, r' < d \end{cases} \tag{6.54}$$

Therefore, the total acoustic field velocity potentials outside the medium particles in the cylindrical coordinate systems (r,θ) and $\left(r',\theta'\right)$ can be obtained separately:

$$\begin{aligned} \Phi_{tot}(r,\theta)|_{r<d} &= \Phi_{inc}(r,\theta) + \Phi_{sca,1}(r,\theta) + \Phi_{sca,2}(r,\theta) \\ &= \phi_0 \left[\sum_{n=-\infty}^{\infty} \left(\Lambda_n J_n(kr) + C_{n,1} H_n^{(1)}(kr)\right) e^{in\theta} \right. \\ &\quad \left. + \sum_{n=-\infty}^{\infty} \left(C_{n,2} \sum_{m=-\infty}^{\infty} J_m(kr) H_{m-n}^{(1)}(kd) e^{im\theta} \right) \right] \end{aligned} \tag{6.55}$$

$$\Phi_{tot}(r,\theta)|_{r>d} = \phi_0 \left[\sum_{n=-\infty}^{\infty} \left(\Lambda_n J_n(kr) + C_{n,1} H_n^{(1)}(kr)\right) e^{in\theta} + \sum_{n=-\infty}^{\infty} \left(C_{n,2} \sum_{m=-\infty}^{\infty} J_{m-n}(kd) H_m^{(1)}(kr) e^{im\theta} \right) \right] \tag{6.56}$$

$$\begin{aligned}\Phi_{tot}(r',\theta')\big|_{r'<d} &= \Phi_{inc}(r',\theta') + \Phi_{sca,1}(r',\theta') + \Phi_{sca,2}(r',\theta') \\ &= \phi_0 \left[\sum_{n=-\infty}^{\infty} \left(\Lambda_n e^{-ikd\cos\beta} J_n(kr') + C_{n,2} H_n^{(1)}(kr\prime)\right) e^{in\theta'} + \sum_{n=-\infty}^{\infty} \left(C_{n,1} \sum_{m=-\infty}^{\infty} J_m(kr\prime) H_{n-m}^{(1)}(kd) e^{im\theta'} \right) \right]\end{aligned} \tag{6.57}$$

$$\Phi_{tot}(r',\theta')\big|_{r'>d} = \phi_0 \left[\sum_{n=-\infty}^{\infty} \left(\Lambda_n e^{-ikd\cos\beta} J_n(kr\prime) + C_{n,2} H_n^{(1)}(kr\prime)\right) e^{in\theta'} + \sum_{n=-\infty}^{\infty} \left(C_{n,1} \sum_{m=-\infty}^{\infty} J_{n-m}(kd) H_m^{(1)}(kr') e^{im\theta'} \right) \right] \tag{6.58}$$

The scattering coefficients are calculated through the boundary conditions. The boundary conditions for particles 1 and 2 are the continuity of sound pressure and normal sound speed at the boundary, namely:

$$P_{tot}(r,\theta)|_{r=A_{1,\theta}} = P_{int,1}(r,\theta)\big|_{r=A_{1,\theta}} \tag{6.59}$$

$$\nabla\Phi_{tot}(r,\theta)\cdot \mathbf{n}_1|_{r=A_{1,\theta}} = \nabla\Phi_{int,1}(r,\theta)\cdot \mathbf{n}_1\big|_{r=A_{1,\theta}} \tag{6.60}$$

$$P_{tot}(r',\theta')\big|_{r'=A_{2,\theta}} = P_{int,2}(r',\theta')\big|_{r'=A_{2,\theta}} \tag{6.61}$$

$$\nabla\Phi_{tot}(r',\theta')\cdot \mathbf{n}_2\big|_{r'=A_{2,\theta}} = \nabla\Phi_{int,2}(r',\theta')\cdot \mathbf{n}_2\big|_{r'=A_{2,\theta}} \tag{6.62}$$

where, $P_{tot} = i\omega\rho\Phi_{tot}$, $P_{int,1} = i\omega\rho_1\Phi_{int,1}$, $P_{int,2} = i\omega\rho_2\Phi_{int,2}$, ρ_1, ρ_2 are the mass densities of particles 1 and 2, $\mathbf{n}_{\{1,2\}}$ is the normal vector of the surface of particles 1 and 2, $A_{\{1,2\},\theta}$ is the surface shape function of particles 1 and 2 [70, 71].

Next, we calculate the acoustic radiation force acting on the particle. The known formula for calculating the acoustic radiation force is [72] as follows:

$$\langle \mathbf{F}_{\{1,2\}} \rangle \underset{kr\to\infty}{=} \frac{1}{2}\rho k^2 \int_0^{2\pi} \Re\{\Phi_{is,\{1,2\}}\} d\mathbf{S} \tag{6.63}$$

where, $\Re\{\cdot\}$ denotes the real part of a complex number, $\Phi_{is,\{1,2\}} \underset{kr\to\infty}{=} \Phi^*_{sca,\{1,2\}}\left[\left(\frac{i}{k}\right)\partial_r\Phi^{eff}_{inc,\{1,2\}} - \Phi^{eff}_{inc,\{1,2\}} - \Phi_{sca,\{1,2\}}\right]$, and $d\mathbf{S} = dS\cdot\mathbf{e}_r$, $dS = rd\theta$ denotes the cross-sectional area of a unit length cylindrical particle, $\mathbf{e}_r = \cos\theta\mathbf{e}_x + \sin\theta\mathbf{e}_y$, where $\mathbf{e}_x$, $\mathbf{e}_y$ are the unit vectors of the Cartesian coordinate system, the symbol $\cdot$ denotes time averaging, and the superscript $*$ denotes the conjugate of a complex number.

Equations (6.55)–(6.58) above represent the total velocity potential around the particles in the medium, while in the equation In Eq. (6.63), the effective velocity potential of each particle is required to calculate the acoustic radiation force. It should be noted that the waves reflected from the surface of particle 1 cannot be incident on the surface of particle 1 again, so this part of the sound wave does not need to be included in the effective sound field of particle 1. The same applies to particle 2.

$$\begin{aligned}\Phi^{eff}_{inc,1}(r,\theta) &= \Phi_{tot}(r,\theta)|_{r<d} - \Phi_{sca,1}(r,\theta) = \Phi_{inc}(r,\theta) + \Phi_{sca,2}(r,\theta)\\ &= \phi_0\left[\sum_{n=-\infty}^{\infty}\Lambda_n J_n(kr)e^{in\theta} + \sum_{n=-\infty}^{\infty}\left(C_{n,2}\sum_{m=-\infty}^{\infty}J_m(kr)H^{(1)}_{m-n}(kd)e^{im\theta}\right)\right]\end{aligned} \tag{6.64}$$

$$\begin{aligned}\Phi^{eff}_{inc,2}(r',\theta') &= \Phi_{tot}(r',\theta')|_{r'<d} - \Phi_{sca,2}(r',\theta') = \Phi_{inc}(r',\theta') + \Phi_{sca,1}(r',\theta')\\ &= \phi_0\left[\sum_{n=-\infty}^{\infty}\Lambda_n e^{-ikd\cos\beta}J_n(kr\prime)e^{in\theta'} + \sum_{n=-\infty}^{\infty}\left(C_{n,1}\sum_{m=-\infty}^{\infty}J_m(kr\prime)H^{(1)}_{n-m}(kd)e^{im\theta'}\right)\right]\\ &= \phi_0\left[\sum_{n=-\infty}^{\infty}\Lambda_n e^{-ikd\cos\beta}J_n(kr\prime)e^{in\theta'} + \sum_{n=-\infty}^{\infty}\left(C_{n,1}\sum_{m=-\infty}^{\infty}J_m(kr\prime)H^{(1)}_{n-m}(kd)e^{im\theta'}\right)\right]\end{aligned} \tag{6.65}$$

Since the effective velocity potentials and the formulas for calculating the acoustic radiation force of particles 1 and 2 have been given, the expressions for the transverse (Y_x) and longitudinal (Y_y) acoustic radiation forces of particles 1 and 2 can be obtained:

$$Y_{x,1} = \frac{2}{kS_{c,1}}\Im\left\{\sum_{n=-\infty}^{\infty}\left(\Lambda_n + C_{n,1} + \sum_{m=-\infty}^{\infty}C_{m,2}H^{(1)}_{n-m}(kd)\right)\left(C^*_{n+1,1} - C^*_{n-1,1}\right)\right\} \tag{6.66}$$

$$Y_{y,1} = -\frac{2}{kS_{c,1}}\Re\left\{\sum_{n=-\infty}^{\infty}\left(\Lambda_n + C_{n,1} + \sum_{m=-\infty}^{\infty}C_{m,2}H^{(1)}_{n-m}(kd)\right)\left(C^*_{n+1,1} + C^*_{n-1,1}\right)\right\} \tag{6.67}$$

$$Y_{x,2} = \frac{2}{kS_{c,2}}\Im\left\{\sum_{n=-\infty}^{\infty}\left(\Lambda_n e^{-ikd\cos\beta} + C_{n,2} + \sum_{m=-\infty}^{\infty}C_{m,1}H^{(1)}_{m-n}(kd)\right)\left(C^*_{n+1,2} - C^*_{n-1,2}\right)\right\} \tag{6.68}$$

$$Y_{y,2} = -\frac{2}{kS_{c,2}}\Re\left\{\sum_{n=-\infty}^{\infty}\left(\Lambda_n e^{-ikd\cos\beta} + C_{n,2} + \sum_{m=-\infty}^{\infty} C_{m,1}H_{m-n}^{(1)}(kd)\right)\left(C_{n+1,2}^{*} + C_{n-1,2}^{*}\right)\right\} \tag{6.69}$$

Both particles are liquid particles. For particle 1, substituting Eqs. (6.55) and (6.64) into boundary conditions (6.59) and (6.60), and using the orthogonality of complex exponential functions, the following expressions can be obtained:

$$\Lambda_l J_l(ka) + C_{l,1}H_l^{(1)}(ka) + J_l(ka)\sum_{n=-\infty}^{n=\infty} C_{n.2}H_{l-n}^{(1)}(kd) = C_{l,1}^{int}J_l(k_1a) \tag{6.70}$$

$$\Lambda_l J_l'(ka) + C_{l,1}H_l^{(1)'}(ka) + J_l'(ka)\sum_{n=-\infty}^{n=\infty} C_{n.2}H_{l-n}^{(1)}(kd) = \left(\frac{\rho c}{\rho_1 c_1}\right)C_{l,1}^{int}J_l'(k_1a) \tag{6.71}$$

where, c_1 is the longitudinal sound speed of the medium inside particle 1, ρ_1 is the mass density of the medium inside cylindrical particle 1. Similarly, for particle 2, substituting Eqs. (6.57) and (6.65) into boundary conditions (6.616.61) and (6.62), and using the orthogonality of complex exponential functions, the following expressions can be obtained:

$$\Lambda_l J_l(kb)e^{-ikd\cos\beta} + C_{l,2}H_l^{(1)}(kb) + J_l(kb)\sum_{n=-\infty}^{n=\infty} C_{n.1}H_{l-n}^{(1)}(kd) = C_{l,2}^{int}J_l(k_2b) \tag{6.72}$$

$$\Lambda_l J_l'(kb)e^{-ikd\cos\beta} + C_{l,2}H_l^{(1)'}(kb) + J_l'(kb)\sum_{n=-\infty}^{n=\infty} C_{n.1}H_{l-n}^{(1)}(kd) = \left(\frac{\rho c}{\rho_2 c_2}\right)C_{l,2}^{int}J_l'(k_2b) \tag{6.73}$$

where, c_2 is the longitudinal sound speed of the medium inside cylindrical particle 2 with radius b, ρ_2 is the mass density of the medium inside cylindrical particle 2.

Equations (6.70)–(6.73) form a set of equations. By representing this set of equations in the form of matrices and vectors, the scattering coefficients $C_{n,1}$ and $C_{n,2}$ can be obtained using matrix inverse transformation. The dimensions of the matrix are $(2N + 1) \times (2N + 1)$ and the dimensions of the vector are $1 \times (2N + 1)$, $N\max(\max(ka, kb)kd)_{\max}$. After obtaining the scattering coefficients, substitute them into Eqs. (6.66)–(6.69) to obtain the longitudinal and transverse acoustic radiation forces of particles 1 and 2.

6.5.2 Simulation Analysis and Discussion

As shown in Fig. 6.22, the distance between particles 1 and 2 is d, the radii are a and b respectively, the particles are made of the same material, the mass density of the medium inside the particles is $\rho_{int} = 100\ \text{kg/m}^3$, the sound speed is $c_{int} = 1631$ m/s, the density of the external fluid water is $\rho_0 = 1000\,\text{kg/m}^3$, and the sound speed is $c_0 = 1500$ m/s. The longitudinal wave number inside particles 1 and 2 is $k_c = w/c_{int}(1 + i\gamma)$, where, $\gamma\left(= 10^{-3}\right)$ is the dimensionless absorption coefficient. The acoustic radiation in Eqs. (6.66)–(6.69) is

Figure 6.26a–d show the transverse and longitudinal acoustic radiation forces of particles 1 and 2 in Eqs. (6.68)–(6.71), with the radius-related parameters of particles 1 and 2 set to $ka = 0.1$, $kb = 1$, the incident angle range of the Ailey wave is: $-90° \leq \beta \leq 90°$, and $(ka + kb) < kd \leq 30$. As can be seen from figure (a), the transverse acoustic radiation force $\left(Y_{x,1}\right)$ of particle 1 is symmetrically distributed with the incident angle β of the Ailey wave, showing certain harmonic characteristics with the change of angle β, and the transverse acoustic radiation force changes between positive and negative, with the absolute value decaying as the distance between particles kd increases. This indicates that when an appropriate incident angle β is chosen, particle 1 can be pushed away from $\left(Y_{x,1} < 0\right)$ or pulled towards $\left(Y_{x,1} > 0\right)$ particle 2, or at the intersection point $\left(Y_{x,1} = 0\right)$, with particle 1 remaining stationary. Figure (b) shows the longitudinal acoustic radiation force $\left(Y_{y,1}\right)$ of particle 1. Like the transverse acoustic radiation force, it also decays with the increase of the distance kd, is symmetrically distributed with the incident angle β, and exhibits certain harmonic characteristics with the increase of the angle β. However, the amplitude of $Y_{y,1}$ remains positive. From figures (a) and (b), it can be found that when the particle spacing kd is close to the particle size and $(ka + kb)$, the absolute value of the acoustic radiation force amplitude is the largest. Figures (c) and (d) show the acoustic radiation force $\left(Y_{x,2},\ Y_{y,2}\right)$ of particle 2. As can be seen from the figure, the transverse and longitudinal acoustic radiation forces of particle 2 are both fan-shaped, and the amplitudes are all positive, decaying with the increase of particle spacing kd. As can be seen from the figure, at the angle $50° \leq |\beta| \leq 90°$, $Y_{x,2}$, $Y_{y,2}$ exhibits certain fluctuations, resembling the propagation pattern of Airey waves.

Figure 6.27 shows the acoustic radiation force when the sizes of the two particles are the same, i.e., $ka = 1$, $kb = 1$, and Fig. 6.28 shows the acoustic radiation force when the size of particle 1 is larger than that of particle 2, i.e., $ka = 5$, $kb = 0.1$. By comparing Figs. 6.26a, b, 6.27a, b, and 6.28a, b, it can be seen that the distribution of the acoustic radiation force of particle 1 basically remains unchanged. The slight difference is that when the radius of particle 1 is equal to or larger than the size of particle 2, the acoustic radiation force of particle 1 is positive, and the oscillations at the peaks and valleys in the distribution diagram of the acoustic radiation force are more pronounced. Similarly, the distribution of the acoustic radiation force of particle 2 remains basically unchanged. Contrary to the oscillation situation of the acoustic radiation force of particle 1, as the ratio $\frac{\text{ka}}{\text{kb}}$ increases, $Y_{x,2}$, $Y_{y,2}$ the fluctuations at

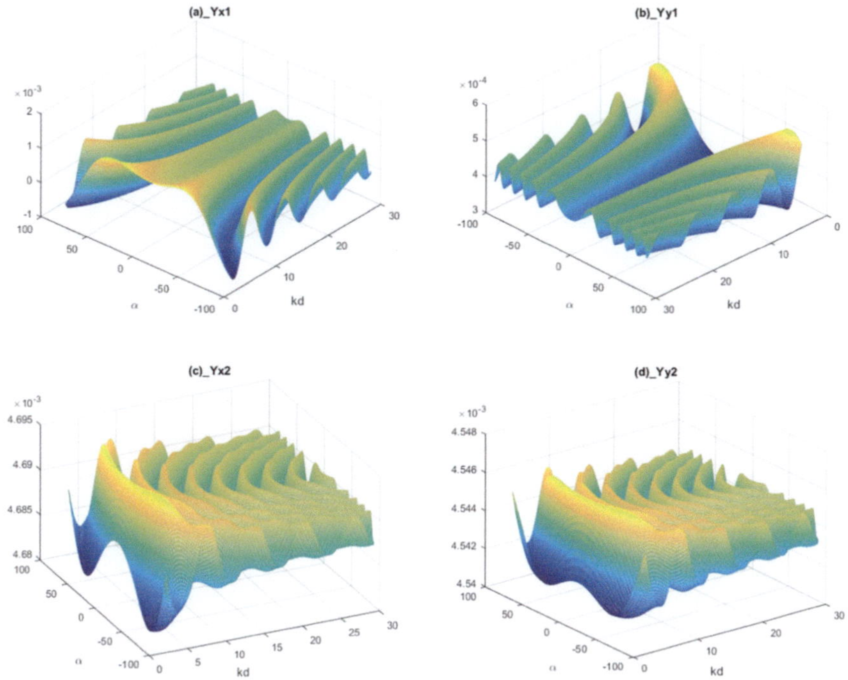

Fig. 6.26 The transverse and longitudinal acoustic radiation forces of particles 1 and 2, $ka = 0.1$, $kb = 1$

the angle $50° \leq |\beta| \leq 90°$ gradually disappear. This is because when the size of particle 1 is larger than that of particle 2, it interferes with the action of Airey waves on particle 2, causing most of the Airey waves to act on particle 1. The Airey wave characteristics of the acoustic radiation force of particle 1 are more pronounced, while those of particle 2 gradually disappear. From Figs. 6.27 and 6.28, it can be seen that when the radius size of particle 1 is equal to or larger than that of particle 2, due to the self-bending characteristic of Airey wave propagation, even at the incident angle $\beta = 0^{\circ}$, the acoustic radiation force on particle 2 does not disappear due to the obstruction of particle 1; similarly, at the incident angle $\beta = -90°(\text{or } 90°)$, the transverse acoustic radiation force of particle 2 is not zero.

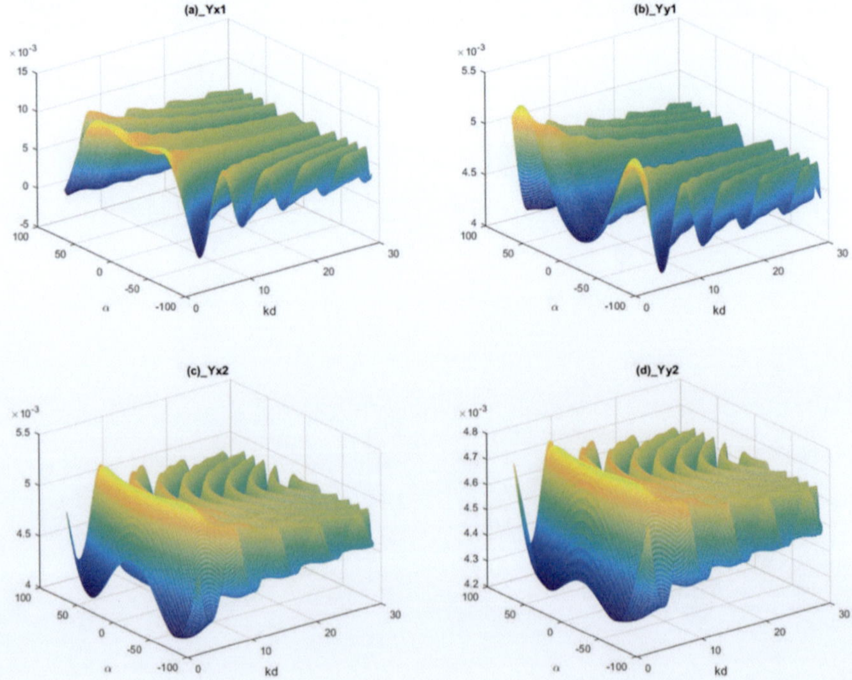

Fig. 6.27 Acoustic radiation forces in the transverse and longitudinal directions of particles 1 and 2, $ka = 1, kb = 1$

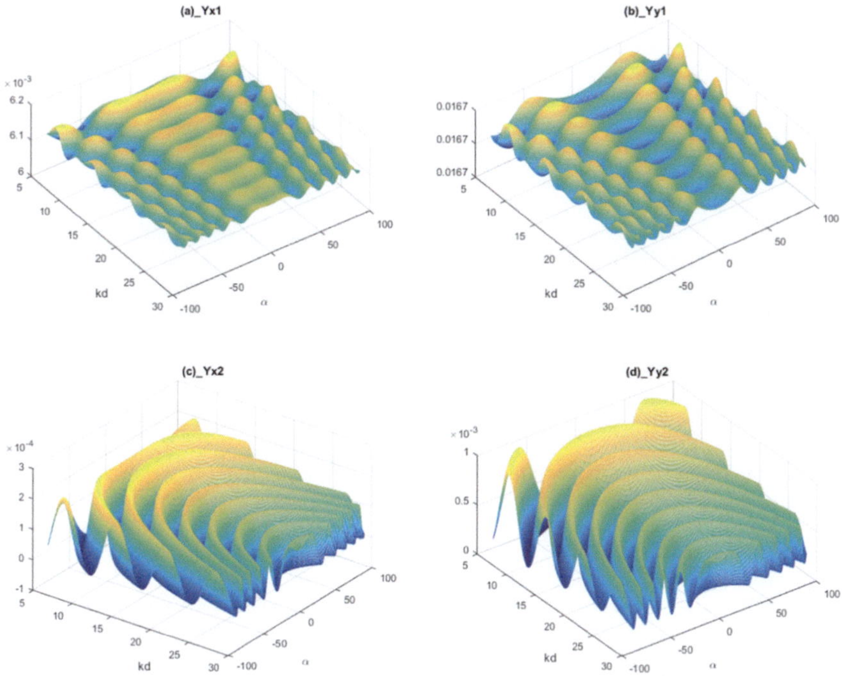

Fig. 6.28 Acoustic radiation forces in the transverse and longitudinal directions of particles 1 and 2, $ka = 5$, $kb = 0.1$

References

1. O'Neil HT (1949) Theory of focusing radiators. J Acoust Soc Am 21:516–526
2. Filipczynski L, Etienne J (1973) Theoretical study and experiments on spherical focusing transducers with Gaussian surface velocity distribution. Acta Acust Acust 28:121–128
3. Lucas BG, Muir TG (1982) The field of a focusing source. J Acoust Soc Am 72:1289–1296
4. Thompson RB, Lopes EF (1984) The effects of focusing and refraction on Gaussian ultrasonic beams. J Nondestr Eval 4:107–123
5. Coulouvrat F (1993) Continuous field radiated by a geometrically focused transducer: numerical investigation and comparison with an approximate model. J Acoust Soc Am 94:1663–1675
6. Stratton JA (1941) Electromagnetic theory. McGraw-Hill, New York, pp 1–631
7. Arfken GB, Weber HJ Mathematical methods for physicists. Academic Press, California
8. Hasegawa T, Inoue N, Matsuzawa K (1987) A new rigorous expansion for the velocity potential of a circular piston source. J Acoust Soc Am 82:706–708
9. Beissner K (2013) On the lateral resolution of focused ultrasonic fields from spherically curved transducers. J Acoust Soc Am 134:943–3947
10. Hasegawa T, Matsuzawa K, Inoue N (1986) A new expansion for the velocity potential of a circular concave piston. J Acoust Soc Am 79:927–931
11. Mair HD, Hutchins DA (1987) In progress in underwater acoustics. Springer, US 619–626
12. Chen X, Apfel R (1996) Radiation force on a spherical object in an axisymmetric wave field and its application to the calibration of high-frequency transducers. J Acoust Soc Am 99:713–724

13. Mitri FG (2009) Acoustic radiation force of high-order Bessel beam standing wave tweezers on a rigid sphere. Ultrasonics 49:794–798
14. Dekker DL, Piziali RL, Dong JE (1974) Spherical wave decomposition approach to ultrasonic field calculations. J Acoust Soc Am 56:87–93
15. Greenspan M (1979) Piston radiator: some extensions of the theory. J Acoust Soc Am 65:608–621
16. Mast TD, Yu FF (2005) Simplified expansions for radiation from a baffled circular piston. J Acoust Soc Am 118:3457–3464
17. Kunmiao L (1998) Methods of mathematical physics, 2nd edn. Higher Education Press, Beijing
18. Gonghuan D, Zhemin Z, Xiufen G (2001) Fundamentals of acoustics, 2nd edn: Nanjing University Press
19. Hasegawa T, Inoue N, Matsuzawa K (1983) A new rigorous expansion for the velocity potential of a circular piston source. J Acoust Soc Am 74:1044
20. Hasegawa T, Inoue N, Matsuzawa K (1984) Ultrasonic scattering by a rigid sphere in the nearfield of a circular. J Acoust Soc Am 75:1048–1051
21. Kino GS (1987) Acoustic waves: devices, imaging, and analog signal processing. Prentice-Hall, NanJing
22. Gutiérrez-Vega JC, Rodríguez-Masegosa R, Chávez-Cerda S (2003) Bessel–Gauss resonator with spherical output mirror: geometrical-and wave-optics analysis. J Opt Soc Am A-Opt Image Sci Vis 20:2113–2122
23. Hakola A, Buchter SC, Kajava T, Turunen J (2004) Bessel–Gauss output beam from a diode-pumped Nd:YAG laser. Opt Commun 238:335–340
24. Mitri FG (2014) Acoustics of a finite-aperture laguerre-Gaussian vortex beam. IEEE Ultrasonics Symp (IUS)
25. Wang SJ (1998) Another approximately analytical description for the Bessel-Gaussian beam fields of a finite aperture. J Northwest Univ 34:22–25
26. Mitri FG (2009) Langevin acoustic radiation force of a high-order Bessel beam on a rigid sphere. IEEE Trans Ultrason Ferroelectr Freq Control 56:1059–1064
27. Mitri FG, Lobo TP, Silva GT (2012) Axial. Acoustic radiation torque of a Bessel vortex beam on spherical shells. Phys Rev E, 85:026602
28. Mitri FG (2015) Acoustical pulling force of a limited-diffracting annular beam centered on a sphere. IEEE Trans Ultrason Ferroelectr Freq Control 62:1827–1834
29. Mitri FG (2015) Acoustical pulling force on rigid spheroids in single Bessel vortex tractor beams. Europhys Lett 112:34002
30. Marston PL (2006) Axial radiation force of a Bessel beam on a sphere and direction reversal of the force. J Acoust Soc Am 120:3518–3524
31. Marston PL (2009) Radiation force of a helicoidal Bessel beam on a sphere. J Acoust Soc Am 125:3539–3547
32. Zhang L, Marston PL (2011) Angular momentum flux of nonparaxial acoustic vortex beams and torques on axisymmetric objects. Phys Rev E 84:065601
33. Zhang XF, Zhang GB (2012) Acoustic radiation force of a Gaussian beam incident on spherical particles in water. Ultrasound Med Biol 38:2007–2017
34. Wu RR, Cheng KX, Liu XZ et al (2016) Study of axial acoustic radiation force on a sphere in a Gaussian quasi-standing field. Wave Motion 62:63–74
35. Gouesbet G, Grehan G, Maheu B (1998) Computations of the gn coefficients in the generalized Lorenz-Mie theory using three different methods. Appl Opt 24:4874–4883
36. Whittaker ET, Watson GN (1966) A course of modern analysis, 4th edn. Cambridge University Press, Cambridge
37. Leighton TG (1995) Bubble population phenomena in acoustic cavitation. Ultrason Sonochem 2:123–136
38. Bandres MA, Gutiérrezvega JC (2007) Airy-Gauss beams and their transformation by paraxial optical systems. Opt Express 15(25):16719–16728
39. Berry MV, Balazs NL (1998) Nonspreading wave packets. Am J Phys 47(3):264–267

40. Siviloglou GA, Christodoulides DN (2007) Accelerating finite energy Airy beams. Opt Lett 32(8):979–981
41. Huang J, Liang Z, Deng F et al (2015) Propagation properties of right-hand circularly polarized Airy-Gaussian beams through slabs of right-handed materials and left-handed materials. J Opt Soc Am A: 32(11):2104–2109
42. Deng DM (2011) Propagation of Airy-Gaussian beams in a quadratic-index medium. Eur Phys J D 65(3):553–556
43. Zhou ML, Chen CD, Chen B et al (2015) Propagation of an Airy-Gaussian beam in uniaxial crystals. Chin Phys B 24(12):309–312
44. Chen C, Chen B, Peng X et al (2015) Propagation of Airy-Gaussian beam in Kerr medium. J Opt 17(3)
45. Peng Y, Peng X, Chen B et al (2016) Interaction of Airy-Gaussian beams in Kerr media. Opt Commun 359:116–122
46. Zhang X (2016) Bound states of breathing Airy–Gaussian beams in nonlocal nonlinear medium. Opt Commun; Mitri FG (2015) Acoustic radiation force on a rigid elliptical cylinder in plane (quasi)standing waves. J Appl Phys 118(21):170–171
47. Mitri FG (2006) Theoretical and experimental determination of the acoustic radiation force acting on an elastic cylinder in a plane progressive wave—far-field derivation approach. New J Phys 8(8):138
48. Mitri FG (2016) Extinction efficiency of "elastic–sheet" beams by a cylindrical (viscous) fluid inclusion embedded in an elastic medium and mode conversion—Examples of nonparaxial Gaussian and Airy beams. J Appl Phys 120(14):463–469
49. Goodman JW (1968) Introduction to Fourier optics. McGraw-Hill, New York, pp 55–62
50. Vallee O, Soares M (2004) Airy functions and applications to physics. London: World Scientific, pp 74–84
51. Colton D, Kress R (2013) Inverse acoustic and electromagnetic scattering theory. Springer Science + Business Media, New York, 33
52. Mitri FG (2015) Interaction of an acoustical 2D-beam with an elastic cylinder with arbitrary location in a non-viscous fluid. Ultrasonics 62:244–252
53. Mitri FG (2005) Theoretical calculation of the acoustic radiation force acting on elastic and viscoelastic cylinders placed in a plane standing or quasistanding wave field. Eur Phys JB - Condensed Matter Complex Syst 44(1):71–78
54. Foldy LL (1945) The multiple scattering of waves. I. General theory of isotropic scattering by randomly distributed scatterers. Phys Rev 67(3–4):107–119
55. Zitron N (1961) Higher-order approximations in multiple scattering. I. Three-dimensional scalar case. J Math Phys 2(3):394–402
56. Zitron N, Karp SN (1961) Higher-order approximations in multiple scattering. II. three-dimensional scalar case. J Math Phys 2(3):402–406
57. Martin PA, Conoir JM (2007) Multiple scattering, interaction of time-harmonic waves with N obstacles. J Acoust Soc Am 121(5):279–282
58. Mitri FG (2017) Acoustic radiation force and torque on a viscous fluid cylindrical particle nearby a planar rigid wall. https://arxiv.org/abs/1801.04863
59. Lierke EG (1996) Acoustic levitation-A comprehensive survey of principles and applications. Acustica 82(2):220–237
60. Trinh EH (1985) Compact acoustic levitation device for studies in fluid dynamics and material science in the laboratory and microgravity. Rev Sci Instrum 56(11):2059–2065
61. Roumeliotis JA, Ziotopoulos AG, Kokkorakis GC (2001) Acoustic scattering by a circular cylinder parallel with another of small radius. J Acoust Soc Am 109(3):870–877
62. Mitri FG (2017) Acoustic attraction, repulsion and radiation force cancellation on a pair of rigid particles with arbitrary cross-sections in 2D: circular cylinders example. Ann Phys 386
63. Mitri FG (2017) Acoustic radiation torques on a pair of fluid viscous cylindrical particles with arbitrary cross-sections: Circular cylinders example. J Appl Phys 121(14):620–623
64. Mitri FG (2017) Extrinsic extinction cross-section in the multiple acoustic scattering by fluid particles. J Appl Phys 121(14):220–237

65. Unnikrishnan K, Rau ARP (1996) Uniqueness of the Airy packet in quantum mechanics. Am J Phys 64(8):1034–1035
66. Goodman JW (1968) Introduction to fourier optics. Springer Science + Business Media, New York, pp 55–62
67. Vallee O, Soares M (2004) Airy functions and applications to physics. world scientific, London, pp 74–84
68. Colton D, Kress R (2013) Inverse acoustic and electromagnetic scattering theory. Springer Science + Business Media, New York, p 33
69. Abramowitz M, Stegun IA (1965) Handbook of mathematical functions: with formulas, graphs, and mathematical tables. Dover Publications, USA
70. Mitri FG (2016) Radiation forces and torque on a rigid elliptical cylinder in acoustical plane progressive and (quasi)standing waves with arbitrary incidence. Phys Fluids 28(7):220–244
71. Mitri FG (2015) Acoustic radiation force on a rigid elliptical cylinder in plane (quasi)standing waves. J Appl Phys 118(21):170–171
72. Mitri FG (2006) Theoretical and experimental determination of the acoustic radiation force acting on an elastic cylinder in a plane progressive wave—far-field derivation approach. New J Phys 8(8):138.0

Chapter 7
The Impact of Boundaries on Acoustic Radiation Force

The key to accurately manipulating particles in an acoustic field lies in the control of the acoustic radiation force. Therefore, in practical applications, we need to consider the acoustic radiation force generated under various circumstances. In previous work, much research has focused on the acoustic radiation force when particles are in a boundary-free field. However, in reality, these particles are often in a bounded environment, and we need to consider the impact of these boundaries. This impact includes the reflection of sound waves by impedance boundaries, which can cause changes in the surrounding acoustic field. These changes, in turn, can affect the magnitude and direction of the acoustic radiation force experienced by the particles, thereby affecting the precise manipulation of the particles.

7.1 Acoustic Radiation Force on Cylindrical Particles Under a Single Boundary by Plane Waves

7.1.1 Vertically Incident Plane Waves

Plane waves are usually the simplest waveforms in acoustics, with their wavefronts being planar. The complex form of a one-dimensional plane sound wave is expressed as:

$$p = p_a e^{i(kx-\omega t)} \tag{7.1}$$

where, p_a is the complex amplitude of the plane wave, k is the wave number, ω is the angular frequency.

Consider the acoustic scattering of cylindrical particles under a single impedance boundary. The physical model is shown in Fig. 7.1.

X. Liu and H. Zheng, *Acoustic Radiation Force*, Advances in Acoustics,
https://doi.org/10.1007/978-981-95-3152-3_7

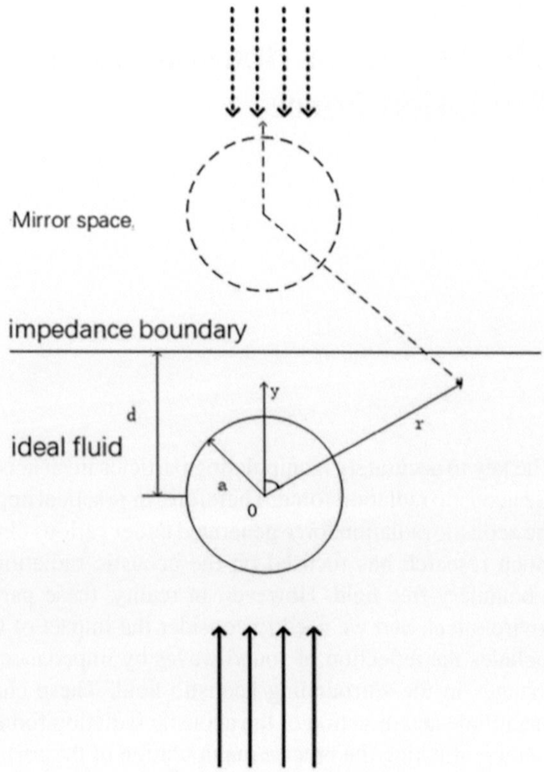

Fig. 7.1 Acoustic scattering of cylindrical particles under a single impedance boundary by plane waves

The plane wave is incident perpendicular to the boundary, and a cylindrical particle with a radius of a and infinite length is placed in the incident sound field. The distance between the boundary and the particle is d. The origin O is located at the center of the cylindrical particle in this cross-section to establish a cylindrical coordinate system. The impact of the boundary on this problem can be solved by the method of images. From the theory of acoustic radiation in theoretical acoustics [1], the effect of the impedance boundary can be equivalent to an image particle at a symmetrical position above the boundary with the particle, with the same size as the original particle and a distance of d from the boundary. Correspondingly, we can also establish an image cylindrical coordinate system, with the origin being the center of the image particle, denoted as $\overline{O}$. After such equivalent processing, the reflected sound wave after the incident sound wave hits the boundary can be regarded as the image incident wave in the opposite direction, and the reflected sound wave after the scattered sound wave hits the boundary can be regarded as the scattered wave of the image particle.

Regarding the fluid medium where the particles are located, for simplicity, let's assume it is an ideal fluid without viscous effects. The density of the medium is ρ_0, and the speed of sound in the medium is c_0.

In an ideal fluid, in the absence of sources and external forces, the sound wave equation in the form of velocity potential is

$$\nabla^2 \Phi - \frac{1}{c_0^2}\frac{\partial^2 \Phi}{\partial t^2} = 0 \tag{7.2}$$

where, ∇^2 is the Laplacian operator, Φ is the velocity potential function of the sound wave.

Its steady-state form is

$$(\nabla^2 + k^2)\Phi = 0 \tag{7.3}$$

where, k is the wave number in the fluid medium.

We know that according to the relevant theory of mathematical physics equations [2], any plane wave can be expanded into an infinite series form using cylindrical Bessel functions, that is

$$\Phi_{inc} = \Phi_0 \sum_{n=-\infty}^{\infty} i^n J_n(kr) e^{in\theta} e^{-i\omega t} \tag{7.4}$$

where, Φ_0 is the amplitude of the incident sound wave, $J_n(\cdot)$ is the first kind of n-th order cylindrical Bessel function.

The reflection wave of the boundary to the incident sound wave can be represented in the following form

$$\Phi_{ref} = \Phi_0 R_s e^{i2kd} \sum_{n=-\infty}^{\infty} (-1)^n i^n J_n(kr) e^{in\theta} e^{-i\omega t} \tag{7.5}$$

where, R_s is the sound pressure reflection coefficient of the impedance interface, its calculation formula should be

$$R_s = \frac{1-\beta}{1+\beta} \tag{7.6}$$

Here, $\beta = \rho_0 c_0 / Z$, is the ratio of the normal acoustic impedance of the interface, Z is the normal acoustic impedance of the interface.

In addition to these two types of sound waves, there are also scattered sound waves of particles for incident sound waves and the corresponding scattered sound waves of image particles, which can be represented as

$$\Phi_{scat} = \Phi_0 \sum_{n=-\infty}^{\infty} a_n i^n H_n^{(1)}(kr) e^{in\theta} e^{-i\omega t} \tag{7.7}$$

$$\Phi_{scat,ref} = \Phi_0 R_s \sum_{n=-\infty}^{\infty} (-1)^n a_n i^n H_n^{(1)}(k\overline{r}) e^{-in\overline{\theta}} e^{-i\omega t} \tag{7.8}$$

where, $H_n^{(1)}(\cdot)$ is the first kind of n-th order cylindrical Hankel function, a_n is the scattering coefficient of the particle for the incident sound wave, determined by the corresponding boundary conditions.

The sound field outside the particle is obviously composed of the above four velocity potentials. The current question is whether there is still the presence of sound waves inside the cylinder, that is, whether there are refracted sound waves. Here we need to discuss two situations: rigid and non-rigid.

7.1.1.1 Scattering Coefficient of Rigid Cylinder

First, consider a simple situation where the cylinder can be regarded as rigid, with its surface being a completely hard boundary. Of course, this is an ideal situation that does not exist in reality. However, when the acoustic impedance inside the particle is much greater than the acoustic impedance of the fluid medium, it can be regarded as a rigid cylinder, and there is no need to consider the issue of refracted waves. Reflected in the scattering coefficient, that is, at this time R_s is close to 1, and the sound wave is almost completely reflected at the surface.

The sound field in the whole space should be the superposition of the above four sound fields, that is,

$$\Phi_{total} = \Phi_{inc} + \Phi_{ref} + \Phi_{scat} + \Phi_{scat,\,ref} \tag{7.9}$$

Before substituting each specific term, we use the addition formula of the cylindrical Bessel function [3, 4] to transform the $\Phi_{scat,ref}$ term mathematically and rewrite it as

$$\Phi_{scat,ref} = \Phi_0 R_s \sum_{n=-\infty}^{\infty} \sum_{m=-\infty}^{\infty} (-1)^m a_m i^m H_{m-n}^{(1)}(2kd) J_n(kr) e^{in\theta} e^{-i\omega t} \tag{7.10}$$

At this time, if you substitute the specific expressions above, you can get

$$\Phi_{total} = \Phi_0 e^{-i\omega t} \left\{ \sum_{n=0}^{\infty} \varepsilon_n A_n i^n J_n(kr) \cos n\theta + \sum_{n=0}^{\infty} \varepsilon_n a_n i^n H_n^{(1)}(kr) \cos n\theta \right\} \tag{7.11}$$

where, A_n is a function introduced purely for computational simplicity, its expression is

$$A_n = 1 + R_s e^{i2kd}(-1)^n + \frac{R_s}{\varepsilon_n i^n} \sum_{m=0}^{\infty} (-1)^m \varepsilon_m a_m i^m H_{m-n}^{(1)}(2kd) J_n(kr) \tag{7.12}$$

In addition, ε_n can be regarded as a symbol, its value is

$$\varepsilon_n = \begin{cases} 1, n = 0 \\ 2, n > 0 \end{cases}$$

According to the relationship between the radial velocity of the particle vibration and the velocity potential, the radial velocity can be obtained

$$v_r = -\Phi_0 e^{-i\omega t} \sum_{n=0}^{\infty} \varepsilon_n A_n i^n k J_n'(ka) \cos n\theta + \sum_{n=0}^{\infty} \varepsilon_n a_n i^n k H_n^{(1)'}(ka) \cos n\theta\} \quad (7.13)$$

Then, according to the relationship between the harmonic sound pressure and the velocity potential, we get

$$p = -i\omega\rho_0 \Phi_{total} \quad (7.14)$$

Since the surface of the cylinder can be regarded as a completely hard boundary at this time, the radial velocity of the particles on the interface is zero.

Let

$$v_r|_{r=a} = 0 \quad (7.15)$$

That is

$$\sum_{n=0}^{\infty} \varepsilon_n A_n i^n k J_n'(ka) \cos n\theta + \sum_{n=0}^{\infty} \varepsilon_n a_n i^n k H_n^{(1)'}(ka) \cos n\theta\} = 0 \quad (7.16)$$

Substitute the specific form of A_n to get the implicit equation about the scattering coefficient a_n. Although it is impossible to obtain its analytical solution, its numerical solution can be obtained by iterative methods.

7.1.1.2 Scattering Coefficient of Non-rigid Cylinder

In reality, the acoustic impedance of the medium inside the cylinder may not be much greater than the acoustic impedance of the surrounding fluid medium,

It may even be smaller. At this time, the surface of the cylinder can no longer be regarded as completely rigid, and the refracted waves inside the cylinder must be considered, which is slightly more complicated than the rigid cylinder.

Still express the refracted wave in the following form through the method of spherical Bessel function expansion

$$\Phi_l = \Phi_0 e^{-i\omega t} \sum_{n=0}^{\infty} i^n B_n J_n(k_l r)\cos n\theta \quad (7.17)$$

where, k_l is the wave number in the medium inside the cylinder, B_n is the scattering coefficient of the elastic wave inside the cylinder.

At this time, the boundary condition of zero radial velocity on the surface of the cylinder naturally no longer holds, and it is replaced by the condition of continuity of sound pressure and particle velocity on the surface of the cylinder. Expressed in formulas as

$$p|_{r=a} = p_l|_{r=a} \quad v|_{r=a} = v_l|_{r=a} \quad (7.18)$$

Similarly, we can easily derive the expressions for sound pressure and particle velocity inside the cylinder

$$p_l = -i\omega\rho_0\Phi_0 e^{-i\omega t} \sum_{n=-\infty}^{\infty} i^n \varepsilon_n B_n J_n(k_l r)\cos n\theta \quad (7.19)$$

$$v_l = -\Phi_0 e^{-i\omega t} \sum_{n=0}^{\infty} i^n \varepsilon_n k_l B_n J_n'(kr)\cos n\theta \quad (7.20)$$

Substitute the above two formulas into Eq. (7.18), and the corresponding scattering coefficients a_n, B_n and function A_n can be solved. Of course, we can only obtain the so-called numerical solution through the iterative method.

7.1.1.3 Solution of Acoustic Radiation Force

So far, we have obtained the scattering coefficient of cylindrical particles through certain boundary conditions. Next, we proceed to the calculation of acoustic radiation force. For cylindrical particles in an ideal fluid, the calculation formula for the acoustic radiation force has been given in many literatures [3–5], which is directly listed as follows

$$\langle F \rangle = -\left\langle \iint_{S_0} \rho_0(v_n n + v_t t)ds \right\rangle + \left\langle \iint_{S_0} \frac{\rho_0}{2}|v|^2 n ds \right\rangle - \left\langle \iint_{S_0} \frac{\rho_0}{2c_0^2}\dot{\Phi}^2 n ds \right\rangle \quad (7.21)$$

where, S_0 is the cross-section surrounding the scattering body, v_n and v_t are the normal and tangential components of the particle vibration velocity, respectively.

The above formula is a vector formula. In practical situations, we are more concerned about the axial component of the acoustic radiation force experienced by the particles. This component is

$$\langle F\rangle = \langle F_r\rangle + \langle F_\theta\rangle + \langle F_{r\theta}\rangle + \langle F_t\rangle \tag{7.22}$$

In related literature, the calculation result has been given, we omit the specific calculation process and directly list the result as follows [4]

$$\langle F\rangle = Y_p S\langle E\rangle \tag{7.23}$$

where,

$$\langle E\rangle = \frac{1}{2}\rho_0 k^2 \Phi_0^2$$

can be regarded as the energy density of the incident wave,

$$S = 2a$$

is the cross-sectional area of the scattering.

And in the formula, Y_p is the so-called acoustic radiation force function, its expression is

$$\begin{aligned} Y_p = -\frac{2}{ka}\sum_{n=0}^{\infty}\Big(&\xi_{n+1}^{(1)}\xi_n^{(2)} + \xi_n^{(1)}\xi_{n+1}^{(2)} + \eta_{n+1}^{(1)}\eta_n^{(2)} \\ &+ \eta_n^{(1)}\eta_{n+1}^{(2)} + 2\xi_{n+1}^{(2)}\xi_n^{(2)} + 2\eta_n^{(2)}\eta_{n+1}^{(2)}\Big) \end{aligned} \tag{7.24}$$

where

$$\xi_n^{(1)} = Re(A_n),\ \eta_n^{(1)} = Im(A_n)\cdot\xi_n^{(2)} = Re(a_n),\ \eta_n^{(2)} = Im(a_n)$$

So far, we have calculated the acoustic radiation force of cylindrical particles under plane wave incidence. Next, we will study the characteristics of the acoustic radiation force under different conditions through specific simulation examples.

7.1.1.4 Simulation Example

Next, we will conduct a more in-depth study through numerical simulation. In this section's simulation, we assume that the cylindrical particles are all located in water, and the water is regarded as an ideal fluid. All discussions are divided into rigid columns and non-rigid columns. In the case of non-rigidity, it is assumed that the column is filled with oleic acid (hereinafter referred to as oleic acid column). Table 7.1 gives the relevant acoustic parameters of water and oleic acid. It can be seen that the acoustic impedance of oleic acid is even smaller than that of water, so it cannot be regarded as rigid.

Table 7.1 Relevant acoustic parameters of water and oleic acid [4]

Material	Density (kg/m^3)	Longitudinal sound speed (m/s)
Water	1000	1500
Oleic acid	938	1450

First, we study the relationship between the acoustic radiation force function of particles under a single boundary and ka. In the simulation, it is assumed that the distance between the particle and the boundary is twice the radius of the particle, i.e., d = 2a. For boundaries with different reflection coefficients, we simulate under both rigid column and oleic acid column conditions. The results are shown in Fig. 7.2.

In Fig. 7.2, the running results are consistent with the results in the reference [4]. The introduction of the boundary exacerbates the oscillation characteristics of the acoustic radiation force function, which stems from the complex interaction of the scattering wave of the column, the scattering wave of the mirror column, and the reflection wave of the boundary [4]. From the simulation diagrams of the rigid column and the oleic acid column, it can be found that the existence of the impedance boundary makes the negative acoustic radiation force possible, which is the so-called "acoustic trapping force". This negative force causes the particles to move away from the boundary. Of course, the appearance of the "acoustic trapping force" depends on the choice of ka value. In the case of a rigid column, it is almost only when $ka \approx 0 \sim 0.8$ The larger acoustic radiation force can only be obtained at low frequencies. Under the conditions of oleic acid columns, the ka range that can produce "acoustic trapping force" is significantly larger, and even at medium and high frequencies, it is possible to generate a larger negative acoustic radiation force. Under rigid column conditions, the increase in the reflection coefficient will reduce the positive acoustic radiation force and increase the negative force. For oleic acid columns, the peaks of both will increase. This reflects the huge impact of the impedance boundary we introduced on the acoustic radiation force of particles. However, under the conditions of rigid columns, the peak value of the acoustic radiation force function is much larger than that of non-rigid columns.

Interestingly, whether it is a rigid column or an oleic acid column, the change in the sound pressure reflection coefficient only changes the peak value of the acoustic radiation force and does not change the ka value at which it takes the extreme value [4]. At the same time, it can be found that both rigid columns and oleic acid columns will have several equilibrium positions, that is, points where the acoustic radiation force is zero. Specifically, at the node of sound pressure or the antinode of velocity, there is a balance point for particles, at this time $ka = \frac{\pi}{4}, \frac{3\pi}{4} \ldots$. At the antinode of sound pressure or the node of velocity, the acoustic radiation force received by the particles reaches its maximum [4].

In actual operation, what can usually be directly changed is only the transmission frequency of the transducer, and the radius of the particles is often fixed. Therefore, we need to further explore the relationship between the acoustic radiation force function and the frequency of the incident sound wave. Assuming that the sound pressure reflection coefficient is 0.5, the distance between the particle and the boundary is

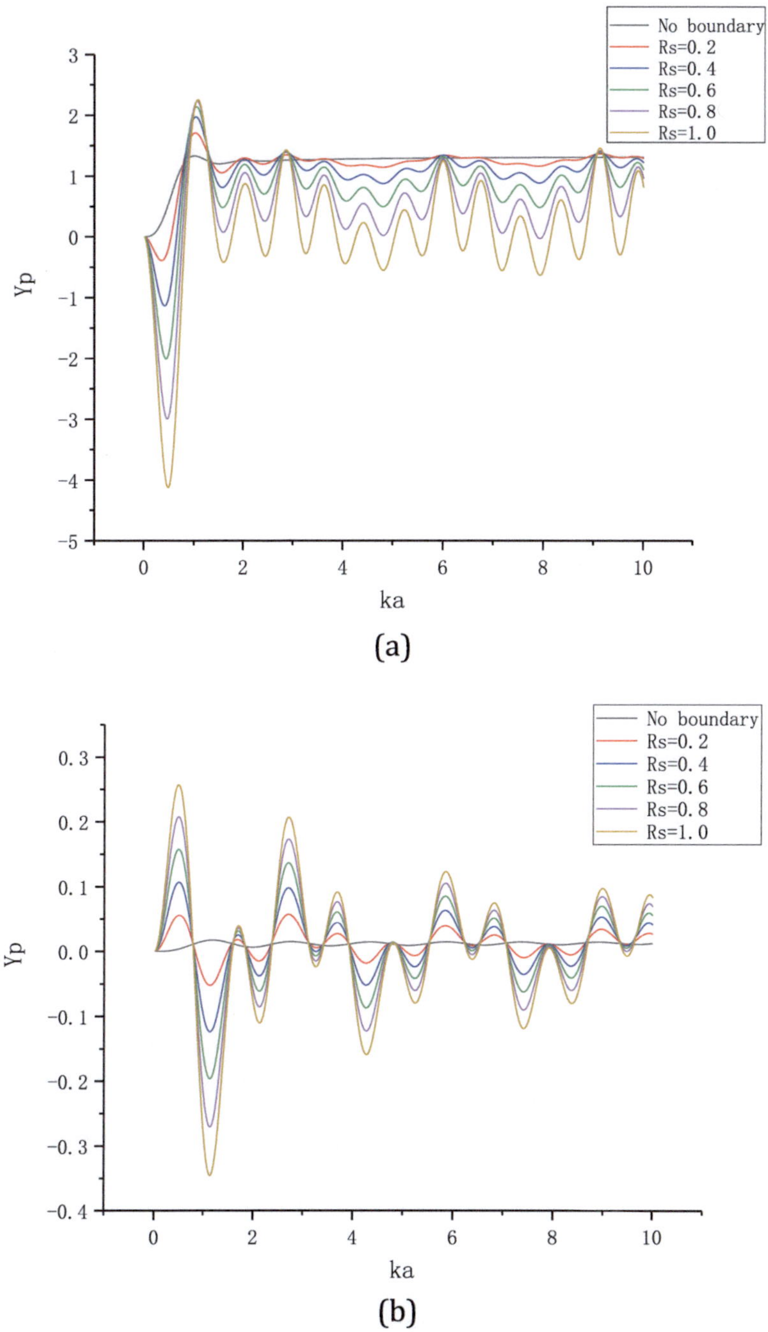

Fig. 7.2 Variation of the acoustic radiation force function of rigid column and oleic acid column with ka when plane waves are incident. **a** Rigid column; **b** Oleic acid column

still twice the radius, simulations are carried out for cylindrical particles with radii of 1, 5, and 9 μm over a frequency range of 0–500 MHz. The results of rigid columns and oleic acid columns are shown in Fig. 7.3.

In the curve of the change of the acoustic radiation force function with the frequency of the incident sound wave, it is not difficult to find that as the frequency increases, the peak values of both the positive and negative acoustic radiation forces will move to the low frequency [4], and it will become more dense. For rigid columns, when the radius is larger, the "acoustic trapping force" can be achieved at low frequencies, but the frequency range is small, and when the radius is smaller, the "acoustic trapping force" will be achieved at medium and high frequencies, and the frequency range is larger. The situation of oleic acid columns is just the opposite. But no matter what the particle radius is, the first peak values of the acoustic radiation force they receive are always equal.

So far, we have not considered the impact of the size of the distance d between the particle and the boundary on the characteristics of the acoustic radiation force, and always assumed that d = 2a. Now we fix ka = 0.063, and study the relationship between the acoustic radiation force function and kd under different boundary reflection coefficients. Still divided into rigid columns and oleic acid columns for discussion. The simulation results are shown in Fig. 7.4.

Figure 7.4 tells us that whether it is a rigid column or an oleic acid column, under the condition of fixed ka, the acoustic radiation force function is a periodic function of kd, and its relationship is similar to a sine curve. As the sound pressure reflection coefficient increases, the peak of the curve also increases. Similarly, the reflection coefficient does not change the kd value when it takes the extreme value. Under the conditions of oleic acid columns, the value of the acoustic radiation force function is slightly smaller for the same sound pressure reflection coefficient and kd value. This is consistent with the previous discussion.

7.1.2 Oblique Incidence of Plane Waves

This section expands on previous research theories [6–10]. The incidence of the beam in one direction perpendicular to the interface has changed to any angle of incidence. When considering multiple reflections of the impedance boundary, the acoustic radiation force on viscoelastic cylindrical objects of different sizes in a non-viscous fluid is studied. This includes the effective sound field (mainly including the incident sound field, boundary scattering field and image scattering), the use of the translation addition theorem, systematic analysis, and the derivation of the acoustic radiation force expression. Numerical simulations give the changes in the acoustic radiation force received by viscoelastic cylindrical particles under different size radii, boundary distances, impedance boundaries, and different incident angles of plane sound waves. The research results can provide effective theoretical support for the micro-manipulation of biological particles, and this theory can also be extended to more complex multi-layer structure particle models.

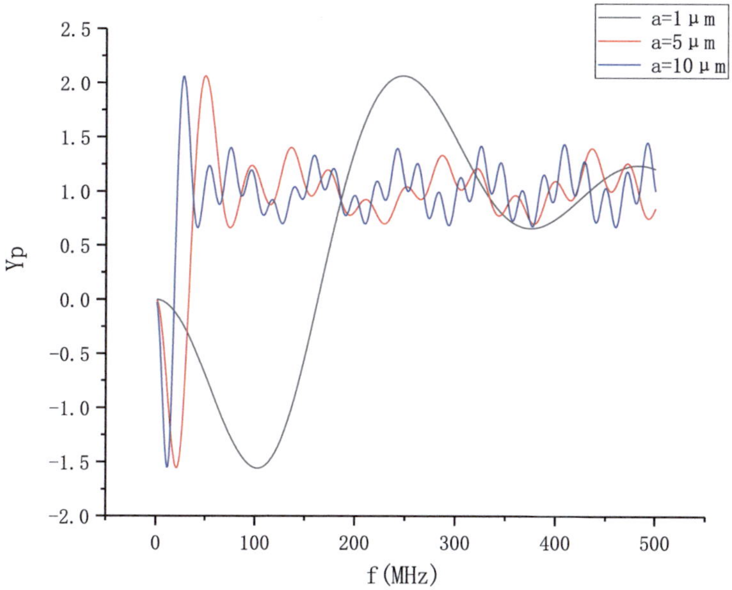

(a)

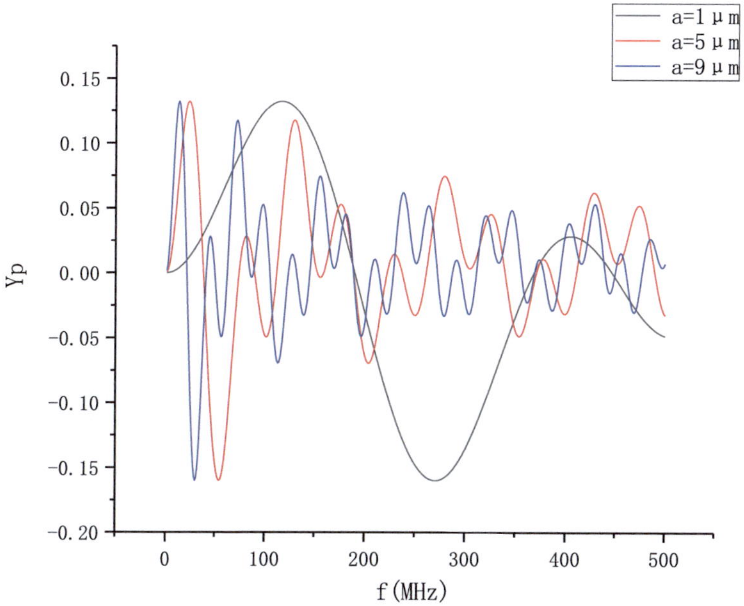

(b)

Fig. 7.3 Changes in the acoustic radiation force function of rigid column and oleic acid column with frequency when plane waves are incident. **a** Rigid column. **b** Oleic acid column

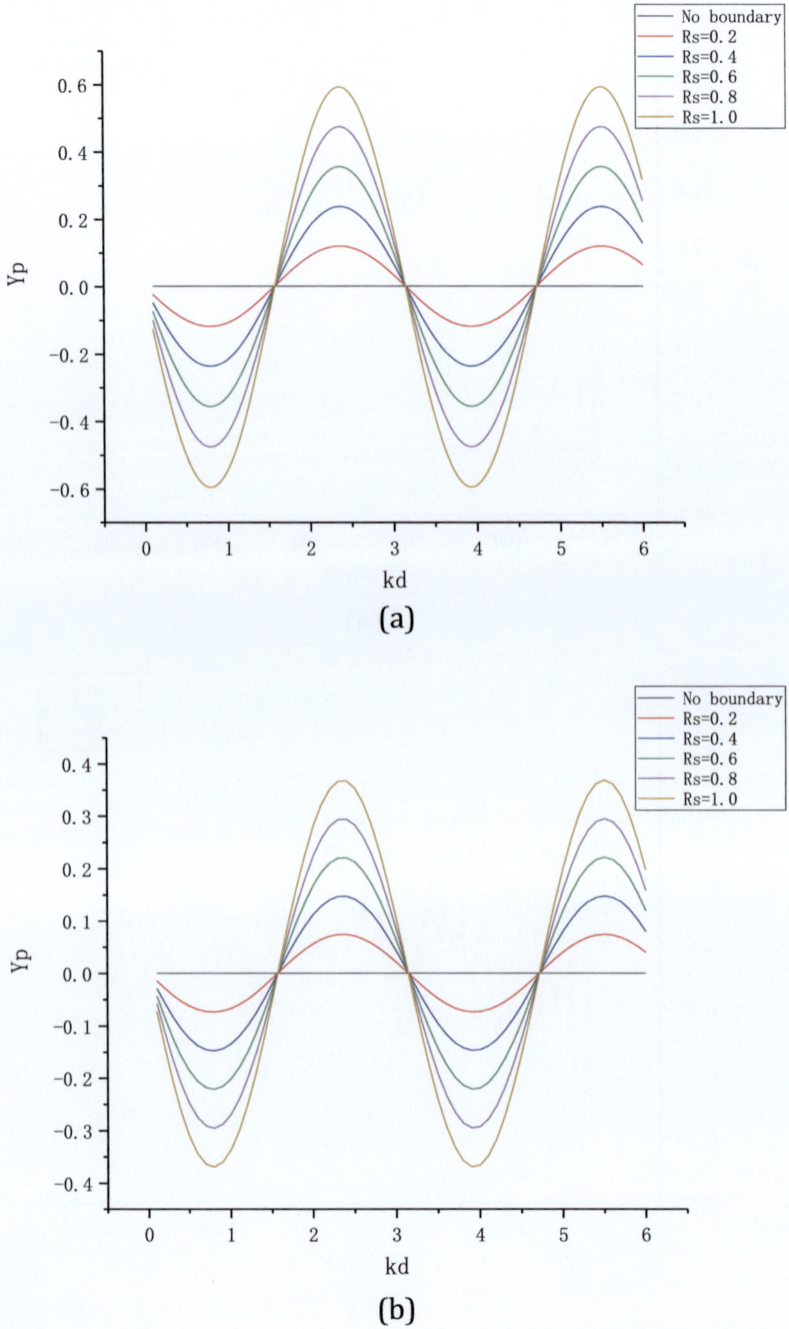

Fig. 7.4 Changes in the acoustic radiation force function of rigid column and oleic acid column with kd when plane waves are incident. **a** Rigid column. **b** Oleic acid column

7.1.2.1 Theoretical Analysis

As shown in Fig. 7.5, suppose xoy There is a cylindrical particle 1 at the origin of the plane, the particle radius is a, a plane wave is perpendicular to z with an incidence angle of β incident, at a distance of d there is an impedance boundary. Using the mirror method, there is a cylindrical particle 2 with a radius of a on the opposite side, located on the plane $x'o'y'$, with a distance between them of $2d$. The velocity potential of the plane wave incident can be expressed as:

$$\Phi_{inc} = \phi_0 e^{i(\boldsymbol{k}\cdot\boldsymbol{r}-\omega t)} \tag{7.25}$$

where, ϕ_0 is the amplitude of the velocity potential of the incident wave, ω is the angular frequency of the incident wave, $\boldsymbol{k}$ is the wave number of the incident wave.

For convenience of calculation, the incident velocity potential in Eq. (7.25) can be expanded into a finite series in the cylindrical coordinate system (r, θ) as follows:

$$\Phi_{inc}(r,\theta,t) = \phi_0 e^{-i\omega t} e^{ikr\cos(\theta-\beta)} = \phi_0 e^{-i\omega t} \sum_{n=-\infty}^{n=\infty} i^n e^{-in\beta} J_n(kr) e^{in\theta}, \tag{7.26}$$

where, $J_n(\cdot)$ is the Bessel function of the first kind, wave number $k = \frac{\omega}{c}$, c is the speed of sound in a non-viscous fluid.

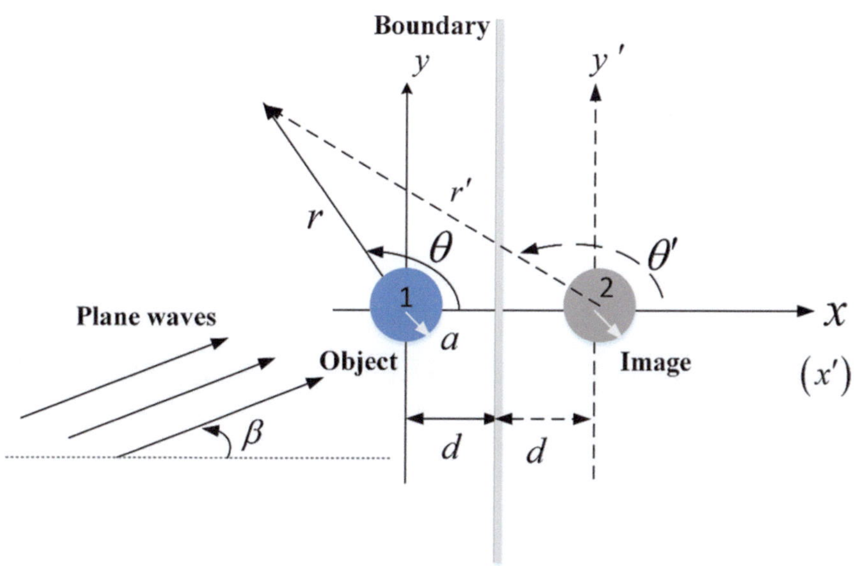

Fig. 7.5 Schematic diagram of the interaction between a plane wave at any incident angle and an elastic cylindrical particle

Due to the presence of an impedance boundary, the sound wave incident on the boundary will be reflected, and will eventually be incident on the cylindrical object again, affecting the distribution of the original sound field. Therefore, the velocity potential of the reflected wave in the cylindrical coordinate system (r, θ) can be expressed as:

$$\begin{aligned}\Phi_R(r,\theta) &= \phi_0 R_s e^{-i\omega t} e^{ikr\cos(\theta-\pi+\beta)} e^{2ikd\cos\beta} \\ &= \phi_0 R_s e^{-i\omega t} e^{2ikd\cos\beta} \sum_{n=-\infty}^{n=\infty} i^n e^{-in(\pi-\beta)} J_n(kr) e^{in\theta}\end{aligned} \tag{7.27}$$

where, R_s is the reflection coefficient of the impedance boundary, which can be approximated as $R_s = \frac{(1-\alpha)}{(1+\alpha)}$ [11, 12], and $\alpha = \frac{\rho c}{Z}$ is the specific impedance of the interface, Z is the normal impedance of the boundary. For simplicity, we use R_s to represent the physical characteristics of the impedance interface.

In addition, the velocity potential of the scattered wave of the object (mainly the scattering after the interaction of the incident plane wave with the object) can be expressed as:

$$\Phi_{sca}^{object}(r,\theta) = \phi_0 e^{-i\omega t} \sum_{n=-\infty}^{n=\infty} C_{n,1} H_n{}^{(1)}(kr) e^{in\theta} \tag{7.28}$$

where, $H_n^{(1)}(\cdot)$ is the Hankel function of the first kind of order n, $C_{n,1}$ is the scattering coefficient, which depends on the boundary conditions.

In addition, there is scattering caused by the reflection from the boundary incident on the object. This part of the scattered wave can be analyzed using the mirror principle, equivalent to the scattered wave from the mirror object (i.e., the scattered wave of the mirror particle). Its velocity potential in the cylindrical coordinate system (r', θ') can be expressed as:

$$\Phi_{sca}^{image}(r',\theta') = \phi_0 R_s e^{-i\omega t} \sum_{n=-\infty}^{n=\infty} C_{n,2} H_n{}^{(1)}(kr') e^{in\theta'} \tag{7.29}$$

where, $C_{n,2}$ It is the scattering coefficient from the mirror particle, which depends on the boundary conditions of the mirror object.

For an elastic cylinder, when a plane wave impinges on the cylinder (including the mirror particle), longitudinal and transverse waves will be excited within the object, i.e., the transmission wave velocity potential inside the particle (longitudinal wave velocity potential Φ, transverse wave velocity potential Ψ) are represented as follows:

$$\Phi_{int}^{object}(r,\theta) = \phi_0 e^{-i\omega t} \sum_{n=-\infty}^{n=\infty} E_n^{int} J_n(k_1 r) e^{in\theta} \tag{7.30}$$

$$\Psi_{int}^{object}(r,\theta) = \phi_0 e^{-i\omega t} \sum_{n=-\infty}^{n=\infty} F_n^{int} J_n(k_2 r) e^{in\theta} \tag{7.31}$$

$$\Phi_{int}^{\text{image}}(r',\theta') = \phi_0 e^{-i\omega t} \sum_{n=-\infty}^{n=\infty} H_n^{int} J_n(k_1 r') e^{in\theta'} \tag{7.32}$$

$$\Psi_{int}^{\text{image}}(r',\theta') = \phi_0 e^{-i\omega t} \sum_{n=-\infty}^{n=\infty} I_n^{int} J_n\left(k_2 r'\right) e^{in\theta'} \tag{7.33}$$

where, k_1 (k_2)is the particle's longitudinal (transverse) wave number, $E_n^{int}, F_n^{int}, H_n^{int}, I_n^{int}$ are the undetermined coefficients of the transmission wave inside the particle (mirror particle), all determined by the corresponding boundary conditions.

Next, we use the boundary conditions to solve the scattering coefficients $C_{n,1}$ and $C_{n,2}$, considering the boundary conditions of the particle when calculating the scattering coefficients. Since the particle and the mirror particle are in different reference frames, when operating in one reference frame, the formula in the other reference frame needs to be converted to the same reference frame. Therefore, using the addition theorem of cylinders [13–16], the Hankel functions in different cylindrical coordinate systems can be converted to each other as follows:

$$H_n^{(1)}(kr')e^{in\theta'} = \sum_{m=-\infty}^{\infty} J_m(kr) H_{m-n}^{(1)}(2kd) e^{im\theta}, r < 2d \tag{7.34}$$

And

$$H_n^{(1)}(kr)e^{in\theta} = \sum_{m=-\infty}^{\infty} J_m(kr') H_{n-m}^{(1)}(2kd) e^{im\theta'}, r' < 2d \tag{7.35}$$

Therefore, the total sound field velocity potential outside the medium particle in the cylindrical coordinate systems (r,θ) and $\left(r',\theta'\right)$ can be obtained respectively:

$$\begin{aligned} \Phi_{tot}(r,\theta)|_{r<2d} &= \Phi_{inc}(r,\theta) + \Phi_R(r,\theta) + \Phi_{sca}^{object}(r,\theta) + \Phi_{sca}^{image}(r,\theta)\Big|_{r<2d} \\ &= \phi_0 \left[\begin{array}{c} \sum\limits_{n=-\infty}^{\infty} \left(\Lambda_{1,n} J_n(kr) + C_{n,1} H_n^{(1)}(kr)\right) e^{in\theta} \\ + \sum\limits_{n=-\infty}^{\infty} \sum\limits_{m=-\infty}^{\infty} R_s C_{n,2} J_m(kr) H_{n-m}^{(1)}(2kd) e^{im\theta} \end{array} \right] \end{aligned} \tag{7.36}$$

where, $\Lambda_{1,n} = i^n(e^{-in\beta} + R_s e^{i2kd\cos\beta} e^{-in(\pi-\beta)})$.

Similarly, the total velocity potential in the cylindrical cooinate system $\left(r',\theta'\right)$ is:

$$\Phi_{tot}\left(r',\theta'\right)\Big|_{r<2d} = \Phi_{inc}\left(r',\theta'\right) + \Phi_R\left(r',\theta'\right) + \Phi_{sca}^{object}\left(r',\theta'\right)\Big|_{r'<2d} + \Phi_{sca}^{image}\left(r',\theta'\right)$$

$$= \phi_0 \left[\begin{array}{l} \sum\limits_{n=-\infty}^{\infty} \left(\Lambda_{2,n} J_n(kr') + C_{n,1} H_n^{(1)}(kr')\right) e^{in\theta'} \\ + \sum\limits_{n=-\infty}^{\infty} \sum\limits_{m=-\infty}^{\infty} R_s C_{n,2} J_m(kr') H_{m-n}^{(1)}(2kd) e^{im\theta'} \end{array} \right] \tag{7.37}$$

where, $\Lambda_{2,n} = i^n (R_s e^{i2kd\cos\beta} e^{-in\beta} + e^{-in(\pi-\beta)})$

Of course, we can also consider multiple reflection situations, in which case the corresponding reflection wave velocity potential can be represented as:

$$\begin{aligned} \Phi_R(r,\theta) &= \phi_0 e^{-i\omega t} \sum_{q=1}^{p} R_s^q e^{ikr\cos(\theta-\pi+\beta)} e^{2^q ikd\cos\beta} \\ &= \phi_0 e^{-i\omega t} \sum_{q=1}^{p} R_s^q e^{2^q ikd\cos\beta} \sum_{n=-\infty}^{n=\infty} i^n e^{-in(\pi-\beta)} J_n(kr) e^{in\theta} \end{aligned} \tag{7.38}$$

In the cylindrical coordinate system (r, θ) and (r', θ'), the scattering wave velocity potentials are respectively represented as:

$$\Phi_{sca}^{image}(r,\theta)\big|_{r<2d} = \sum_{q=1}^{p} \sum_{n=-\infty}^{\infty} \sum_{m=-\infty}^{\infty} R_s^q C_{n,2} J_m(kr) H_{n-m}^{(1)}(2^q kd) e^{im\theta} \tag{7.39}$$

$$\Phi_{sca}^{object}(r',\theta')\big|_{r'<2d} = \sum_{q=1}^{p} \sum_{n=-\infty}^{\infty} \sum_{m=-\infty}^{\infty} R_s^q C_{n,1} J_m(kr') H_{n-m}^{(1)}(2^q kd) e^{im\theta'} \tag{7.40}$$

Similarly, in the cylindrical coordinate system (r, θ) and (r', θ'), the total sound field velocity potentials outside the medium particles can be respectively represented as:

$$\Phi_{tot}(r,\theta)|_{r<2d} = \phi_0 \left[\begin{array}{l} \sum\limits_{-\infty}^{\infty} \left(\Pi_{1,n} J_n(kr) + C_{n,1} H_n^{(1)}(kr)\right) e^{in\theta} \\ + \sum\limits_{q=1}^{p} \sum\limits_{n=-\infty}^{\infty} \sum\limits_{m=-\infty}^{\infty} R_s^q C_{n,2} J_m(kr) H_{n-m}^{(1)}(2^q kd) e^{im\theta} \end{array} \right] \tag{7.41}$$

where, $\Pi_{1,n} = i^n \left(e^{-in\beta} + \sum\limits_{q=1}^{p} R_s^q e^{i2^q kd\cos\beta} e^{-in(\pi-\beta)} \right)$.

$$\Phi_{tot}(r',\theta')\big|_{r'<2d} = \phi_0 \left[\begin{array}{l} \sum\limits_{n=-\infty}^{\infty} \left(\Pi_{2,n} J_n(kr') + C_{n,1} H_n^{(1)}(kr')\right) e^{in\theta'} \\ + \sum\limits_{q=1}^{p} \sum\limits_{n=-\infty}^{\infty} \sum\limits_{m=-\infty}^{\infty} R_s^q C_{n,2} J_m(kr') H_{n-m}^{(1)}(2^q kd) e^{im\theta'} \end{array} \right] \tag{7.42}$$

where, $\Pi_{2,n} = i^n\left(\sum_{q=1}^{p} R_s^q e^{i2^q kd\cos\beta} e^{-in\beta} + e^{-in(\pi-\beta)}\right)$.

The scattering coefficient is calculated through boundary conditions. The boundary conditions for the particle and the image particle are: at the boundary ($r = a$), the normal displacement, normal stress, and tangential stress are continuous, that is:

$$U_{2r} = U_{1r},\quad \sigma_{2rr} = \sigma_{1rr},\quad \sigma_{2r\theta} = 0 \tag{7.43}$$

$$U_{2r'} = U_{1r'},\quad \sigma_{2r'r'} = \sigma_{1r'r'},\quad \sigma_{2r'\theta'} = 0 \tag{7.44}$$

where, $U_{1r}(U_{1r'})$is the normal displacement of the particle outside the cylinder, $U_{2r}(U_{2r'})$ is the normal displacement of the particle inside the cylinder, $\sigma_{1rr}(\sigma_{1r'r'})$ is the normal stress of the particle outside the cylinder, $\sigma_{2rr}(\sigma_{2r'r'})$ is the normal stress of the particle inside the cylinder, $\sigma_{2r\theta}(\sigma_{2r'\theta'})$ is the tangential stress inside the cylinder.

Next, we calculate the acoustic radiation force acting on the particle. The known formula for calculating the acoustic radiation force [10] is as follows:

$$\langle \boldsymbol{F}_{\{1,2\}}\rangle \underset{kr\to\infty}{=} \frac{1}{2}\rho k^2 \int_0^{2\pi} \Re\{\Phi_{is,\{1,2\}}\} d\boldsymbol{S} \tag{7.45}$$

where, $\Re\{\cdot\}$ represents taking the real part of the complex number, $\Phi_{is,\{1,2\}} \underset{kr\to\infty}{=} \Phi^*_{sca,\{1,2\}}[(\frac{i}{k})\partial_r\Phi^{eff}_{inc,\{1,2\}} - \Phi^{eff}_{inc,\{1,2\}} - \Phi_{sca,\{1,2\}}]$, and $d\boldsymbol{S} = dS\cdot\boldsymbol{e}_r$, $dS = rd\theta$ represents the cross-sectional area of the cylindrical particle of unit length, $\boldsymbol{e}_r = \cos\theta\boldsymbol{e}_x + \sin\theta\boldsymbol{e}_y$, where $\boldsymbol{e}_x$, $\boldsymbol{e}_y$ are the unit vectors of the Cartesian coordinate system, the symbol $\langle\cdot\rangle$ indicates time averaging, and the superscript $*$ denotes the complex conjugate.

Equations (7.36) and (7.41) represent the total velocity potential around the particle in the medium. It should be noted that the waves reflected from the particle surface are not considered to re-impinge on the particle surface, so this part of the sound wave is not included in the effective sound field of the particle [9].

$$\begin{aligned}
\Phi^{eff}_{inc}(r,\theta) &= \Phi_{tot}(r,\theta)\big|_{r<2d} - \Phi^{object}_{sca}(r,\theta)\\
&= \Phi_{inc}(r,\theta) + \Phi_R(r,\theta) + \Phi^{image}_{sca}(r,\theta)\\
&= \phi_0\left[\sum_{n=-\infty}^{\infty}(\Lambda_{1,n}J_n(kr)e^{in\theta} + R_s\sum_{m=-\infty}^{\infty}\left(C_{n,2}\sum_{m=-\infty}^{\infty}J_n(kr)H^{(1)}_{n-m}(2kd)e^{in\theta}\right)\right]
\end{aligned} \tag{7.46}$$

If multiple reflections are considered, Eq. (7.47) is expressed as:

$$\Phi^{eff}_{inc}(r,\theta) = \Phi_{tot}(r,\theta)\big|_{r<2d} - \Phi^{object}_{sca}(r,\theta)$$

$$= \Phi_{inc}(r,\theta) + \Phi_R(r,\theta) + \Phi_{sca}^{image}(r,\theta)$$

$$= \phi_0\left[\sum_{n=-\infty}^{\infty}(\Pi_{1,n}J_n(kr)e^{in\theta} + \sum_{q=1}^{p}\sum_{n=-\infty}^{\infty}\sum_{m=-\infty}^{\infty}R_s^q C_{n,2}J_m(kr)H_{n-m}^{(1)}(2^q kd)e^{im\theta}\right] \tag{7.47}$$

Based on the effective velocity potential of the particle and the formula for calculating the acoustic radiation force, the expression for the acoustic radiation force on the particle can be obtained:

$$Y_x = \frac{2}{kS_{c,1}}\Im\left\{\sum_{n=-\infty}^{\infty}\left(\Lambda_{1,n} + C_{n,1} + R_s\sum_{m=-\infty}^{\infty}C_{m,2}H_{n-m}^{(1)}(2kd)\right)\left(C_{n+1,1}^{*} - C_{n-1,1}^{*}\right)\right\} \tag{7.48}$$

$$Y_y = -\frac{2}{kS_{c,1}}\Re\left\{\sum_{n=-\infty}^{\infty}\left(\Lambda_{1,n} + C_{n,1} + R_s\sum_{m=-\infty}^{\infty}C_{m,2}H_{n-m}^{(1)}(2kd\right)\left(C_{n+1,1}^{*} + C_{n-1,1}^{*}\right)\right\} \tag{7.49}$$

If multiple reflections are considered, Eqs. (7.48) and (7.49) are expressed as:

$$Y_x = \frac{2}{kS_{c,1}}\Im\left\{\sum_{n=-\infty}^{\infty}\left(\Pi_{1,n} + C_{n,1} + \sum_{q=1}^{p}\sum_{m=-\infty}^{\infty}R_s^q C_{m,2}H_{n-m}^{(1)}(2^q kd)\right)\right. \left.\left(C_{n+1,1}^{*} - C_{n-1,1}^{*}\right)\right\} \tag{7.50}$$

$$Y_y = -\frac{2}{kS_{c,1}}\Re\left\{\sum_{n=-\infty}^{\infty}\left(\Pi_{1,n} + C_{n,1} + \sum_{q=1}^{p}\sum_{m=-\infty}^{\infty}R_s^q C_{m,2}H_{n-m}^{(1)}(2^q kd)\right)\right. \left.\left(C_{n+1,1}^{*} + C_{n-1,1}^{*}\right)\right\} \tag{7.51}$$

For physical particles, by substituting Eq. (7.36) into the boundary condition Eq. (7.43) and using the orthogonality of the complex exponential function, the following expressions can be obtained:

$$\Pi_{1,n}kaJ_n'(ka) + C_{n,1}kaH_n^{(1)\prime}(ka) + kaJ_n'(ka)\sum_{m=-\infty}^{\infty}R_s C_{m,2}H_{n-m}^{(1)\prime}(2kd)$$
$$= E_n^{\text{int}}k_1 aJ_n'(k_1 a) + F_n^{\text{int}}nJ_n(k_2 a) \tag{7.52}$$

$$-\lambda(\text{ka})^2\Pi_{1,n}J_n(\text{ka}) - \lambda(\text{ka})^2 C_{n,1}H_n^{(1)}(\text{ka})$$

$$- \lambda(\mathrm{ka})^2 J_n(\mathrm{ka}) \sum_{m=-\infty}^{\infty} R_s C_{m,2} H_{n-m}^{(1)'}(2\mathrm{kd})$$
$$= 2\mu_1 (k_1 a)^2 E_n^{\mathrm{int}} J_n''(k_1 a) + 2\mu_1 n k_2 a F_n^{\mathrm{int}} J_n'(k_2 a)$$
$$- 2\mu_1 n F_n^{\mathrm{int}} J_n(k_2 a) - \lambda (k_1 a)^2 E_n^{\mathrm{int}} J_n(k_1 a) \quad (7.53)$$

$$- 2E_n^{\mathrm{int}} n k_1 a J_n'(k_1 a) + 2nE_n^{\mathrm{int}} J_n(k_1 a) - n^2 F_n^{\mathrm{int}} J_n(k_2 a)$$
$$- (k_2 a)^2 F_n^{\mathrm{int}} J_n''(k_2 a) + k_2 a F_n^{\mathrm{int}} J_n'(k_2 a) = 0 \quad (7.54)$$

Similarly, for image particles, by substituting Eq. (7.37) into the boundary condition Eq. (7.44) and using the orthogonality of the complex exponential function, the following expressions can be obtained:

$$\Pi_{2,n} ka J_n\prime(ka) + C_{n,2} ka H_n^{(1)}\prime(ka) + ka J_n\prime(ka)$$
$$\sum_{m=-\infty}^{\infty} R_s C_{m,1} H_{m-n}^{(1)}\prime(2kd) = H_n^{\mathrm{int}} k_1 a J_n\prime(k_1 a) + I_n^{\mathrm{int}} n J_n(k_2 a) \quad (7.55)$$

$$- \lambda (ka)^2 \Pi_{2,n} J_n(ka) - \lambda (ka)^2 C_{n,2} H_n^{(1)}(ka)$$
$$- \lambda (ka)^2 J_n(ka) \sum_{m=-\infty}^{\infty} R_s C_{m,1} H_{n-m'}^{(1)}(2kd)$$
$$= 2\mu_1 (k_1 a)^2 H_n^{\mathrm{int}} J_n''(k_1 a) + 2\mu_1 n k_2 a I_n^{\mathrm{int}} J_n'(k_2 a)$$
$$- 2\mu_1 n I_n^{\mathrm{int}} J_n(k_2 a) - \lambda (k_1 a)^2 H_n^{\mathrm{int}} J_n(k_1 a) \quad (7.56)$$

$$- 2H_n^{\mathrm{int}} n k_1 a J_n'(k_1 a) + 2nH_n^{\mathrm{int}} J_n(k_1 a) - n^2 I_n^{\mathrm{int}} J_n(k_2 a)$$
$$- (k_2 a)^2 I_n^{\mathrm{int}} J_n''(k_2 a) + k_2 a I_n^{\mathrm{int}} J_n'(k_2 a) = 0 \quad (7.57)$$

Equations (7.52)–(7.57) form a set of equations. By re-expressing this set of equations in the form of matrices and vectors, the scattering coefficients can be obtained using matrix inverse transformation. $C_{n,1}$ and $C_{n,2}$, where the dimensions of the matrix are $(2N+1)\times(2N+1)$, and the dimensions of the vector are $1\times(2N+1)$, $N_{max} = [max(max(ka, kb), kd)] + 25$. After obtaining the scattering coefficients, substitute them into Eqs. (7.50) and (7.51) to calculate the longitudinal and transverse acoustic radiation forces on the particles. Equations (7.52)–(7.57) can be further simplified and organized as follows:

$$\Pi_{1,n} x_{1n} \left[x_{5n} ka J_n'(ka) + x_{6n} J_n(ka) \right]$$
$$+ C_{n,1} x_{1n} \left[x_{5n} ka H_n^{(1)'}(ka) + x_{6n} H_n^{(1)}(ka) \right]$$
$$+ x_{1n} \left[x_{5n} ka J_n'(ka) + x_{6n} J_n(ka) \right] \sum_{m=-\infty}^{\infty} R_s C_{m,2} H_{n-m}^{(1)\,'}(2kd) = 0 \quad (7.58)$$

$$\begin{aligned}
&\Pi_{2,n} y_{1n}\left[y_{6n} ka J_n'(ka) + y_{5n} J_n(ka)\right] \\
&+ C_{n,2} y_{1n}\left[y_{6n} ka H_n^{(1)\prime}(ka) + y_{5n} H_n^{(1)}(ka)\right] \\
&+ y_{1n}\left[y_{6n} ka J_n{}'(ka) + y_{5n} J_n(ka)\right] \sum_{m=-\infty}^{\infty} R_s C_{m,1} H_{m-n}^{(1)}{}'(2kd) = 0
\end{aligned} \tag{7.59}$$

where,

$$\begin{aligned}
x_{1n} &= -2nk_1 a J_n'(k_1 a) + 2nE_n^{\text{int}} J_n(k_1 a), \\
x_{2n} &= n^2 J_n(k_2 a) + (k_2 a)^2 J_n''(k_2 a) - k_2 a J_n'(k_2 a), \\
x_{3n} &= [2\mu_1 (k_1 a)^2 J_n''(k_1 a) - \lambda (k_1 a)^2 J_n(k_1 a)]/[\lambda (ka)^2], \\
x_{4n} &= [2\mu_1 n k_2 a J_n'(k_2 a) - 2\mu_1 n J_n(k_2 a)]/[\lambda (ka)^2], \\
x_{5n} &= x_{2n} x_{3n} + x_{1n} x_{4n}, \\
x_{6n} &= x_{2n} k_1 a J_n'(k_1 a) + x_{1n} n J_n(k_2 a), \\
y_{1n} &= -2nk_1 a J_n'(k_1 a) + 2nE_n^{\text{int}} J_n(k_1 a), \\
y_{2n} &= n^2 J_n(k_2 a) + (k_2 a)^2 J_n''(k_2 a) - k_2 a J_n'(k_2 a), \\
y_{3n} &= [2\mu_1 (k_1 a)^2 J_n''(k_1 a) - \lambda (k_1 a)^2 J_n(k_1 a)]/[\lambda (ka)^2], \\
y_{4n} &= [2\mu_1 n J_n(k_2 a) - 2\mu_1 n k_2 a J_n'(k_2 a)]/[\lambda (ka)^2], \\
y_{5n} &= y_{2n} k_1 a J_n'(k_1 a) + y_{1n} n J_n(k_2 a), \\
y_{6n} &= y_{2n} y_{3n} - y_{1n} y_{4n},
\end{aligned}$$

The system of Eqs. (7.58) and (7.59) can be represented in matrix form as follows:

$$\boldsymbol{a} + \boldsymbol{C}_1 * \boldsymbol{A} + R_s \boldsymbol{C}_2 * \boldsymbol{C} = 0 \tag{7.60}$$

$$\boldsymbol{b} + \boldsymbol{C}_2 * \boldsymbol{B} + R_s \boldsymbol{C}_1 * \boldsymbol{D} = 0 \tag{7.61}$$

$C_{n,1}$ and $C_{n,2}$ can be represented as,

$$\boldsymbol{C}_1 = (-\boldsymbol{a} - \boldsymbol{C}_2 * \boldsymbol{C}) * \boldsymbol{A}^{-1} \tag{7.62}$$

$$\boldsymbol{C}_2 = (\boldsymbol{a} * \boldsymbol{A}^{-1} * \boldsymbol{D} - \boldsymbol{b}) * (\boldsymbol{B} - \boldsymbol{C} * \boldsymbol{A}^{-1} * \boldsymbol{D})^{-1} \tag{7.63}$$

where,

$$\begin{aligned}
\boldsymbol{a} &= \left[a_{-N}, a_{-N+1}, \cdots, a_{N-1}, a_N\right]_{1\times(2N+1)} \\
\boldsymbol{b} &= \left[b_{-N}, b_{-N+1}, \cdots, b_{N-1}, b_N\right]_{1\times(2N+1)} \\
\boldsymbol{C}_1 &= \left[C_{-N,1}, C_{-N+1,1}, \cdots, C_{N-1,1}, C_{N,1}\right]_{1\times(2N+1)}
\end{aligned}$$

$$\boldsymbol{C}_2 = \left[C_{-N,2}, C_{-N+1,2}, \cdots, C_{N-1,2}, C_{N,2}\right]_{1\times(2N+1)}$$

$$\boldsymbol{A} = \begin{bmatrix} A_{-N} & 0 & \cdots & 0 & 0 \\ 0 & A_{-N+1} & 0 & \cdots & 0 \\ \vdots & 0 & \ddots & 0 & \vdots \\ 0 & \cdots & 0 & A_{N-1} & 0 \\ 0 & 0 & \cdots & 0 & A_N \end{bmatrix}_{(2N+1)\times(2N+1)}$$

$$\boldsymbol{B} = \begin{bmatrix} B_{-N} & 0 & \cdots & 0 & 0 \\ 0 & B_{-N+1} & 0 & \cdots & 0 \\ \vdots & 0 & \ddots & 0 & \vdots \\ 0 & \cdots & 0 & B_{N-1} & 0 \\ 0 & 0 & \cdots & 0 & B_N \end{bmatrix}_{(2N+1)\times(2N+1)}$$

$$\boldsymbol{C} = \begin{bmatrix} E_{-N}H^{(1)}_{-N-(-N)}(kd) & E_{-N+1}H^{(1)}_{-N+1-(-N)}(kd) & \cdots & E_{N-1}H^{(1)}_{N-1-(-N)}(kd) & E_N H^{(1)}_{N-(-N)}(kd) \\ E_{-N}H^{(1)}_{-N-(-N+1)}(kd) & E_{-N+1}H^{(1)}_{-N+1-(-N+1)}(kd) & \cdots & E_{N-1}H^{(1)}_{N-1-(-N+1)}(kd) & E_N H^{(1)}_{N-(-N+1)}(kd) \\ \vdots & \vdots & \ddots & \vdots & \vdots \\ E_{-N}H^{(1)}_{-N-(N-1)}(kd) & E_{-N+1}H^{(1)}_{-N+1-(N-1)}(kd) & \cdots & E_{N-1}H^{(1)}_{N-1-(N-1)}(kd) & E_N H^{(1)}_{N-(N-1)}(kd) \\ E_{-N}H^{(1)}_{-N-(N)}(kd) & E_{-N+1}H^{(1)}_{-N+1-(N)}(kd) & \cdots & E_{N-1}H^{(1)}_{N-1-(N)}(kd) & E_N H^{(1)}_{N-(N)}(kd) \end{bmatrix}_{(2N+1)\times(2N+1)}$$

$$\boldsymbol{D} = \begin{bmatrix} F_{-N}H^{(1)}_{-N-(-N)}(kd) & F_{-N+1}H^{(1)}_{-N-(-N+1)}(kd) & \cdots & F_{N-1}H^{(1)}_{-N-(N-1)}(kd) & F_N H^{(1)}_{-N-N}(kd) \\ F_{-N}H^{(1)}_{-N+1-(-N)}(kd) & F_{-N+1}H^{(1)}_{-N+1-(-N+1)}(kd) & \cdots & F_{N-1}H^{(1)}_{-N+1-(N-1)}(kd) & F_N H^{(1)}_{-N+1-N}(kd) \\ \vdots & \vdots & \ddots & \vdots & \vdots \\ F_{-N}H^{(1)}_{N-1-(-N)}(kd) & F_{-N+1}H^{(1)}_{N-1-(-N+1)}(kd) & \cdots & F_{N-1}H^{(1)}_{N-1-(N-1)}(kd) & F_N H^{(1)}_{N-1-N}(kd) \\ F_{-N}H^{(1)}_{N-(-N)}(kd) & F_{-N+1}H^{(1)}_{N-(-N+1)}(kd) & \cdots & F_{N-1}H^{(1)}_{N-(N-1)}(kd) & F_N H^{(1)}_{N-N}(kd) \end{bmatrix}_{(2N+1)\times(2N+1)} \quad (7.64)$$

Other quantities can be represented as:

$$\begin{aligned} a_n &= \Pi_{1,n}x_{1n}\left[x_{5n}kaJ_n{}'(ka) + x_{6n}J_n(ka)\right] \\ b_n &= \Pi_{2,n}y_{1n}\left[y_{6n}kaJ_n{}'(ka) + y_{5n}J_n(ka)\right] \\ A_n &= x_{1n}\left[x_{5n}kaH_n^{(1)\prime}(ka) + x_{6n}H_n^{(1)}(ka)\right] \\ B_n &= y_{1n}\left[y_{6n}kaH_n^{(1)\prime}(ka) + y_{5n}H_n^{(1)}(ka)\right] \\ E_n &= x_{1n}\left[x_{5n}kaJ_n{}'(ka) + x_{6n}J_n(ka)\right] \\ F_n &= y_{1n}\left[y_{6n}kaJ_n{}'(ka) + y_{5n}J_n(ka)\right], \end{aligned} \quad (7.65)$$

7.1.2.2 Simulation Analysis and Discussion

A simulation analysis was conducted on the acoustic radiation force experienced by particles near the impedance boundary when plane waves incident at any angle. In this chapter, the material of the particles is chosen to be polyethylene. The known external

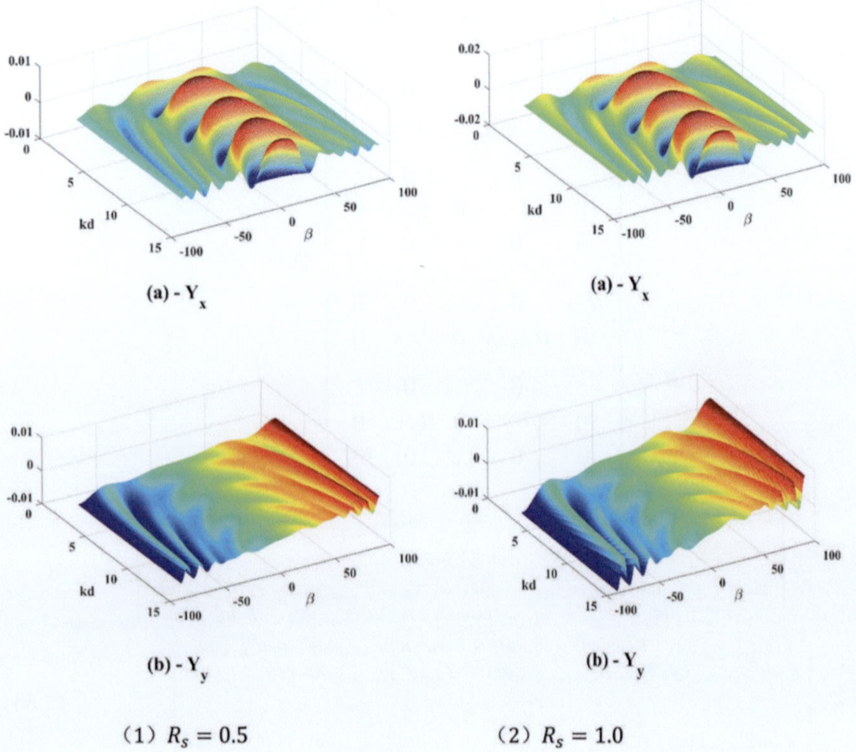

Fig. 7.6 Acoustic radiation force on particle (ka = 0.1) during single reflection. **a** Lateral acoustic radiation force Y_x, **b** Longitudinal acoustic radiation force $\mathrm{Y_y}$

medium (water) has a density of $\rho_1 = 1000$ kg/m^3, and the speed of sound in water is $c_1 = 1500$ m/s The density of polyethylene material is ρ_2=0.957 × 10^3 kg/m^3, the longitudinal wave speed in the material is $c_L = 2430$ m/s, and the transverse wave speed is $c_T = 950$ m/s. In Eqs. (7.48)–(7.51), the expression for acoustic radiation force $S_c(= 2a)$ represents the cross-sectional area per unit length of the particle. In the following numerical simulations, we will analyze and study the changes in the acoustic radiation force from the aspects of the size of the particle cross-section, the distance from the boundary, and different incident angles of the plane wave.

Acoustic Radiation Force Under Specific Particle Size

First, we fixed the size of the particle (taken as $ka = 0.1$), and considered the influence of different incident angles and different boundary distances on the acoustic radiation force of the particle. The specific numerical simulation results are shown in Figs. 7.6 and 7.7.

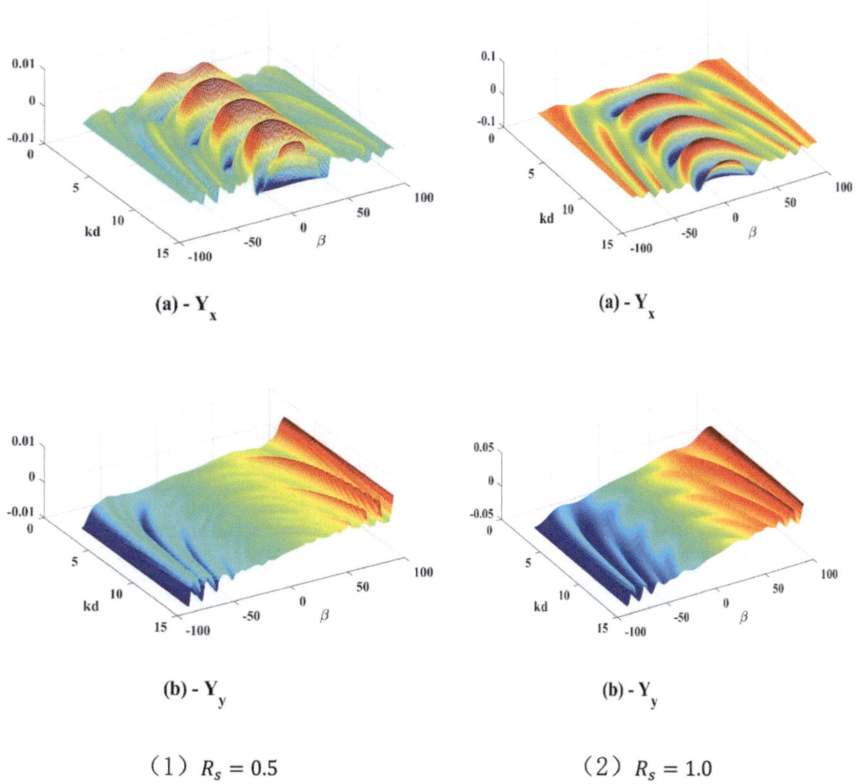

Fig. 7.7 Acoustic radiation force on particle ($ka = 0.1$) during three reflection **a** Lateral acoustic radiation force Y_x, **b** Longitudinal acoustic radiation force $\mathrm{Y_y}$

Figure 7.6 shows the simulation results of the transverse and longitudinal acoustic radiation forces of the particle during a single reflection. In the left side of Fig. 7.6, the reflection rate of the impedance boundary is $R_s = 0.5$, the range of plane wave incident angles is $-90° \leq \beta \leq 90°$, and the normalized distance of the particle from the impedance boundary is $0.1 < kd \leq 15$. As can be seen from Figure (a), the transverse acoustic radiation force function of the particle Y_x is symmetrically distributed with the incident angle β of the plane wave ($Y_x(\beta) = Y_x(-\beta)$), and shows a certain fan-shaped distribution with the change of angleβ. As the incident angle increases, the absolute value of the radiation force gradually decreases. In addition, Y_x changes between positive and negative, and its absolute value decreases as the boundary spacing kd increases. This indicates that at a specific boundary distance, when an appropriate incident angle β is selected, the particle can be pushed away from ($Y_x < 0$) or pulled towards ($Y_x > 0$) the impedance boundary, or at the intersection point ($Y_x = 0$), the particle can remain stationary.

Figure (b) shows the longitudinal acoustic radiation force of the particle Y_y, which, similar to the transverse acoustic radiation force Y_x, also decreases as the distance *kd* increases. Y_y is inversely symmetrical to the incident angle β in a fan-shaped distribution, and increases as the angle β increases. At normal incidence $\beta = 0°$, $Y_y = 0$; at $-90° \leq \beta < 0°$, the amplitude of Y_y always remains negative; and at $0° < \beta \leq 90°$, the amplitude of Y_y is always positive. From Fig. 7.6a, b, it can be found that when the particle spacing *kd* is close to the particle size ($ka = 0.1$), the absolute value of the acoustic radiation force (Y_x, Y_y) is at its maximum.

The right side of Fig. 7.6 takes the boundary reflection rate as $R_s = 1.0$ (i.e., full reflection at the rigid boundary). Comparing the transverse radiation force Y_x and the longitudinal radiation force Y_y in the left side (1), it can be found that they have the same characteristics with the change of incident angle β and boundary distance *kd*. The only difference is that as the boundary reflection rate R_s increases, their values also increase, indicating that the absolute value of Y_x, Y_y increases with the increase of R_s. In actual particle manipulation, these changes should be paid special attention to, as different values will affect the manipulation of particles.

Figure 7.7 shows the numerical simulation results of the lateral and longitudinal acoustic radiation forces (Y_x, Y_y) on particles during three reflections. Compared to Fig. 7.6, which only considers a single reflection, the corresponding values here have increased to a certain extent. Upon careful analysis, it can be found that when the reflection rate of the impedance boundary R_s is low, the number of reflections can be disregarded in the theoretical calculation of the acoustic radiation force. However, when the boundary reflection rate is high, the calculation of the acoustic radiation force should consider multiple reflections.

Acoustic Radiation Force Under Specific Boundary Distance

Next, we fixed the distance from the particle to the impedance boundary (*d* is the distance, *kd* is the normalized distance, here taken as $kd = 6$), and considered the changes in the acoustic radiation force under different incident angles and particle sizes. The specific simulation results are shown in Figs. 7.8 and 7.9.

Figure 7.8 shows the simulation curves of the lateral and longitudinal acoustic radiation forces of the particles during a single reflection. The boundary reflection rate in the left side of Fig. 7.6(1) is taken as $R_s = 0.5$, which gives the variation of the radiation force of the particles with the incident angle under four particle sizes $ka = 0.1$, 0.6, 1.1, and 1.6. The range of the plane wave incident angle is: $-90° \leq \beta \leq 90°$. From Figure (a), it can be seen that the lateral acoustic radiation force function of the particles Y_x is symmetrically distributed with the incident angle β of the plane wave ($Y_x(\beta) = Y_x(-\beta)$), and presents a certain oscillation state with the change of angle β. At $ka = 0.1$, the value of $=Y_x$ is very small, but as the particle size increases, the radiation force value increases, but it fluctuates, and in most cases, Y_x is positive, only at individual incident angles and specific particle sizes, Y_x a negative value appears. This means that under these conditions, when an appropriate incident angle β is selected, the particle can be pulled towards the ($Y_x > 0$) impedance boundary.

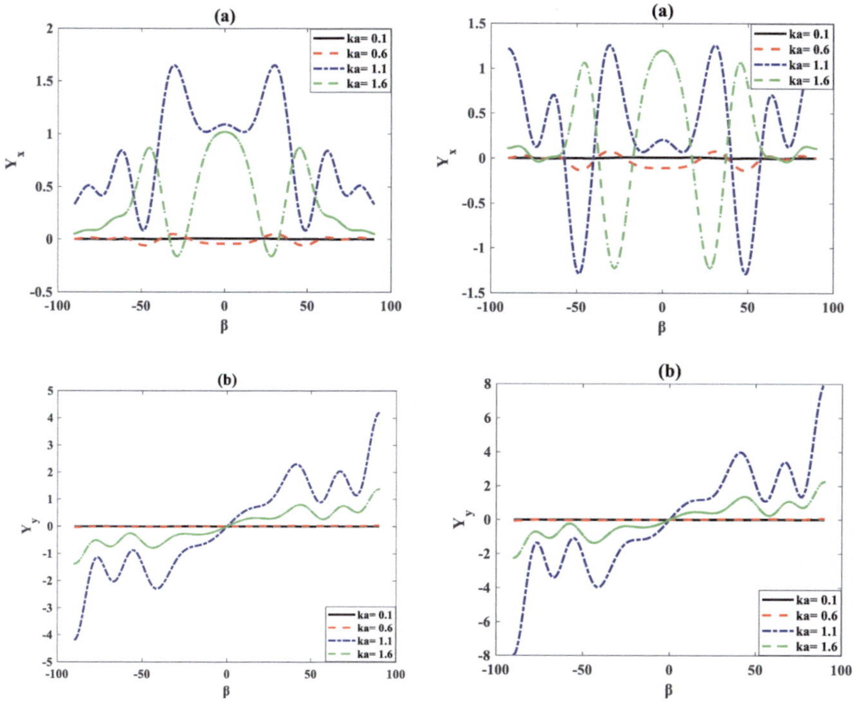

(1) $R_s = 0.5$ (2) $R_s = 1.0$

Fig. 7.8 Acoustic radiation force on particle (kd = 6) during single reflection. **a** Lateral acoustic radiation force Y_x, **b** Longitudinal acoustic radiation force Y_y.

Figure (b) shows the variation curve of the longitudinal acoustic radiation force of the particles Y_y. Unlike the lateral acoustic radiation force Y_x, Y_y is anti-symmetric with the incident angle β ($Y_y(\beta)$=-$Y_y(-\beta)$), the value increases with the increase of angle β, and at normal incidence $\beta = 0°$, $Y_y = 0$; while at $-90° \leq \beta < 0°$ The amplitude of Y_y remains negative at all times; at $0° < \beta \leq 90°$ time, the amplitude of Y_y remains positive. For particles of different sizes, the values of the longitudinal acoustic radiation force Y_y also vary. When the particle size is small, its Y_y value is also relatively small.

The boundary reflectance in Fig. 7.8 on the right (2) is taken as $R_s = 1.0$ (i.e., full reflection at rigid boundaries). Comparing the transverse radiation force Y_x and the longitudinal radiation force Y_y in the left (1), it can be found that they have the same characteristics with the change of incident angle β and boundary distance kd. However, their values increase when the boundary reflectance increases, indicating that the absolute value of Y_x, Y_y increases with the increase of R_s. Among them, the change of the transverse radiation force Y_x is larger, and a larger negative value appears. This means that when an appropriate incident angle β is selected, the particle

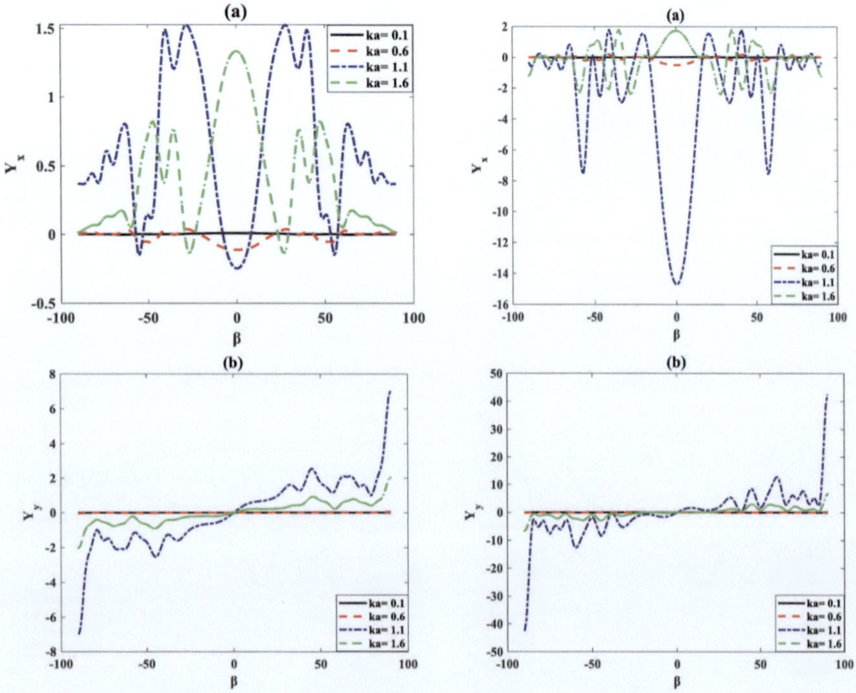

Fig. 7.9 Acoustic radiation force on particle ($kd = 6$) during three reflections. **a** Lateral acoustic radiation force Y_x, **b** Longitudinal acoustic radiation force Y_y

can also be pushed away from the ($Y_x < 0$) impedance boundary. These changes should be noted at rigid boundaries as they will affect the manipulation of particles.

Figure 7.9 shows the variation curves of the transverse and longitudinal acoustic radiation forces of particles considering three reflections. Compared to Fig. 4.4, which only considers a single reflection, the corresponding values have significantly increased. Detailed analysis reveals that when the impedance boundary reflectance is low, the influence on the results is limited if the number of reflections is not considered in the theoretical calculation process; however, when the boundary reflectance is high, the influence of multiple reflections must be considered when calculating the acoustic radiation force, as the transverse and longitudinal acoustic radiation forces of the particles have increased by an order of magnitude compared to before.

Acoustic Radiation Force Under Specific Incident Angles

Finally, we fix the plane wave incident angle β (considering two situations $\beta = 30°$ and $\beta = 60°$), and consider the changes in the acoustic radiation force under different boundary distances, different particle sizes, and multiple (three) reflections. The specific simulation results are shown in Figs. 7.10–7.11.

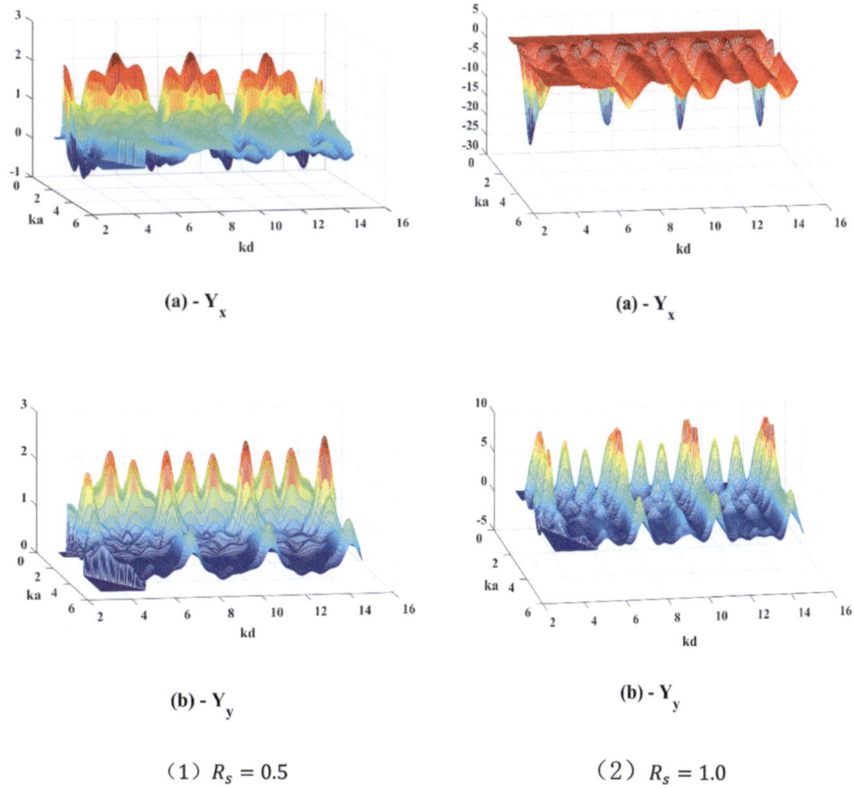

Fig. 7.10 Acoustic radiation force of particle at incident angle$\beta = 30°$. **a** Lateral acoustic radiation force Y_x, **b** Longitudinal acoustic radiation force Y_y

Figure 7.10 shows the simulation results of the transverse and longitudinal acoustic radiation forces of particles when the incident angle is $\beta = 30°$. The boundary reflectance in the left (1) of the figure is taken as $R_s = 0.5$, which gives the radiation force of particles of different sizes with the boundary distance kd The variation pattern can be observed from Figure (a), where the lateral acoustic radiation force function of the particle Y_x fluctuates with the increase in boundary distance kd. As the particle size increases, the value shows a trend of decreasing fluctuations. Y_x The value can be positive or negative, implying that under this incident angle, at a specific boundary distance, different particles can generate different lateral acoustic radiation forces. When $Y_x > 0$, the particle is pulled towards the impedance boundary, and vice versa when $Y_x > 0$. Figure (b) shows the simulation of the particle's longitudinal acoustic radiation force Y_y, which draws a similar conclusion to the lateral acoustic radiation force Y_x.

In Fig. 7.10, the boundary reflectance on the right side (2) is taken as $R_s = 1.0$ (i.e., full reflection of the rigid boundary). Comparing the lateral radiation force Y_x and the longitudinal radiation force Y_y in the left side (1), it can be found that they have

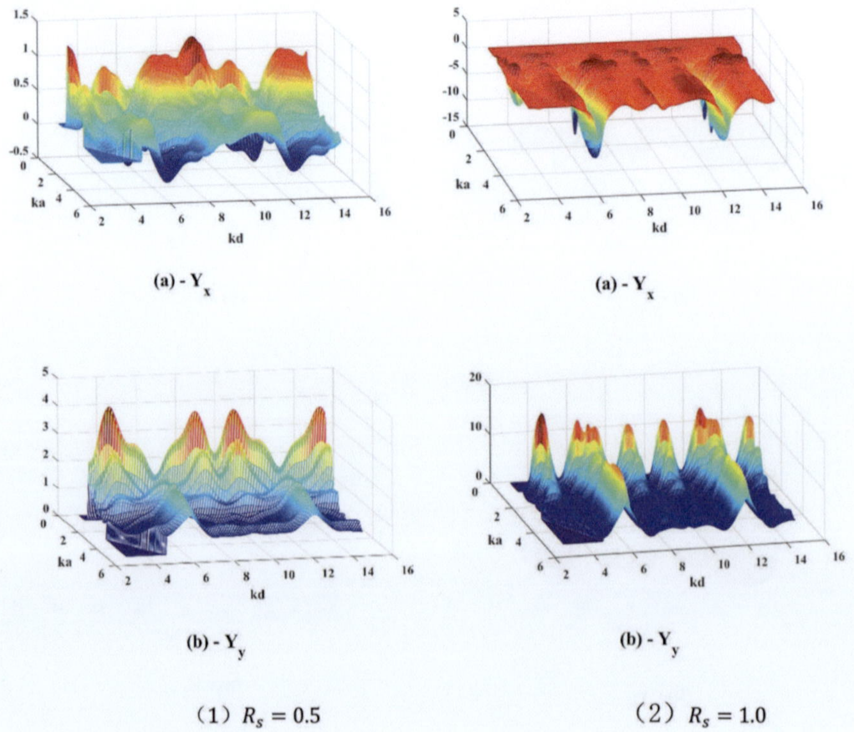

Fig. 7.11 Acoustic radiation force of particle at incident angle $\beta = 60°$. **a** Lateral acoustic radiation force Y_x, **b** Longitudinal acoustic radiation force Y_y

similar characteristics with the change of incident angle β and boundary distance kd. The difference is that besides the increase in their values with the increase of boundary reflectance R_s, the negative value (pulling force) of the lateral radiation force Y_x has increased significantly.

Figure 7.11 shows the simulation results of the lateral and longitudinal acoustic radiation forces of particles at the incident angle $\beta = 60°$. Compared to Fig. 7.10, they have similar conclusions, and the analysis process will not be repeated here.

7.2 Acoustic Radiation Force on Cylindrical Particles under Single Boundary by Gaussian Wave

In addition to the most basic plane wave, the Gaussian wave is also a common beam, especially in the field of biomedical ultrasound, it has important applications [18]. Here we attempt to expand the Gaussian beam through the finite series method, thereby calculating the acoustic radiation force it exerts on the cylindrical particle

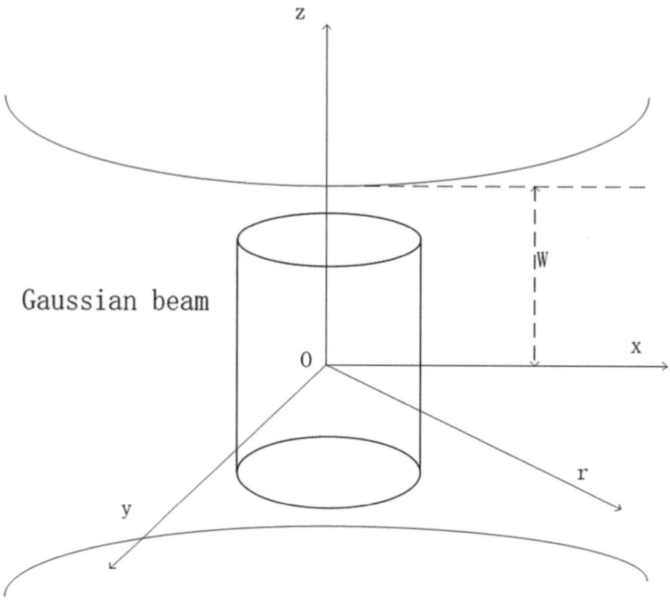

Fig. 7.12 Gaussian beam incident on cylindrical particles [19]

under a single boundary [19]. Unlike plane waves, the wavefront of a Gaussian beam is not a plane when it propagates in space, but it can be approximated as a plane in the waist of the beam. When the Gaussian beam is incident on the cylindrical particle, its physical model is shown in Fig. 7.12. For simplicity, we did not draw the impedance boundary and the image source.

The Gaussian beam propagates in the positive direction of the x-axis, taking its center as the origin of the coordinates, the line where the cylindrical particle is located as the z-axis, and the center of the particle is exactly the origin of the coordinates. The angle between the vector r in the xOy plane and the x-axis is θ. In this case, the incident Gaussian beam can be written as [19]

$$p_i = p_0 e^{-\frac{z^2+y^2}{w^2}} e^{ikx} \tag{7.66}$$

Here, w is the waist width of the Gaussian beam.

As can be seen, compared to plane waves, there is an additional exponential adjustment factor, so its amplitude is no longer equal everywhere. Due to its similarity to plane waves, we naturally wonder if it can also be expanded using a finite series method like plane waves. In fact, this has been achieved in related literature [19]. We will not elaborate on the specific derivation process here, but only provide the result of the expansion.

$$\Phi_{inc} = \Phi_0 \sum_{n=-\infty}^{\infty} i^n g_n J_n(kr) e^{in\theta} e^{-i\omega t} \tag{7.67}$$

Compared to the expansion of plane waves Eq. (7.4), there is an additional term g_n, called the beam factor. Apart from this, the remaining parameters have the same meaning as in the case of plane waves.

Naturally, we can also represent the remaining sound waves in this form, as shown in Eqs. (7.68)–(7.70).

$$\Phi_{ref} = \Phi_0 R_{s1} e^{i2kd1} \sum_{n=0}^{\infty} (-1)^n i^n J_n(kr) e^{in\theta} e^{-i\omega t} \tag{7.68}$$

$$\Phi_{scat} = \Phi_0 \sum_{n=0}^{\infty} g_n a_n i^n H_n(kr) e^{in\theta} e^{-i\omega t} \tag{7.69}$$

$$\Phi_{scat,ref\,1} \Phi_0 R_{s1} \sum_{n=-\infty}^{\infty} (-1)^n a_n i^n g_n H_n(k\overline{r}) e^{-in\overline{\theta}} e^{-i\omega t} \tag{7.70}$$

These correspond to the plane wave cases of Eqs. (7.5), (7.7) and 7.8.

The current problem is obviously about the calculation of the beam factor. This task can be solved by referring to related literature [18]. We cite its results as follows:

$$g_n = \begin{cases} 1, n = 0 \\ \sum_{q=0}^{n/2} 2^{n-2q} \frac{n(n-q-1)!(-1)^{-q} s^{n-2q}}{q!\left[\frac{n-2q}{2}\right]! \varepsilon_n i^n}, n \text{ is even} \\ \sum_{q=0}^{n/2} 2^{n-2q} \frac{n(n-q-1)!(-1)^{-q} s^{n-2q-1}}{nq!\left[\frac{n-1-2q}{2}\right]! \varepsilon_n i^{n-1}}, n \text{ is odd} \end{cases} \tag{7.71}$$

where, $s = 1/kw$, k is the wave number in the fluid medium. The remaining parameters have the same meaning as in the case of plane waves.

In the above calculations, we find that as n increases, the beam factor gradually decreases from 1 to zero. We treat all values less than 0.001 in the calculation results as zero and use it as a sign of convergence. From the table, it can be seen that as the ratio of the waist width to the wavelength increases, the convergence speed of the beam factor becomes slower. We know that plane waves can be regarded as a special "Gaussian wave" with a constant beam factor of 1, so it can be imagined that when the beam factor is large enough, it can be approximated as a plane wave for processing [7].

7.2.1 Solution of Acoustic Radiation Force

Having solved the problem of calculating the beam factor, in principle, we can apply the theory of Sect. 7.1 to solve the acoustic radiation force at this time. The total velocity potential in the fluid medium is still the sum of the four velocity potentials, which can be substituted into Eqs. (7.67)–(7.70) to obtain the specific expression of the velocity potential [20].

$$\Phi_{total} = \Phi_0 e^{-i\omega t}\left\{\sum_{n=0}^{\infty} \varepsilon_n g_n A_n i^n J_n(kr)\cos n\theta + \sum_{n=0}^{\infty} \varepsilon_n a_n g_n i^n H_n^{(1)}(kr)\cos n\theta\right\} \tag{7.72}$$

where

$$A_n = 1 + R_s e^{i2kd}(-1)^n + \frac{R_s}{\varepsilon_n g_n i^n}\sum_{m=0}^{\infty}(-1)^m \varepsilon_m g_m a_m i^m H_{m-n}^{(1)}(2kd)J_n(kr)$$

The meanings of the remaining parameters are the same as mentioned before.

Under the two conditions of particle rigidity and non-rigidity, different boundary conditions are given, and the scattering numbers under the two conditions can be easily obtained.

No matter what wave shape is incident, the original formula of the acoustic radiation force is still in the form of Eq. (7.21). After a series of complicated calculations, we find that the acoustic radiation force of the particle at this time can still be written in the form of Eq. (7.23), that is, the sound wave energy density and the scattering cross-sectional area have not changed. The only thing that needs to be corrected is the so-called acoustic radiation force function Y_p. Its expression is [7].

$$\begin{aligned} Y_p = -\frac{2}{ka}\sum_{n=0}^{n=\infty}\Big\{&\mathrm{Re}\big(g_n g_{n+1}^*\big)\Big(\xi_{n+1}^{(1)}\xi_n^{(2)} + \xi_n^{(1)}\xi_{n+1}^{(2)} + \eta_{n+1}^{(1)}\eta_n^{(2)} \\ &+ \eta_n^{(1)}\eta_{n+1}^{(2)} + 2\xi_{n+1}^{(2)}\xi_n^{(2)} + 2\eta_n^{(2)}\eta_{n+1}^{(2)}\Big) + \mathrm{Im}\big(g_n g_{n+1}^*\big)(\xi_{n+1}^{(2)}\eta_n^{(1)} \\ &+ \xi_{n+1}^{(1)}\eta_n^{(2)} - \xi_n^{(1)}\eta_{n+1}^{(2)} - \xi_n^{(2)}\eta_{n+1}^{(1)} + 2\xi_n^{(2)}\eta_{n+1}^{(2)} - 2\xi_n^{(2)}\eta_{n+1}^{(2)} \end{aligned} \tag{7.73}$$

where

$$\begin{aligned} &\xi_n^{(1)} = \mathrm{Re}(A_n),\ \eta_n^{(1)} = \mathrm{Im}(A_n)\cdot \xi_n^{(2)} = \mathrm{Re}(a_n)\eta_n^{(2)} = \mathrm{Im}(a_n) \\ &g_n^r = \mathrm{Re}(g_n)g_n^i = \mathrm{Im}(g_n) \end{aligned}$$

So far, we have derived the formula for the acoustic radiation force on cylindrical particles under a single boundary, laying the foundation for the next step of simulation.

7.2.2 Simulation Example

In this section, we will explore the characteristics of the acoustic radiation force of Gaussian beams on cylindrical particles under a single impedance boundary through specific examples, in order to have an intuitive understanding of the characteristics of the acoustic radiation force of Gaussian waves.

We continue to first study the variation of the acoustic radiation force function with ka under the condition of a fixed particle-to-boundary distance d. Here, we still assume d = 2a. Compared to the case of plane wave incidence, we also need to consider another parameter, namely the waist width w of the Gaussian wave. As can be deduced from previous discussions, the size of the waist width affects the convergence rate of the beam factor. The larger the waist width, the slower the beam factor converges, and the closer it is to the case of plane wave incidence. In the simulations here, to more intuitively reflect the characteristics of the Gaussian wave acoustic radiation force different from the plane wave, we set the beam factor to be three times the wavelength, i.e., $w = 3\lambda$. The simulation results for rigid spheres and oleic acid spheres are shown in Fig. 7.13.

Similar to the case of plane waves, the presence of an impedance boundary exacerbates the oscillatory characteristics of the acoustic radiation force. Similarly, in the case of Gaussian wave incidence, there is also the phenomenon of "acoustic trapping force", which is quite different from the case of Gaussian wave incidence in free space. Whether it is a rigid column or an oleic acid column, the change in the sound pressure reflection coefficient has a similar impact on the amplitude of the acoustic radiation force as in the case of plane wave incidence.

Similarly, we can also study the variation of the acoustic radiation force function with the sound wave frequency and with kd under Gaussian wave incidence, and obtain curves similar to Figs. 7.3 and 7.4. Here, only two curves for the rigid column are given, as shown in Figs. 7.14 and 7.15. The relevant parameters are the same as previously mentioned.

It is not difficult to see that the trends of these two curves are similar to the case of plane waves, and will not be further elaborated.

In actual operation, the waist width of the Gaussian beam can be arbitrarily chosen, so it is necessary to consider the impact of different waist widths on the acoustic radiation force function curve. At this time, the boundary sound pressure reflection coefficient is fixed at $R_s = 0.5$, and the distance between the particle and the boundary is still twice the radius. The simulation results for the rigid column are shown in Fig. 7.16.

It can be found that the waist width does not have a significant impact on the acoustic radiation force function curve. After the waist width reaches 3λ, it is basically the same as the plane wave incidence, and all curves with a waist width greater than 3λ almost coincide. Only when the beam factor is as small as one wavelength, does it show some differences in the mid-high frequency range, and the value of the acoustic radiation force function is slightly lower than that of the plane wave.

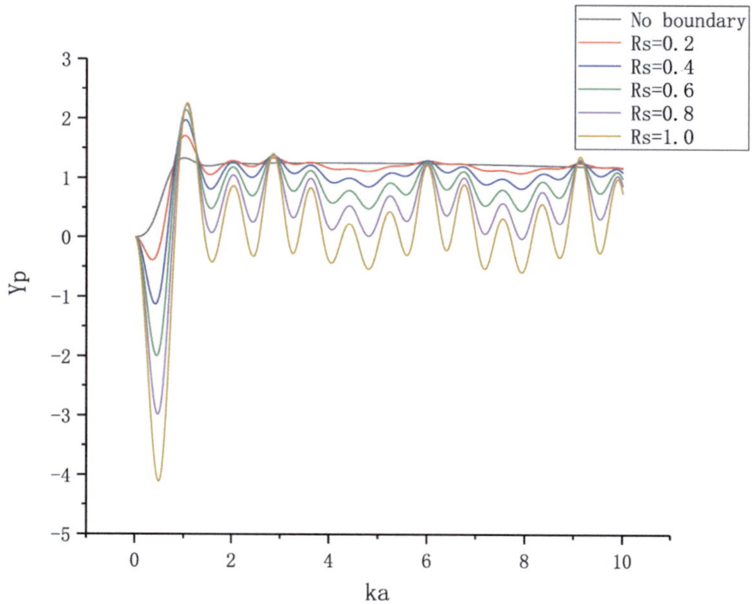

(a)

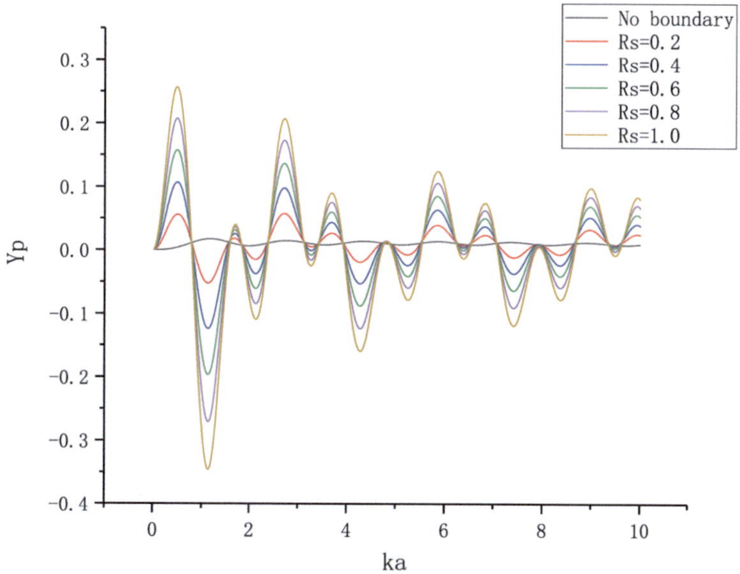

(b)

Fig. 7.13 Variation of the acoustic radiation force function with ka for rigid column and oleic acid column under Gaussian wave incidence. **a** Rigid column. **b** Oleic acid column

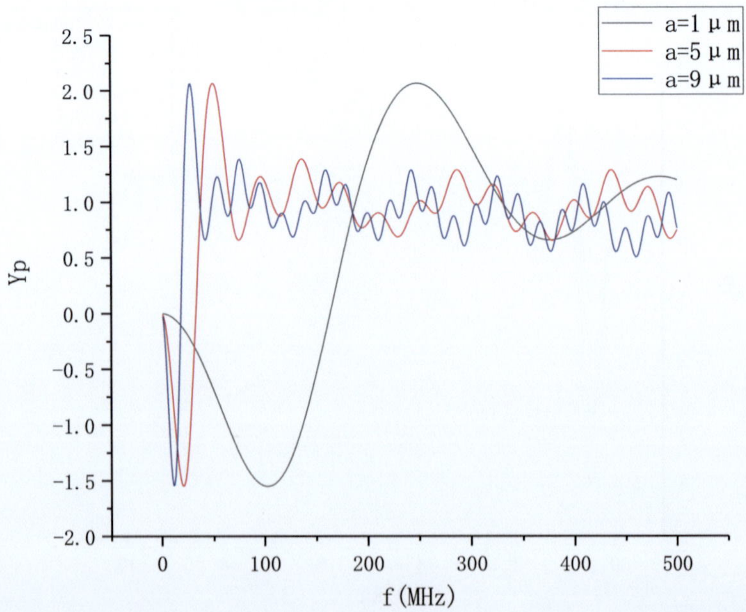

Fig. 7.14 Variation of the acoustic radiation force function with frequency for a rigid column under Gaussian wave incidence

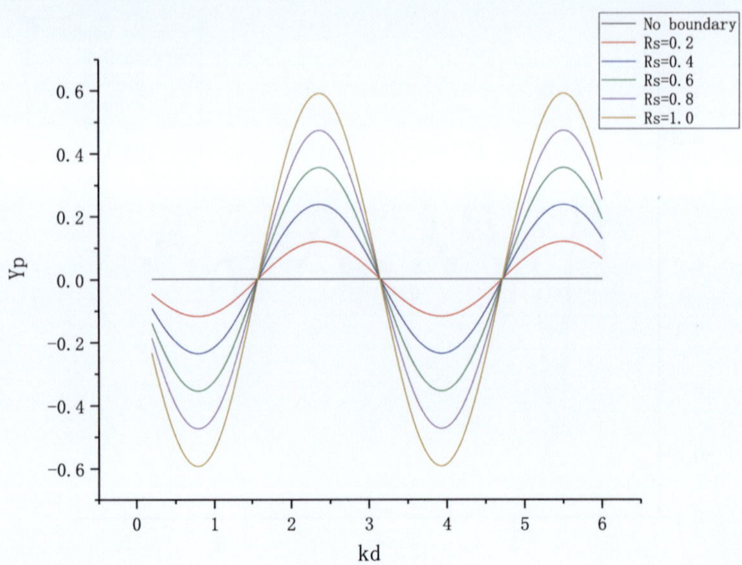

Fig. 7.15 Variation of the acoustic radiation force function with kd for a rigid column under Gaussian wave incidence

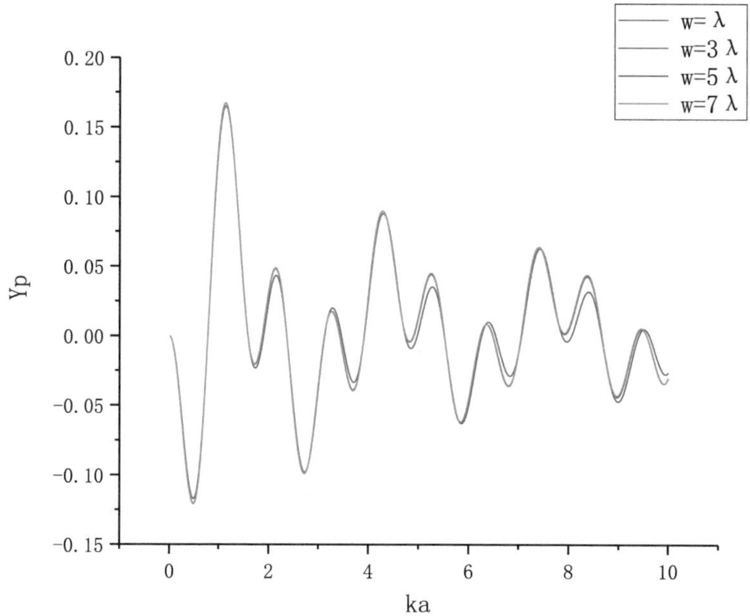

Fig. 7.16 Acoustic radiation force function of Gaussian waves on a rigid column at different waist widths

7.3 Acoustic Radiation Force of Plane Waves on Spherical Particles Under a Single Boundary

As mentioned earlier, plane waves are the most basic and simplest type of waveforms. In this section, we will first consider the case of plane wave incidence.

Consider a plane wave incident perpendicularly on a spherical particle with a radius of a under a single boundary, with the particle at a distance d from the impedance boundary. This physical model is very similar to Fig. 7.1, except that the cylindrical particle in the original figure is replaced with a spherical one. Of course, the coordinate system should also be changed to a spherical coordinate system for convenience, with the center of the sphere as the origin O, and the polar angle and elevation angle denoted as θ and φ respectively.

According to the theory of plane wave expansion into spherical functions in mathematical physics [21], the incident sound wave can be expressed as

$$\Phi_{inc} = \Phi_0 e^{-i\omega t} \sum_{n=0}^{\infty} (2n+1) i^n j_n(kr) P_n(\cos\theta) \tag{7.74}$$

where, $j_n(\cdot)$ is the first kind of n-th order spherical Bessel function, $P_n(\cdot)$ is the n-th order Legendre function.

According to the mirror theory, the reflected wave of the impedance boundary can be expressed as

$$\Phi_{ref} = \Phi_0 e^{-i\omega t} \sum_{n=0}^{\infty} R_s e^{j2kd} (-1)^n (2n+1) i^n j_n(kr) P_n(\cos\theta) \tag{7.75}$$

where, R_s still represents the sound pressure reflection coefficient of the boundary, which can be calculated by Eq. (7.6).

The scattered waves of the spherical particle for the incident sound wave and the mirror source are respectively

$$\Phi_{scat} = \Phi_0 e^{-i\omega t} \sum_{n=0}^{\infty} (2n+1) i^n a_n h_n^{(1)}(kr) P_n(\cos\theta) \tag{7.76}$$

$$\Phi_{scat,ref} = \Phi_0 e^{-i\omega t} \sum_{n=0}^{\infty} R_s (-1)^n (2n+1) i^n a_n(\omega) h_n^{(1)}(k\overline{r}) P_n\left(\cos\overline{\theta}\right) \tag{7.77}$$

where, $h_n^{(1)}(\cdot)$ is the first kind of n-order spherical Hankel function, a_n is the scattering coefficient of the sphere.

Formally, these velocity potential expressions are the same as those for cylindrical particles, except that the expansion function of plane waves is different in spherical coordinates.

7.3.1 Solution for Scattering Coefficients

Having obtained the specific forms of each velocity potential, the next step is to solve for the scattering coefficients. This can be achieved by assigning different boundary conditions to the surface of the sphere. The total velocity potential in the fluid medium is obviously the sum of the above four terms. Using the addition formula of spherical functions [22], it can also be written as [23]

$$\Phi_{total} = \Phi_0 e^{-i\omega t} \left\{ \sum_{n=0}^{\infty} A_n i^n j_n(kr) P_n(\cos\theta) + \sum_{n=0}^{\infty} a_n i^n h_n^{(1)}(kr) P_n(\cos\theta) \right\} \tag{7.78}$$

where,

$$A_n = 1 + R_s e^{i2kd} (-1)^n + \frac{R_s}{(2n+1) i^n} \sum_{m=0}^{\infty} (-1)^m (2m+1) a_m i^m Q_{mn} \tag{7.79}$$

$$Q_{mn} = \sqrt{(2n+1)(2m+1)}i^{m-n} \sum_{\sigma=|m-n|}^{m+n} (-1)^{\sigma} i^{\sigma} b_{\sigma}^{mn} h_{\sigma}^{(1)}(kd) \tag{7.80}$$

When q is an even number

$$b_{\sigma}^{mn} = \frac{(-1)^{q+\sigma} q!}{(q-n)!(q-m)!(q-\sigma)!} \times \sqrt{\frac{2\sigma+1}{(2q+1)!}(2q-2n)!(2q-2m)!(2q-2\sigma)!} \tag{7.81}$$

When q is an odd number

$$b_{\sigma}^{mn} = 0$$

The calculation method of harmonic sound pressure is still Eq. (7.14), as for the radial particle velocity, it can be obtained by deriving the velocity potential

$$v_r = -\Phi_0 e^{-i\omega t} \left\{ \sum_{n=0}^{\infty} A_n i^n k j_n'(ka) P_n(\cos\theta) + \sum_{n=0}^{\infty} a_n i^n k h_n^{(1)\prime}(ka) P_n(\cos\theta) \right\} \tag{7.82}$$

The solution of the scattering coefficient depends on certain boundary conditions. For a rigid sphere, like a rigid cylinder,

The radial velocity of the particle surface is zero, that is, Eq. (7.15) still holds, but the r there is the radial component of the cylindrical coordinates, while here it is the spherical coordinates. The specific form is

$$\sum_{n=0}^{\infty} A_n i^n k j_n'(ka) P_n(\cos\theta) + \sum_{n=0}^{\infty} a_n i^n k h_n^{(1)\prime}(ka) P_n(\cos\theta) = 0 \tag{7.83}$$

Hence, the numerical solution of a_n can be obtained from the iterative method. For non-rigid cylinders, the existence of refracted waves inside the sphere should be considered

$$\Phi_l = \Phi_0 e^{-i\omega t} \sum_{n=0}^{\infty} i^n B_n j_n(k_l r) P_n(\cos\theta) \tag{7.84}$$

where, B_n, k_l has the same meaning as in section two.

The boundary condition at this time becomes Eq. (7.18), and the scattering coefficient can also be obtained by the iterative method.

7.3.2 Solution for Acoustic Radiation Force

Now that we have obtained the sound scattering coefficients under different boundaries, we can calculate the acoustic radiation force. The original formula for calculating the acoustic radiation force is still Eq. (7.21), as for the axial component we are concerned about, it is in the form of Eq. (7.22). After a series of mathematical calculations, we find that the final result is still in the form of Eq. (7.23) [23], that is,

$$\langle F \rangle = Y_p S \langle E \rangle \tag{7.85}$$

where, $\langle E \rangle = \frac{1}{2}\rho_0 k^2 \Phi_0^2$ is the spatial energy density of the sound wave, the same as for cylindrical particles. But the scattering cross-sectional area $S = \pi a^2$ is different. As for the core acoustic radiation force function, its expression is also different. At this time

$$Y_p = -\frac{4}{(ka)^2}\sum_{n=0}^{\infty}(n+1)(\xi_{n+1}^{(1)}\xi_n^{(2)} + \xi_n^{(1)}\xi_{n+1}^{(2)} + \eta_{n+1}^{(1)}\eta_n^{(2)} + \eta_n^{(1)}\eta_{n+1}^{(2)} + 2\xi_{n+1}^{(2)}\xi_n^{(2)} + 2\eta_n^{(2)}\eta_{n+1}^{(2)}) \tag{7.86}$$

where

$$\xi_n^{(1)} = \mathrm{Re}(A_n),\ \eta_n^{(1)} = \mathrm{Im}(A_n) \cdot \xi_n^{(2)} = \mathrm{Re}(a_n)\eta_n^{(2)} = \mathrm{Im}(a_n) \tag{7.87}$$

Thus, the theoretical calculation of the acoustic radiation force of spherical particles under a single boundary is solved, and numerical simulation can be carried out.

7.3.3 Simulation Examples

Having obtained the calculation formula for the acoustic radiation force of spherical particles, we will further explore its acoustic radiation force characteristics through some specific simulation examples.

We still first study the relationship between the acoustic radiation force function and ka under different boundary sound pressure reflection coefficients, divided into rigid cylinders and oleic acid cylinders. The distance between the particle center and the boundary is 2a, and the simulation results are shown in Fig. 7.17

Overall, the acoustic radiation force of rigid spheres is much greater than that of oleic acid spheres. Both rigid and non-rigid spheres exhibit "acoustic trapping force" within a specific ka range, and peaks of positive or negative acoustic radiation force functions are more likely to occur at mid-low frequencies. Similar to cylindrical particles, an increase in the boundary sound pressure reflection coefficient also increases

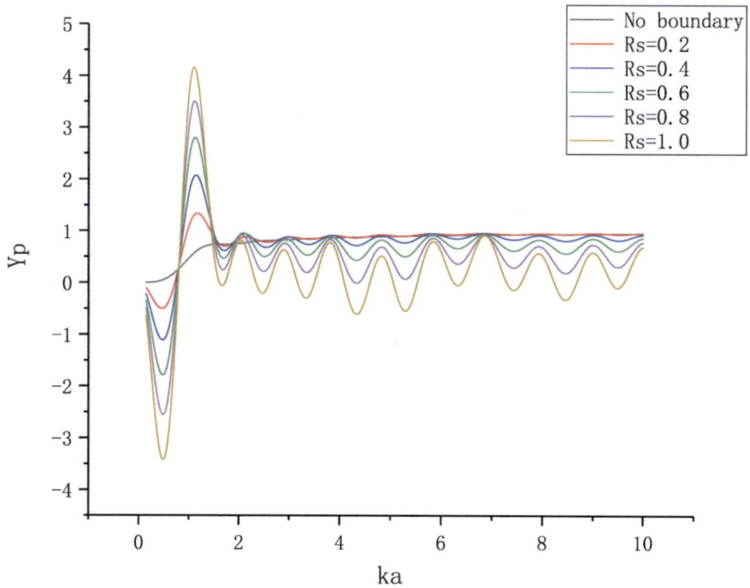

(a)

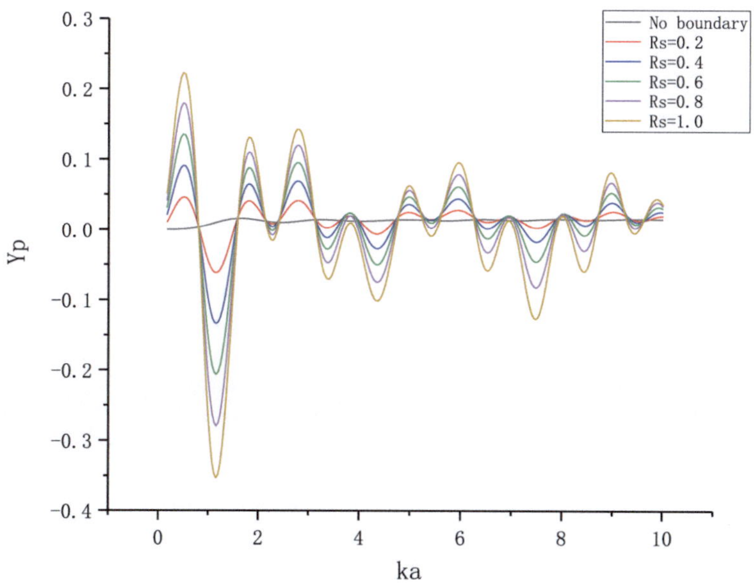

(b)

Fig. 7.17 Variation of the acoustic radiation force function of a rigid sphere and an oleic acid sphere with ka when a plane wave is incident. **a** Rigid sphere. **b** Oleic acid sphere the peak of the acoustic radiation force function, but it hardly changes the ka values at these peaks. Next, we study the relationship between the acoustic radiation force function and the frequency of the incident sound wave under different particle radii, with the boundary sound pressure reflection coefficient Rs = 0.5. The simulation results for rigid spheres and oleic acid spheres are shown in Fig. 7.18.

Similar to the case of cylindrical particles, an increase in particle radius shifts the peaks of the positive and negative acoustic radiation force functions towards lower frequencies, and the curve oscillates more intensely, making it easier to reach extremes. However, the size of the radius does not affect the peak of the acoustic radiation force function. For example, at the first positive peak of the three curves, although the frequencies are different, the values of the acoustic radiation force function are equal. The distance between the particle and the boundary will also obviously affect the acoustic radiation force. With ka fixed at 0.063, the simulation results for rigid spheres and oleic acid spheres under different boundary reflection coefficients are shown in Fig. 7.19. It is not difficult to see that the variation of the acoustic radiation force function of spherical particles with kd is basically consistent with that of cylindrical particles, similar to a sine curve. An increase in the boundary reflection coefficient leads to an increase in the function peak. So far, we have studied the characteristics of the acoustic radiation force of spherical particles under a single impedance boundary when plane waves are incident, laying the foundation for the study of Gaussian beams in the next section.

7.4 Acoustic Radiation Force of Gaussian Waves on Spherical Particles Under a Single Boundary

When a Gaussian beam is incident, its physical model is similar to Fig. 7.12 [24], only the cylindrical particles need to be replaced with spherical particles. The center of the particle coincides with the center of the beam, which is also the origin of the coordinate system. It is still assumed that the Gaussian beam propagates in the positive direction of the x-axis, and the waist of the beam can be approximated as a plane. The Gaussian beam can naturally also be represented in the form of Eq. (7.66), but in order to calculate the scattering coefficient and acoustic radiation force later, it is necessary to rewrite it into a finite series form similar to Eq. (7.67).

$$\Phi_{inc} = \Phi_0 e^{-i\omega t} \sum_{n=0}^{\infty} g_n (2n+1) i^n j_n(kr) P_n(\cos\theta) \tag{7.88}$$

where, the beam factor is still called g_n, but this time it is expanded under spherical coordinates instead of cylindrical coordinates, so it is different from the beam factor in Sect. 7.2.

For the remaining beams, we can also expand them in a similar way.

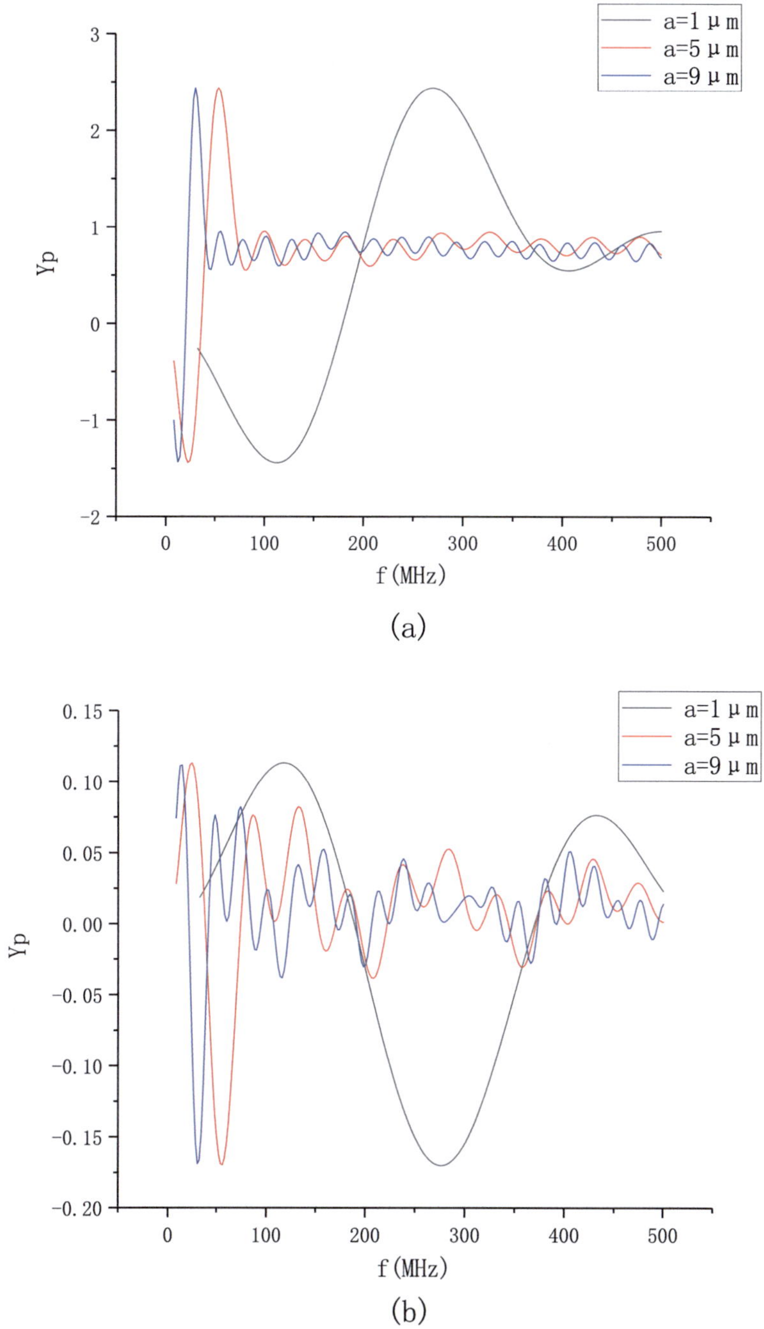

Fig. 7.18 Variation of the acoustic radiation force function of a rigid sphere and an oleic acid sphere with frequency when plane waves are incident. **a** Rigid sphere. **b** Oleic acid sphere

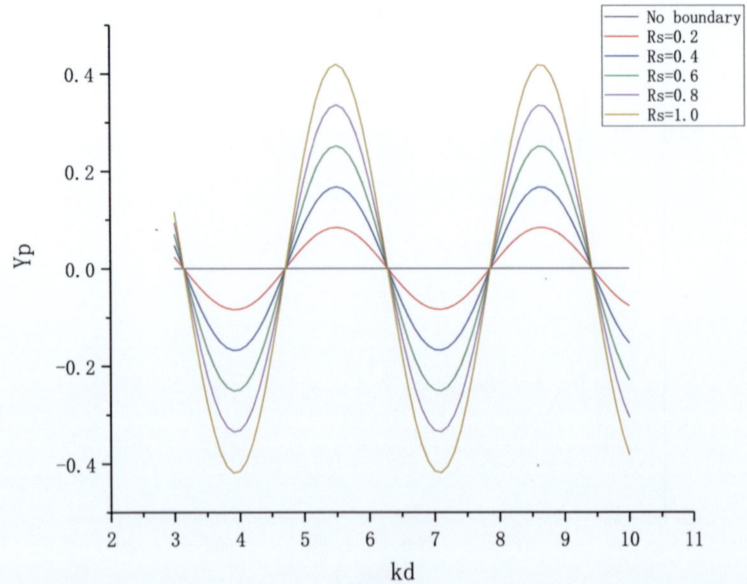

(a)

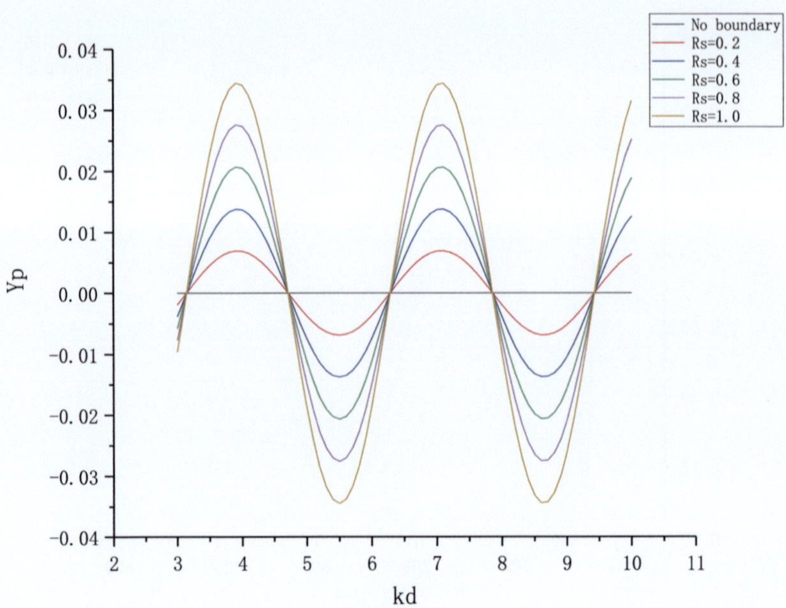

(b)

Fig. 7.19 Variation of the acoustic radiation force function of a rigid sphere and an oleic acid sphere with kd when plane waves are incident. **a** Rigid sphere. **b** Oleic acid sphere

$$\Phi_{ref} = \Phi_0 e^{-i\omega t} \sum_{n=0}^{\infty} R_s e^{j2kd} (-1)^n g_n (2n+1) i^n j_n(kr) P_n(\cos\theta) \tag{7.89}$$

$$\Phi_{scat} = \Phi_0 e^{-i\omega t} \sum_{n=0}^{\infty} g_n (2n+1) i^n a_n h_n^{(1)}(kr) P_n(\cos\theta) \tag{7.90}$$

$$\Phi_{scat,ref} = \Phi_0 e^{-i\omega t} \sum_{n=0}^{\infty} R_s (-1)^n g_n (2n+1) i^n a_n(\omega) h_n^{(1)}(k\overline{r}) P_n\left(\cos\overline{\theta}\right) \tag{7.91}$$

They correspond to Eqs. (7.75)–(7.77) under the condition of plane wave incidence. Next, we calculate the beam factor. In fact, the result has been given in the relevant literature [24, 25]. We list it here.

$$g_n = \begin{cases} \frac{\Gamma(\frac{n}{2}+1)}{\Gamma(\frac{n+1}{2})} \sum_{q=0}^{n/2} \frac{\Gamma(\frac{n}{2}+q+\frac{1}{2})}{(\frac{n}{2}-q)!q!} (-4s^2)^q, \ n \ is \ even \\ \frac{\Gamma(\frac{n}{2}+1)}{\Gamma(\frac{n+3}{2})} \sum_{q=0}^{n/2} \frac{\Gamma(\frac{n}{2}+q+\frac{3}{2})}{(\frac{n}{2}-q)!q!} (-4s^2)^q, \ n \ is \ odd \end{cases} \tag{7.92}$$

From the calculation, it can be seen that no matter how wide the beam waist is, as n increases, the beam factor will gradually decay from 1 to 0. We still use less than 0.001 as the decay mark. The larger the beam waist, the slower the decay, and the closer to the situation of plane waves [4].

7.4.1 Solution of Acoustic Radiation Force

Now that the beam factor has been calculated, we can turn to the solution of the scattering coefficient. The total sound field in the fluid medium can be represented as [23]

$$\Phi_{total} = \Phi_0 e^{-i\omega t} \left\{ \sum_{n=0}^{\infty} A_n g_n i^n j_n(kr) P_n(\cos\theta) + \sum_{n=0}^{\infty} a_n g_n i^n h_n^{(1)}(kr) P_n(\cos\theta) \right\} \tag{7.93}$$

where,

$$A_n = 1 + R_s e^{i2kd} (-1)^n + \frac{R_s}{(2n+1) g_n i^n} \sum_{m=0}^{\infty} g_m (-1)^m (2m+1) a_m i^m Q_{mn} \tag{7.94}$$

For rigid spheres, the boundary condition of zero surface normal velocity must naturally be satisfied. For non-rigid spheres, then to satisfy the condition of continuity between surface sound pressure and radial velocity, the values of various scattering coefficients can always be obtained by iterative methods, regardless of whether the sphere is rigid or non-rigid.

The original formula for axial acoustic radiation force remains unchanged. After a series of mathematical operations, its specific expression still takes the form of Eq. (7.86), and the expressions for $\langle E\rangle$ and S are the same as when plane waves are incident. The only change is the acoustic radiation force function, which at this time is [23]

$$\begin{aligned} Y_p = -\frac{4}{(ka)^2}\sum_{n=0}^{\infty}(n+1)\Big\{Re\big(g_n g_{n+1}^*\big) \\ \Big(\xi_{n+1}^{(1)}\xi_n^{(2)} + \xi_n^{(1)}\xi_{n+1}^{(2)} + \eta_{n+1}^{(1)}\eta_n^{(2)} + \eta_n^{(1)}\eta_{n+1}^{(2)} + 2\xi_{n+1}^{(2)}\xi_n^{(2)} + 2\eta_n^{(2)}\eta_{n+1}^{(2)}\Big) \\ + Im\big(g_n g_{n+1}^*\big) \\ \Big(\xi_{n+1}^{(2)}\eta_n^{(1)} + \xi_{n+1}^{(1)}\eta_n^{(2)} - \xi_n^{(1)}\eta_{n+1}^{(2)} - \xi_n^{(2)}\eta_{n+1}^{(1)} + 2\xi_n^{(2)}\eta_{n+1}^{(2)} - 2\xi_n^{(2)}\eta_{n+1}^{(2)}\Big)\Big\} \end{aligned} \tag{7.95}$$

where,

$$\begin{aligned} \xi_n^{(1)} = \mathrm{Re}(A_n)\eta_n^{(1)} = \mathrm{Im}(A_n) \\ \xi_n^{(2)} = \mathrm{Re}(a_n)\eta_n^{(2)} = \mathrm{Im}(a_n) \\ g_n^r = \mathrm{Re}(g_n)g_n^i = \mathrm{Im}(g_n) \end{aligned}$$

It seems that the method of finite series expansion is still feasible for spherical particles. So far, we have obtained the formula for the acoustic radiation force of a Gaussian beam on spherical particles under a single boundary, and we can proceed to the next step of simulation.

7.4.2 Simulation Example

First plot the curve of the acoustic radiation force function with the change of ka. In the simulation, we still take the waist width of the Gaussian wave as $w = 3\lambda$, and the distance between the particle and the boundary is still twice the radius. The simulation results of rigid spheres and oleic acid spheres are shown in Fig. 7.20.

Compared with the incident plane wave, the overall trend of the curve is similar, and the acoustic radiation force of the rigid sphere is much larger than that of the oleic acid sphere. At the appropriate position in the mid-low frequency range, both can generate larger positive and negative acoustic radiation forces, and the larger the boundary sound pressure reflection coefficient, the larger the peak. This is consistent with the expected results. Next, we study the relationship between the acoustic radiation force function and frequency and the distance between the particle and the boundary. The waist width is still selected as $w = 3\lambda$, and the simulation results for

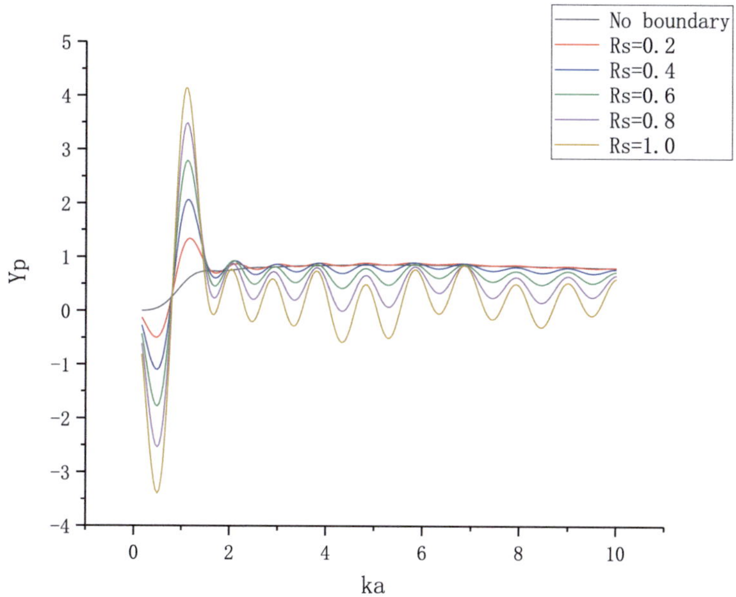

(a)

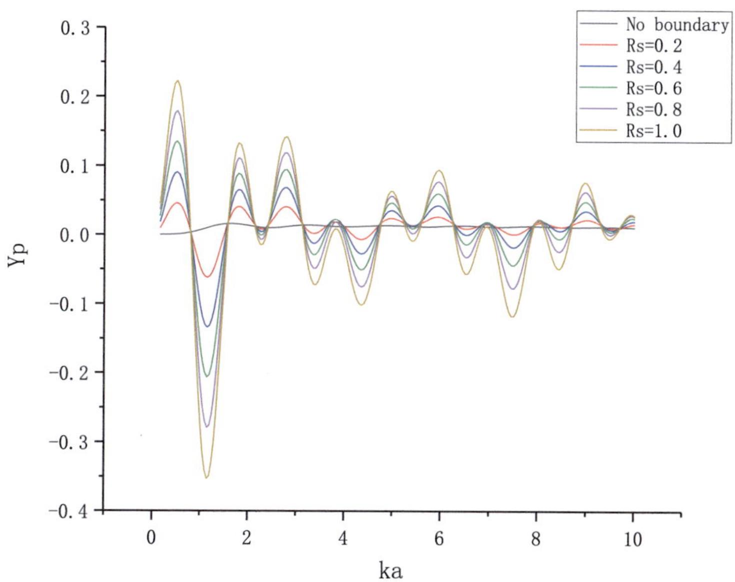

(b)

Fig. 7.20 Change of the acoustic radiation force function of a rigid sphere and an oleic acid sphere with ka when Gaussian waves are incident. **a** Rigid sphere. **b** Oleic acid sphere

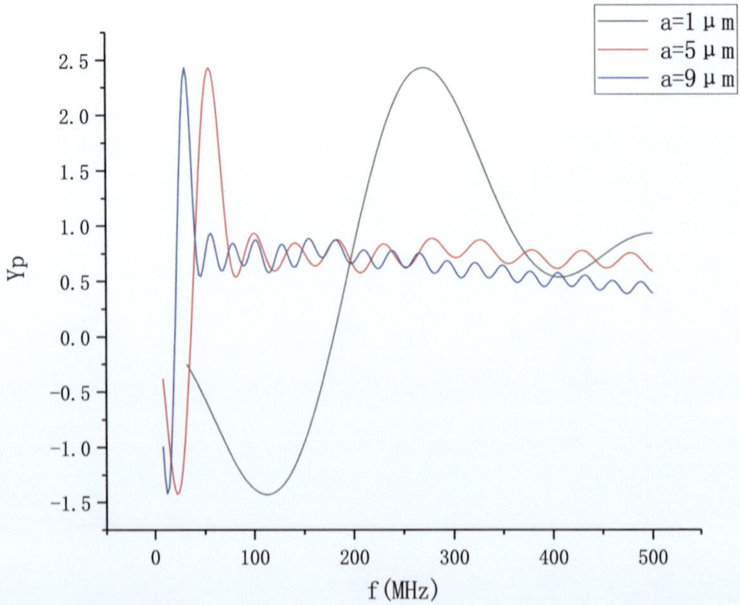

Fig. 7.21 Change of the acoustic radiation force function of a rigid sphere with frequency when Gaussian waves are incident

rigid spheres are shown in Figs. 7.21 and 7.22. In Fig. 7.22, ka = 0.063. It can be found that their rules are similar to those when plane waves are incident, and they are not repeated. Finally, we try to explore the impact of the waist width of the Gaussian beam on the acoustic radiation force function curve. Assuming that the reflection coefficient is $Rs = 0.5$, for rigid spheres, the simulation curves under different waist widths are shown in Fig. 7.23.

When the waist width is small (less than three times the wavelength), compared with the incident plane wave, the acoustic radiation force function curve shows a significant decline in the mid-high frequency range, and the larger the ka, the greater the difference. However, as the waist width increases, the acoustic radiation force function curve becomes closer to the incident situation of the plane wave, almost coinciding with it.

7.5 Acoustic Radiation Force of Plane Waves on Cylindrical Particles Under Double Boundaries

As shown in Fig. 7.24, there are three ideal fluid media in the whole space, forming two impedance boundaries 1 and 2. The cylindrical particle is located in medium 1, the radius of the particle is a, the distance from impedance boundary 1 is d_1, and the distance from impedance boundary 2 is d_2. The plane wave is incident perpendicular

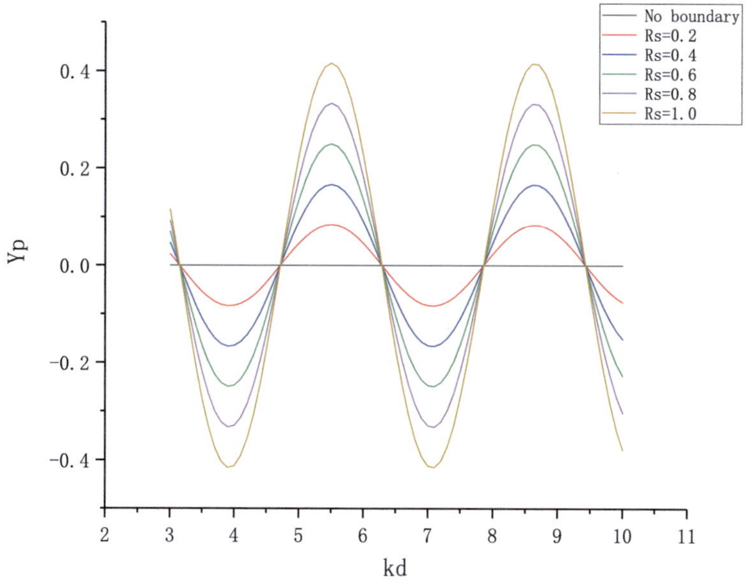

Fig. 7.22 Change of the acoustic radiation force function of a rigid sphere with kd when Gaussian waves are incident

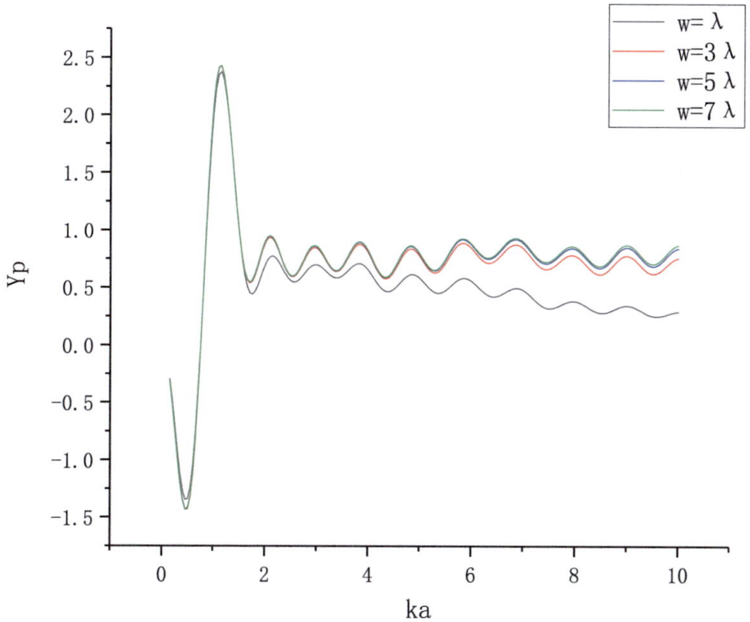

Fig. 7.23 Acoustic radiation force function of Gaussian waves on a rigid sphere under different waist widths

to the boundary. As for the influence of the boundary, in the second section, we have solved this problem with the mirror method. When the boundary increases to two, is it still feasible? We imagine that impedance boundary 1 is equivalent to the first mirror source, its position is at the symmetrical position of the particle relative to boundary 1, and its size is exactly the same as the original particle. At the same time, impedance boundary 2 is equivalent to the second mirror source, its position is at the symmetrical position of the particle relative to boundary 2, and its size is exactly the same. A coordinate system is established with the center of the cylindrical particle in this section as the origin O.

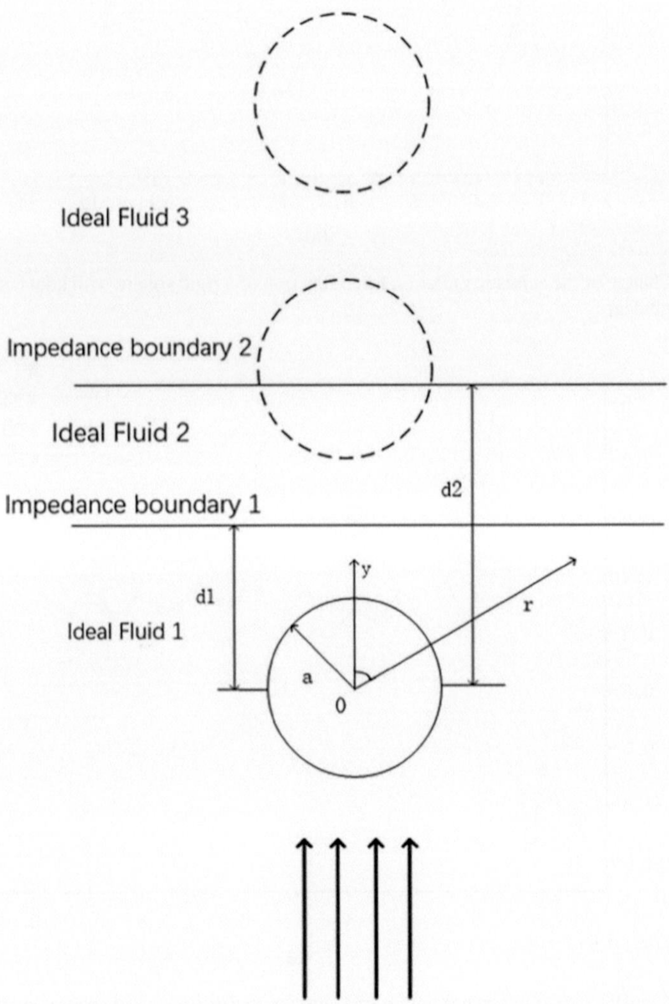

Fig. 7.24 Sound scattering of cylindrical particles under double impedance boundaries against plane waves

Consider all sound wave components in the whole space. For incident plane waves and scattered waves of particles, they can still be represented in the form of Eqs. (7.4) and (7.7), no need to say more. The reflected sound waves after the sound waves are incident on the two impedance boundaries can be equivalent to the incident mirror waves in the opposite direction. What is different from before is that there are now two components of the mirror wave.

$$\Phi_{ref1} = \Phi_0 R_{s1} e^{i2kd1} \sum_{n=-\infty}^{\infty} (-1)^n i^n J_n(kr) e^{in\theta} e^{-i\omega t} \tag{7.96}$$

$$\Phi_{ref2} = \Phi_0 R_{s2} e^{i2kd2} \sum_{n=-\infty}^{\infty} (-1)^n i^n J_n(kr) e^{in\theta} e^{-i\omega t} \tag{7.97}$$

where, R_{s1} and R_{s2} represent the sound pressure reflection coefficients of the two boundaries, respectively.

The reflection waves of the particle scattering waves from the two boundaries can also be equivalently represented as scattering waves from mirror sources. The difference now is that there are two mirror sources.

$$\Phi_{scat,ref1} = \Phi_0 R_{s1} \sum_{n=-\infty}^{\infty} (-1)^n a_n i^n H_n(k\overline{r}) e^{-in\overline{\theta}} e^{-i\omega t} \tag{7.98}$$

$$\Phi_{scat,ref2} = \Phi_0 R_{s2} \sum_{n=-\infty}^{\infty} (-1)^n a_n i^n H_n(k\overline{r}) e^{-in\overline{\theta}} e^{-i\omega t} \tag{7.99}$$

7.5.1 Determination of Scattering Coefficients

Clearly, the total sound field in the fluid medium is the superposition of the above six terms, that is,

$$\Phi_{total} = \Phi_{inc} + \Phi_{ref1} + \Phi_{ref2} + \Phi_{scat,ref1} + \Phi_{scat,ref2} \tag{7.100}$$

Similarly, for the convenience of subsequent calculations, it is rewritten in the form of Eq. (7.101)

$$\Phi_{total} = \Phi_0 e^{-i\omega t} \left\{ \sum_{n=0}^{\infty} \varepsilon_n A_n i^n J_n(kr) \cos n\theta + \sum_{n=0}^{\infty} \varepsilon_n a_n i^n H_n^{(1)}(kr) \cos n\theta \right\} \tag{7.101}$$

where,

$$A_n = 1 + R_{s1}e^{i2kd1}(-1)^n + R_{s2}e^{i2kd2}(-1)^n + \frac{R_{s1}}{\varepsilon_n i^n}\sum_{m=0}^{\infty}(-1)^m \varepsilon_m a_m i^m H_{m-n}^{(1)}(2kd1) + \frac{R_{s2}}{\varepsilon_n i^n}\sum_{m=0}^{\infty}(-1)^m \varepsilon_m a_m i^m H_{m-n}^{(1)}(2kd2) \tag{7.102}$$

Based on the relationship between harmonic sound pressure and velocity potential, the distribution of the total sound pressure field can be obtained. According to the relationship between velocity potential and particle velocity, the radial velocity can be obtained.

$$v_r = \Phi_0 e^{-i\omega t}\left\{\sum_{n=0}^{\infty}\varepsilon_n A_n i^n k J_n'(ka)\cos n\theta + \sum_{n=0}^{\infty}\varepsilon_n a_n i^n k H_n^{(1)\prime}(ka)\cos n\theta\right\} \tag{7.103}$$

Considering that the cylindrical particle is rigid, that is, the surface can be regarded as a completely hard boundary, the radial vibration velocity of the surface particle is zero, which can be obtained as

$$\left\{\sum_{n=0}^{\infty}\varepsilon_n A_n i^n k J_n'(ka)\cos n\theta + \sum_{n=0}^{\infty}\varepsilon_n a_n i^n k H_n^{(1)\prime}(ka)\cos n\theta\right\} = 0 \tag{7.104}$$

Thus, we have obtained the implicit equation for the scattering coefficient a_n, which can be numerically solved by iteration.

If the cylindrical particle is not rigid, the existence of refracted waves inside the cylinder should be considered at this time. The sound pressure and particle velocity of its refracted waves can still be represented in the form of Eqs. (7.19) and (7.20). The boundary conditions should also be changed to Eq. (7.18) at this time. By substituting the sound pressure and particle velocity into the boundary conditions, the numerical solution of the scattering coefficient a_n can also be obtained.

7.5.2 Calculation of Acoustic Radiation Force

In fact, the calculation formula of the acoustic radiation force is exactly the same as that under a single boundary. The addition of a boundary only affects the problem of determining the scattering coefficient. Therefore, the formula for the acoustic radiation force at this time is still in the form of Eq. (7.23), and the physical meanings of each term are exactly the same. As for the acoustic radiation force function, it is still in the form of Eq. (7.24).

7.5.3 Simulation Examples

After completing the theoretical derivation, we can now proceed with some simulation examples. For the convenience of our research, let's assume that fluid 1 and fluid 3 are the same medium (in fact, this is the case in most cases) [22], and fluid 2 is different. According to Eq. (7.6), the sound pressure reflection coefficients of the sound waves incident on the two boundaries should be opposite to each other. Under such assumptions, we conduct the following research.

First, we explore the relationship between the acoustic radiation force function and ka. We set the distance between the particle and boundary 1 to twice the radius, and the distance between the particle and boundary 2 to six times the radius. The simulation results for rigid columns and oleic acid columns under different sound pressure reflection coefficients are shown in Fig. 7.25.

The results obtained are consistent with the reference [22]. Compared with the single boundary situation, whether it is a rigid column or a non-rigid column, the overall trend is the same. A larger "acoustic trapping force" can appear at appropriate frequencies, but the frequency band range is significantly narrower at this time. For each negative peak of the acoustic radiation force under a single boundary, there are two peaks under a double boundary, that is, the change is more complicated than before. Especially when the reflection coefficient is large, this phenomenon is particularly obvious. This "splitting" phenomenon of the peak is caused by the fact that the peak positions of the double boundaries do not coincide when they exist separately, which is a typical feature of the acoustic radiation force function curve of the double boundaries. When the reflection coefficient is small, the peaks and valleys of the curve are not obvious. Therefore, the acoustic radiation force function curve at this time is very similar to that under a single boundary, which is also understandable in physics.

Next, we will further investigate the relationship between the particle acoustic radiation force function and the distance d1 from the particle to the first boundary. The distance between the two boundaries is fixed at four times the radius, with a fixed ka = 0.063. The simulation results for rigid and oleic acid columns are shown in Fig. 7.26.

As can be seen, the acoustic radiation force function still varies sinusoidally with d_1/a, and the acoustic radiation force is much greater for the rigid column than for the non-rigid column, similar to the single boundary case. In fact, since the distance between the two boundaries remains constant, this is equivalent to the two boundaries moving as a whole, and the natural laws are similar.

Unlike the single boundary case, the distance between the two boundaries can also be adjusted. Next, we will specifically study the influence of this "intermediate layer" thickness on the acoustic radiation force function. At this time, the distance between the particle and boundary 1 is fixed at twice the radius, and ka = 0.063 is still set in the simulation. For simplicity, we set the independent variable as the distance d_2 from the particle to boundary 2, rather than the thickness of the intermediate layer. However, since boundary 1 is already fixed, the two are actually changing

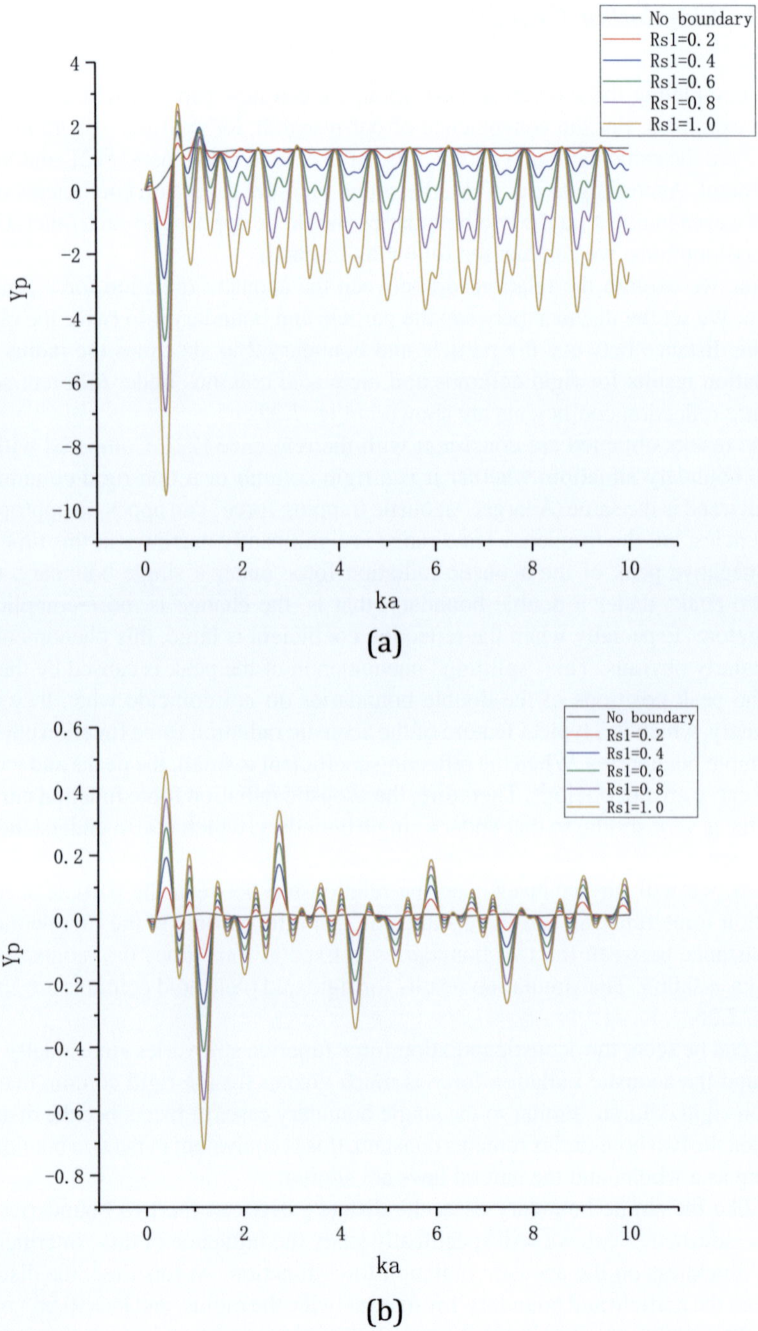

Fig. 7.25 Variation of the acoustic radiation force function of a rigid column and an oleic acid column with ka when plane waves are incident. **a** Rigid column; **b** Oleic acid column

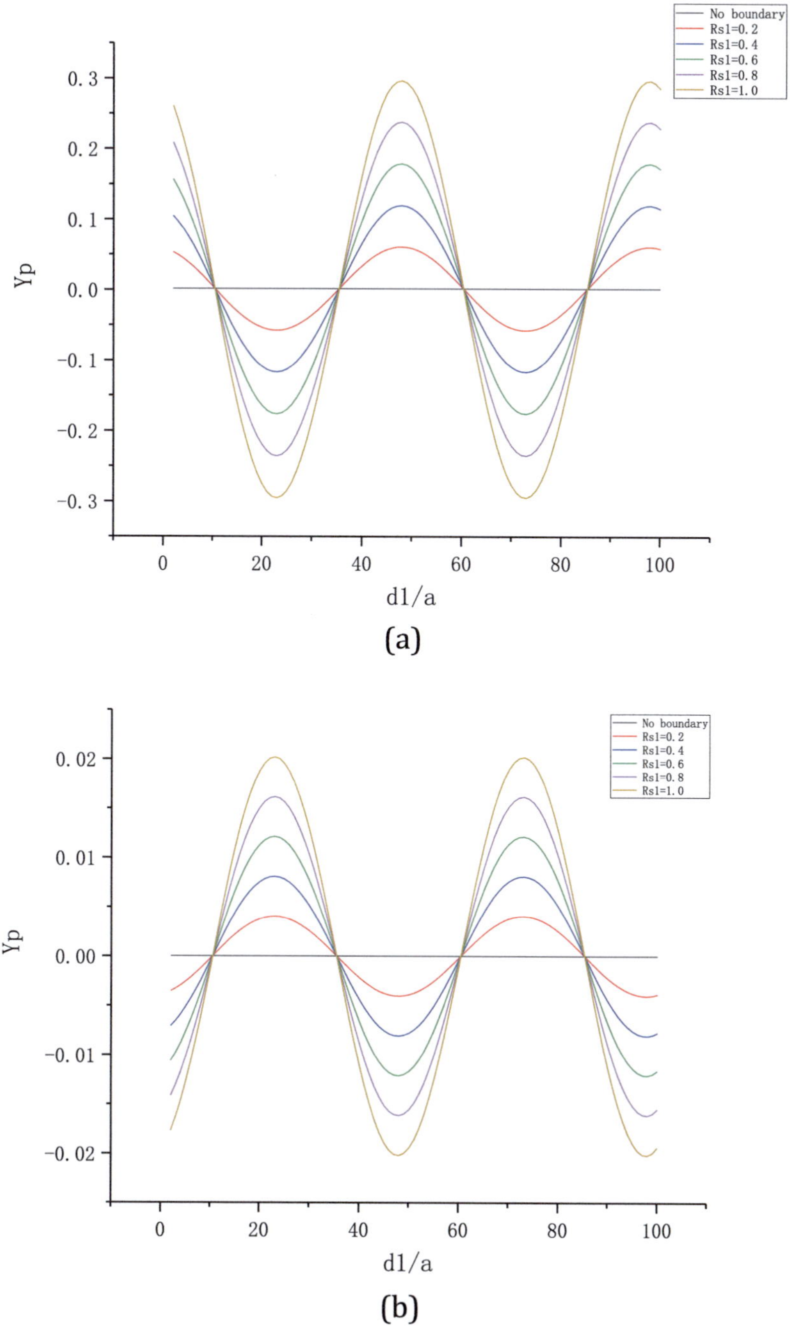

Fig. 7.26 Variation of the acoustic radiation force function for a rigid column and an oleic acid column with d_1/a under plane wave incidence. **a** Rigid column. **b** Oleic acid column

simultaneously. The simulation results for rigid and oleic acid columns are shown in Fig. 7.27.

Surprisingly, whether it is a rigid column or a non-rigid column, although the acoustic radiation force function still varies periodically with d_2, it is no longer a sinusoidal curve! In other words, when boundary 1 is fixed and boundary 2 is moved, the positive and negative peaks of the acoustic radiation force function are not equal. For the rigid column, the negative acoustic radiation force peak is larger, while for the oleic acid column, it is the opposite. The reason for this is that there are two different media in the distance between the particle and boundary 2, resulting in this asymmetry. This provides a theoretical basis for us to adjust the acoustic radiation force on the particles by changing the thickness of the intermediate layer.

7.6 Acoustic Radiation Force on Cylindrical Particles Under Double Boundaries by Gaussian Waves

7.6.1 Solution of the Acoustic Radiation Force

The incident model of Gaussian waves under double impedance boundaries is also simple, just need to change the plane wave incidence in Fig. 7.24 to Gaussian wave incidence. At this time, the incident sound wave can still be expressed in the form of Eq. (7.4). Using the mirror image principle, all other sound waves existing in the whole space can be expressed as

$$\Phi_{ref1} = \Phi_0 R_{s1} e^{i2kd1} \sum_{n=-\infty}^{\infty} (-1)^n g_n i^n J_n(kr) e^{in\theta} e^{-i\omega t} \tag{7.105}$$

$$\Phi_{ref2} = \Phi_0 R_{s2} e^{i2kd2} \sum_{n=-\infty}^{\infty} (-1)^n g_n i^n J_n(kr) e^{in\theta} e^{-i\omega t} \tag{7.106}$$

$$\Phi_{scat,ref1} = \Phi_0 R_{s1} \sum_{n=-\infty}^{\infty} (-1)^n a_n g_n i^n H_n(k\overline{r}) e^{-in\overline{\theta}} e^{-i\omega t} \tag{7.107}$$

$$\Phi_{scat,ref1} = \Phi_0 R_{s1} \sum_{n=-\infty}^{\infty} (-1)^n a_n g_n i^n H_n(k\overline{r}) e^{-in\overline{\theta}} e^{-i\omega t} \tag{7.108}$$

where, g_n acts as the beam factor.

The total sound field in the fluid medium is the superposition of all the above sound waves, which can be expressed as

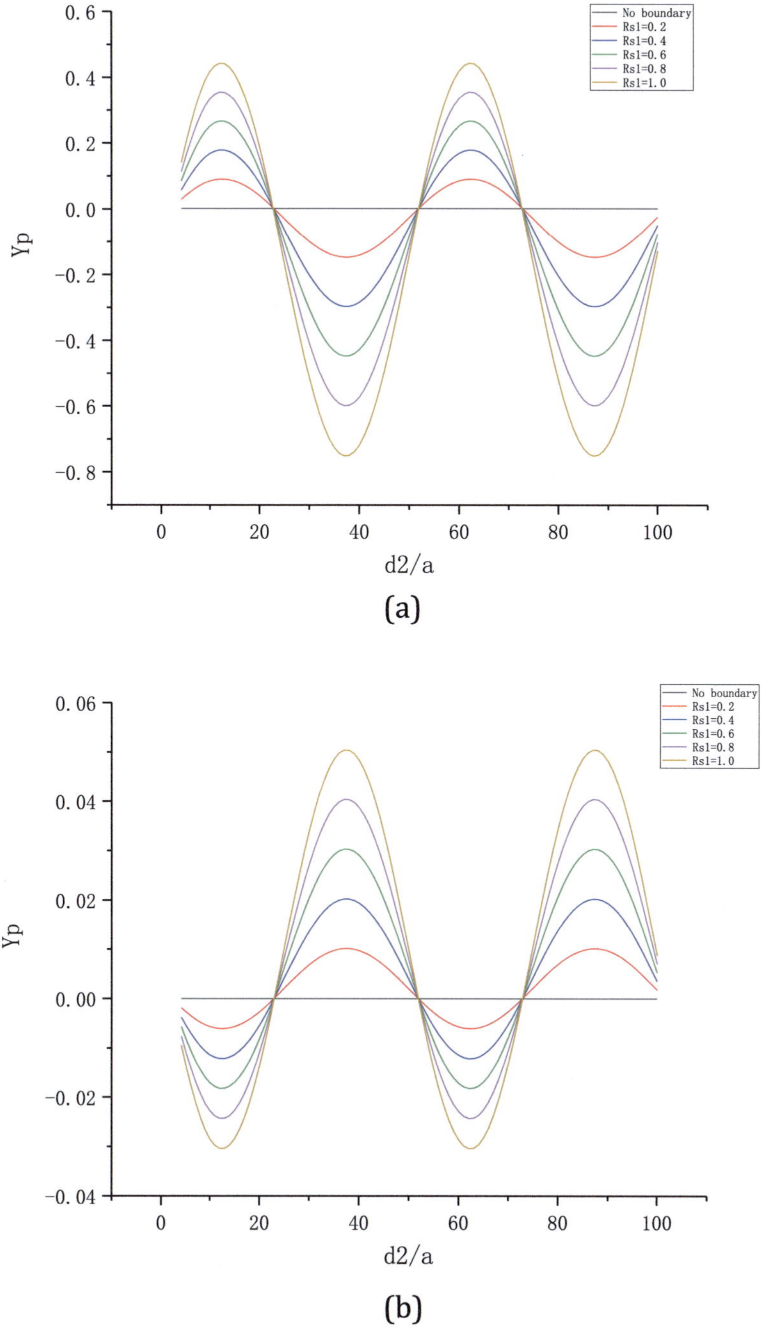

Fig. 7.27 Variation of the acoustic radiation force function for a rigid column and an oleic acid column with d_2/a under plane wave incidence. **a** Rigid column. **b** Oleic acid column

$$\Phi_{total} = \Phi_0 e^{-i\omega t}\left\{\sum_{n=0}^{\infty}\varepsilon_n A_n i^n g_n J_n(kr)\cos n\theta + \sum_{n=0}^{\infty}\varepsilon_n a_n g_n i^n H_n^{(1)}(kr)\cos n\theta\right\} \tag{7.109}$$

where,

$$\begin{aligned} A_n = {} & 1 + R_{s1}e^{i2kd1}(-1)^n + R_{s2}e^{i2kd2}(-1)^n \\ & + \frac{R_{s1}}{\varepsilon_n g_n i^n}\sum_{m=0}^{\infty}(-1)^m \varepsilon_m a_m g_m i^m H_{m-n}^{(1)}(2kd1) \\ & + \frac{R_{s2}}{\varepsilon_n g_n i^n}\sum_{m=0}^{\infty}(-1)^m \varepsilon_m a_m g_m i^m H_{m-n}^{(1)}(2kd2) \end{aligned} \tag{7.110}$$

Once the velocity potential is obtained, it is easy to obtain the distribution of the sound pressure field and the velocity field. With appropriate boundary conditions on the surface of the cylinder, the scattering coefficient of the particle to the sound wave a_m.

As for the formula for solving the acoustic radiation force, Eq. (7.95) can still be used.

7.6.2 Simulation Example

In order to highlight the characteristics of the Gaussian beam, we still choose a Gaussian wave with a beam waist width $w = 3\lambda$. All other parameters remain identical to those for plane wave incidence. For both the rigid cylinder and oleic acid cylinder, the acoustic radiation force functions are illustrated in Fig. 7.28

Just as with plane wave incidence, the addition of a boundary leads to the so-called "peak splitting" phenomenon. Regardless of whether it is a rigid or non-rigid column, each original negative acoustic radiation force peak splits into two peaks, making the curve changes more drastic. At low frequencies, this phenomenon significantly narrows the frequency band range for obtaining "acoustic trapping force". Moreover, an increase in the boundary sound pressure reflection coefficient makes this phenomenon more pronounced, which is expected. Next, we will briefly study the relationship between the acoustic radiation force function under double boundaries and the distance of the particle to the two boundaries. Here, only the results for the rigid column are given, as shown in Figs. 7.29 and 7.30. The change pattern is basically consistent with that during plane wave incidence. The acoustic radiation force function still follows a sine curve with d1/a, while with d2/a, there is a phenomenon of unequal positive and negative peak values.

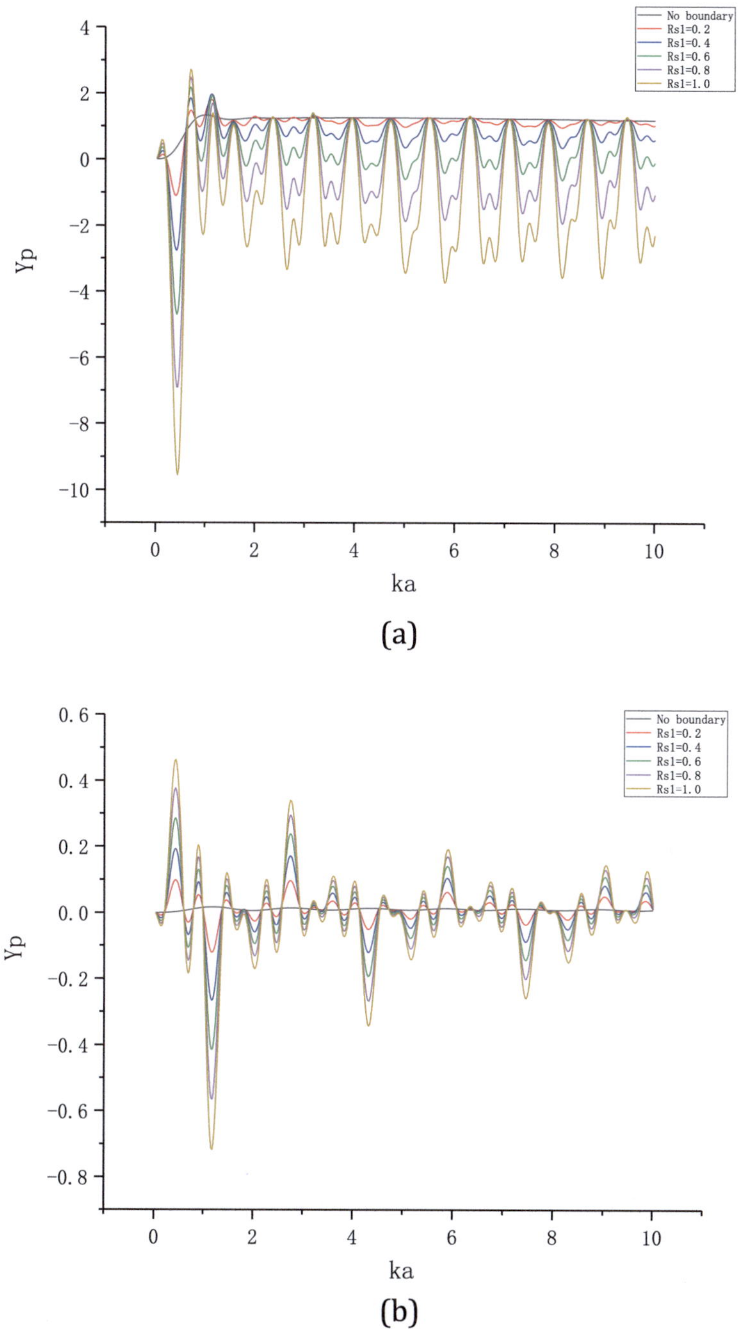

Fig. 7.28 Variation of the acoustic radiation force function for a rigid column and an oleic acid column with ka under Gaussian wave incidence. **a** Rigid column, **b** Oleic acid column

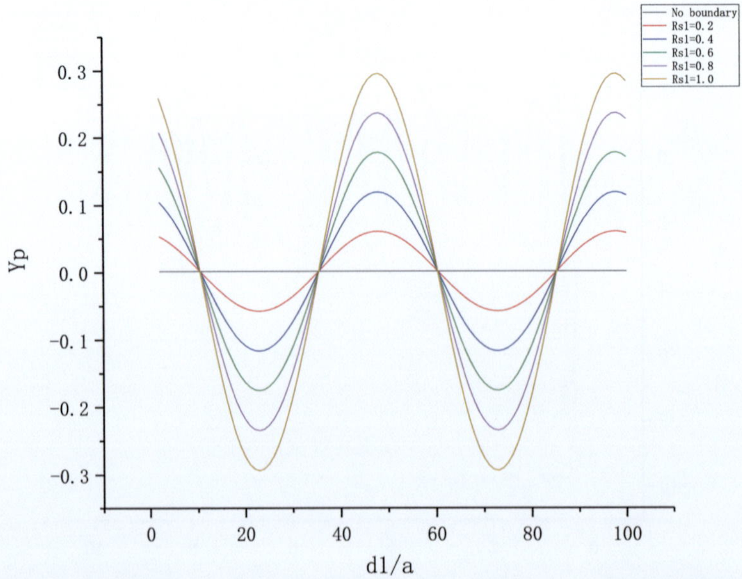

Fig. 7.29 Variation of the acoustic radiation force function of the rigid column with d_1/a during Gaussian wave incidence

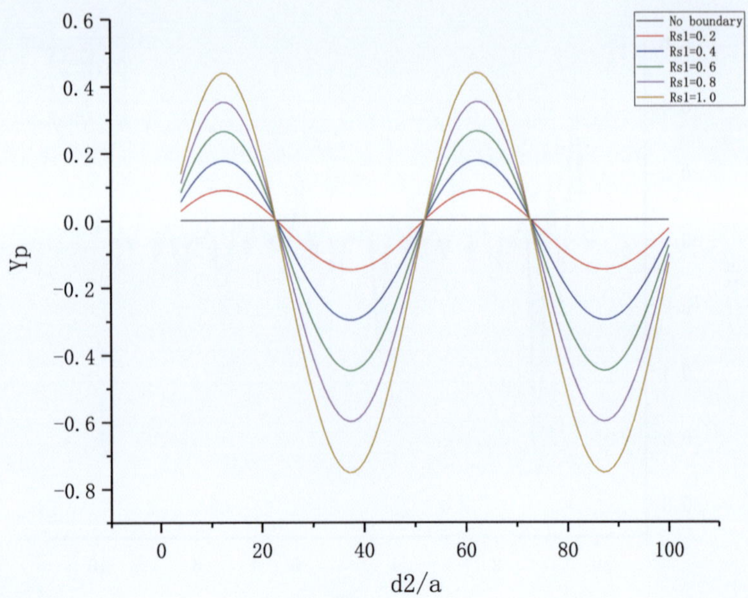

Fig. 7.30 Variation of the acoustic radiation force function of a rigid column with d_2/a during Gaussian wave incidence

7.7 Acoustic Radiation Force of Plane Waves on Spherical Particles Under Double Boundaries

Consider a plane wave incident perpendicularly on a spherical particle with a radius of a under double boundaries. The distance of the particle to the impedance boundary 1 is d1, and to the impedance boundary 2 is d2. This physical model is very similar to Fig. 7.24, except that the cylindrical particle in the original figure is replaced with a spherical one. Of course, the coordinate system should be changed to a spherical coordinate system for convenience, with the center of the sphere as the origin O, and the polar angle and elevation angle denoted as θ and φ.

Considering all acoustic wave components in the unbounded space. The incident plane wave and the scattered waves from the particle can still be expressed in the forms of Eqs. (7.74) and (7.76). The reflected waves from the two boundaries for the incident sound wave, using the method of images, can be represented as

$$\Phi_{ref1} = \Phi_0 e^{-i\omega t} \sum_{n=0}^{\infty} R_{s1} e^{j2kd1} (-1)^n (2n+1) i^n j_n(kr) P_n(\cos\theta) \tag{7.111}$$

$$\Phi_{ref2} = \Phi_0 e^{-i\omega t} \sum_{n=0}^{\infty} R_{s2} e^{j2kd2} (-1)^n (2n+1) i^n j_n(kr) P_n(\cos\theta) \tag{7.112}$$

As for the reflected waves from both boundaries of the scattered waves by the particle, they can be expressed as

$$\Phi_{scat,ref1} = \Phi_0 e^{-i\omega t} \sum_{n=0}^{\infty} R_{s1} (-1)^n (2n+1) i^n a_n h_n^{(1)}(k\overline{r}) P_n\left(\cos\overline{\theta}\right) \tag{7.113}$$

$$\Phi_{scat,ref2} = \Phi_0 e^{-i\omega t} \sum_{n=0}^{\infty} R_{s2} (-1)^n (2n+1) i^n a_n h_n^{(1)}(k\overline{r}) P_n\left(\cos\overline{\theta}\right) \tag{7.114}$$

The meanings of all parameters in the above four expressions are identical to those previously defined and are therefore not restated.

7.7.1 Solution of the Scattering Coefficient

The total sound field in the fluid medium is evidently the superposition of all six components mentioned above, expressed by the formula

$$\Phi_{total} = \Phi_0 e^{-i\omega t} \left\{ \sum_{n=0}^{\infty} A_n i^n j_n(kr) P_n(\cos\theta) + \sum_{n=0}^{\infty} a_n i^n h_n^{(1)}(kr) P_n(\cos\theta) \right\} \tag{7.115}$$

where,

$$
\begin{aligned}
A_n = {} & 1 + R_{s1} e^{i2kd1}(-1)^n + R_{s2} e^{i2kd2}(-1)^n \\
& + \frac{R_{s1}}{(2n+1) g_n i^n} \sum_{m=0}^{\infty} g_m (-1)^m (2m+1) a_m i^m Q_{mn} \\
& + \frac{R_{s2}}{(2n+1) g_n i^n} \sum_{m=0}^{\infty} g_m (-1)^m (2m+1) a_m i^m Q_{mn}
\end{aligned}
\tag{7.116}
$$

Based on the relationship between harmonic sound pressure and velocity potential, the distribution of the total sound pressure field can be obtained. Based on the relationship between velocity potential and particle velocity, the radial velocity can be derived.

$$
v_r = -\Phi_0 e^{-i\omega t} \left\{ \sum_{n=0}^{\infty} A_n i^n k j_n'(ka) P_n(\cos\theta) + \sum_{n=0}^{\infty} a_n i^n k h_n^{(1)\prime}(ka) P_n(\cos\theta) \right\} \tag{7.117}
$$

If the spherical particle can be considered rigid, the radial velocity on its surface vanishes, i.e.

$$
\sum_{n=0}^{\infty} A_n i^n k j_n'(ka) P_n(\cos\theta) + \sum_{n=0}^{\infty} a_n i^n k h_n^{(1)\prime}(ka) P_n(\cos\theta) = 0 \tag{7.118}
$$

Each scattering coefficient can be solved.

If the particle cannot be considered rigid, the presence of refracted waves inside the sphere must be accounted for, expressed by Eq. (7.84). In this case, the boundary conditions requiring continuity of sound pressure and normal particle velocity at the particle surface must be satisfied, and the scattering coefficients are determined by solving these equations jointly.

7.7.2 Solution of the Acoustic Radiation Force

The formula for the acoustic radiation force remains applicable to a single boundary, consistent with Eq. (7.85). As for the acoustic radiation force function, it retains the form given by Eq. (7.86).

7.7.3 Numerical Simulation Example

With the theoretical derivation completed, we now proceed to numerical simulations. For simplicity as Chapter 7.5, assume Fluid 1 and Fluid 3 are identical media, while Fluid 2 differs. The distance between the particle and Boundary 1 is twice its radius, and the distance to Boundary 2 is six times its radius. The acoustic radiation force function curves for both a rigid sphere and an oleic acid sphere are shown in Fig. 7.31.

Compared to the single boundary case, it can be observed that the acoustic radiation force function curve of the rigid sphere under the double boundary condition changes more dramatically. Within the range of ka we simulated, the number of peak points has increased. Especially at low frequencies, the acoustic radiation force function can quickly change from a negative peak to a positive peak, creating very sharp peaks and valleys. For oleic acid spheres, surprisingly, a large negative acoustic radiation force function peak appears in the mid-to-high frequency range of ka 7–10, which is very different from the single boundary case! These are all due to the addition of a boundary, which increases the interaction between different sound waves.

As for the relationship between the particle acoustic radiation force function and the distance from the particle to boundary 1 and boundary 2, the simulation results are shown in Figs. 7.32 and 7.33, respectively, with the same simulation parameters as mentioned above.

Similar to the case of cylindrical particles, the acoustic radiation force function curve still changes with d1 in a sinusoidal pattern with equal positive and negative peaks, identical to the single boundary case. However, when d1 is fixed and d2 is changed, a significant "asymmetry" phenomenon occurs, i.e., the positive and negative peaks are no longer equal. Interestingly, in the case of rigid spheres, positive acoustic radiation forces hardly appear, i.e., the sound wave is a trapping force for most d2 values, while the situation for oleic acid spheres is just the opposite, with "acoustic trapping forces" rarely appearing.

7.8 Acoustic Radiation Force of Gaussian Waves on Spherical Particles with Double Boundaries

7.8.1 Solution of the Acoustic Radiation Force

The incident model of Gaussian waves under double boundaries is also simple, just replacing the plane wave in the previous section with a Gaussian wave. The incident sound wave and scattered sound wave can still be represented in the form of Eqs. (7.93) and (7.95), and the reflections of the incident sound wave from the two boundaries can be represented as

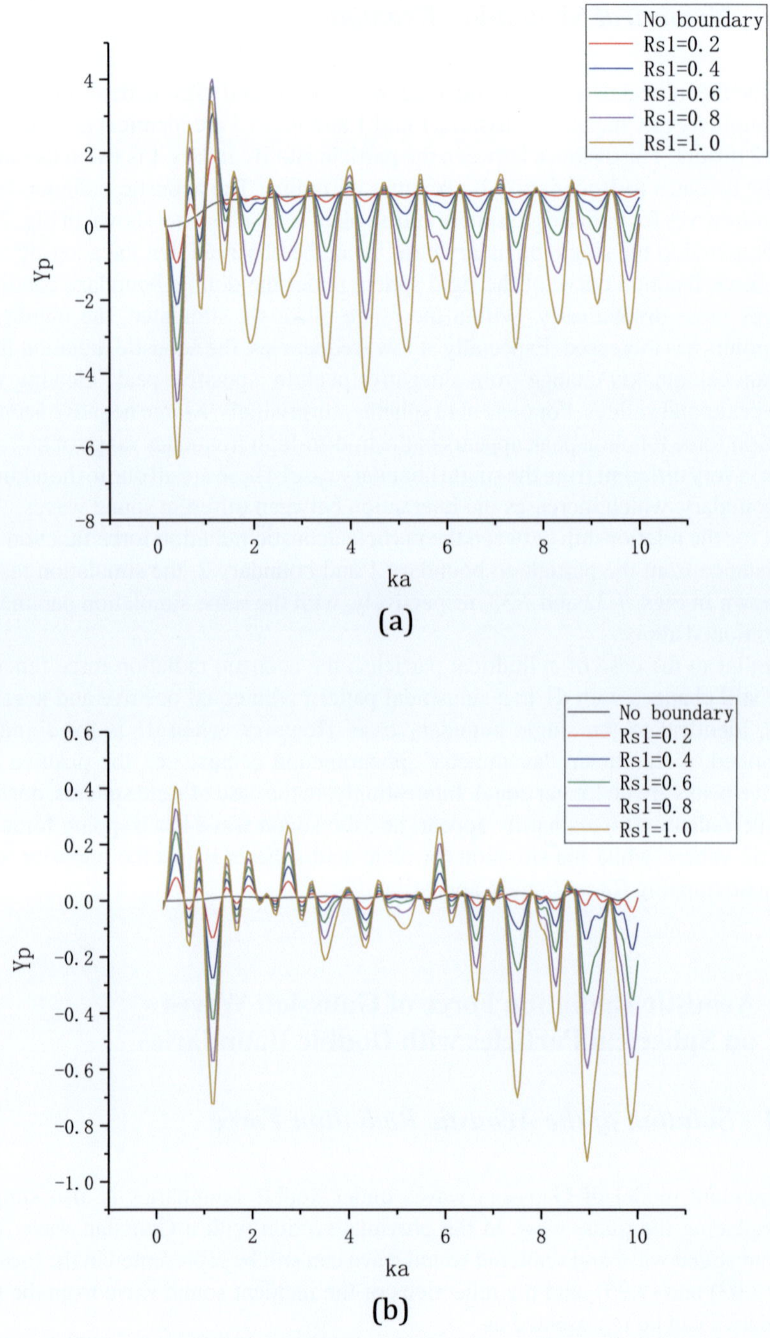

Fig. 7.31 Variation of the acoustic radiation force function with ka for a rigid sphere and an oleic acid sphere under plane wave incidence. **a** Rigid sphere. **b** Oleic acid sphere

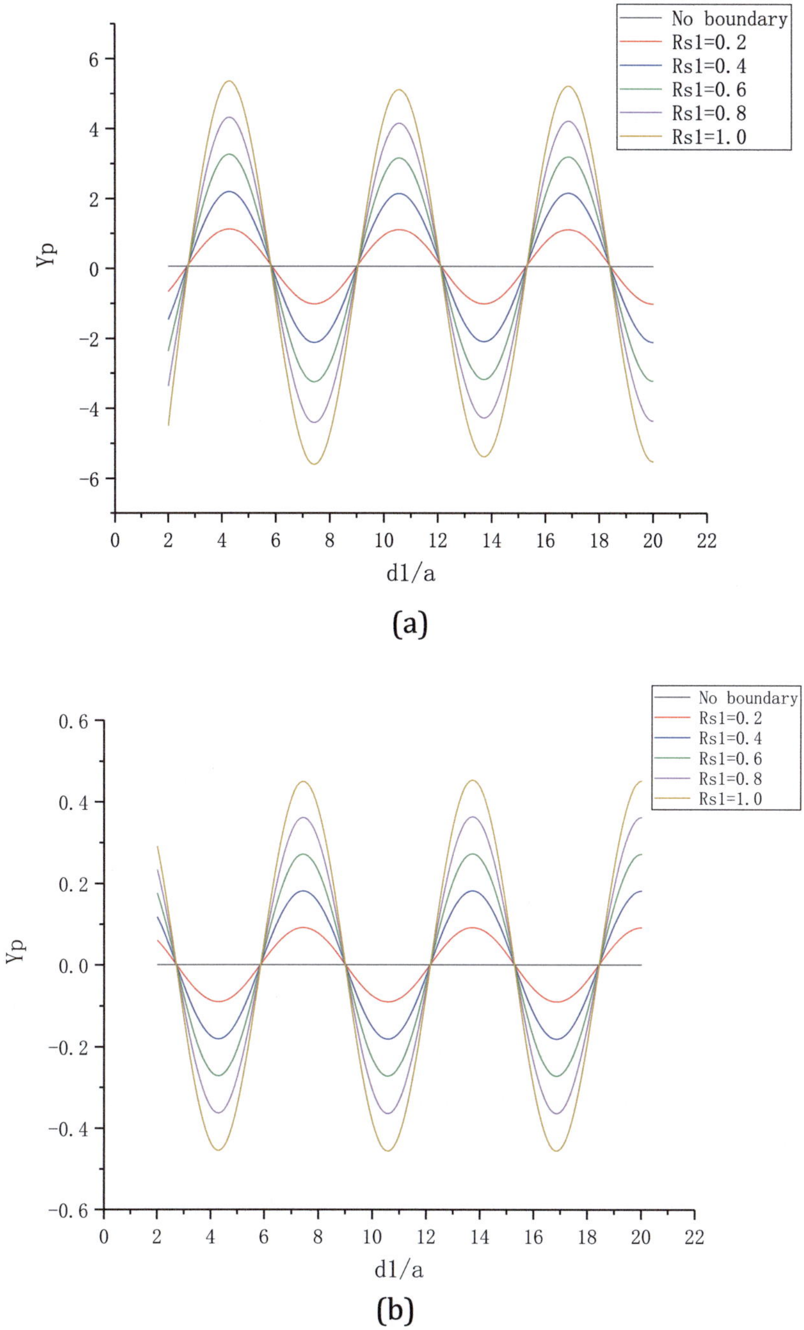

Fig. 7.32 Variation of the acoustic radiation force function of a rigid sphere and an oleic acid sphere with d_1/a under plane wave incidence. **a** Rigid sphere. **b** Oleic acid sphere

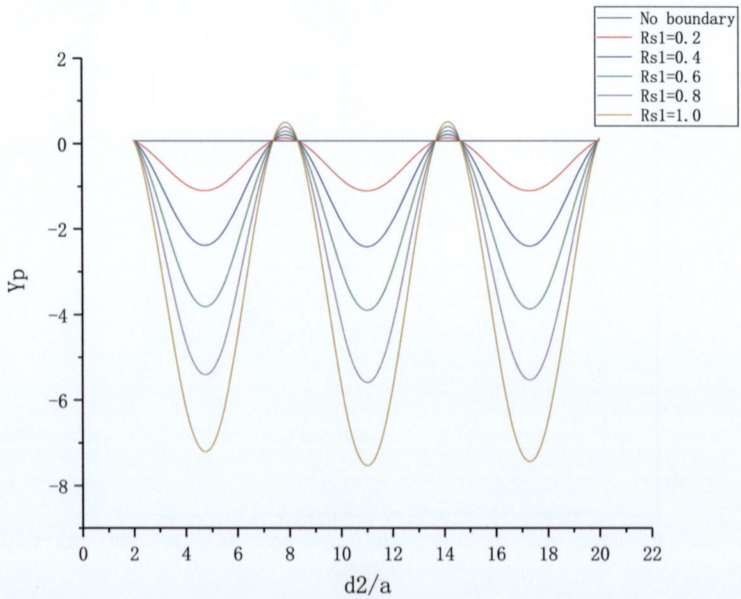

(a)

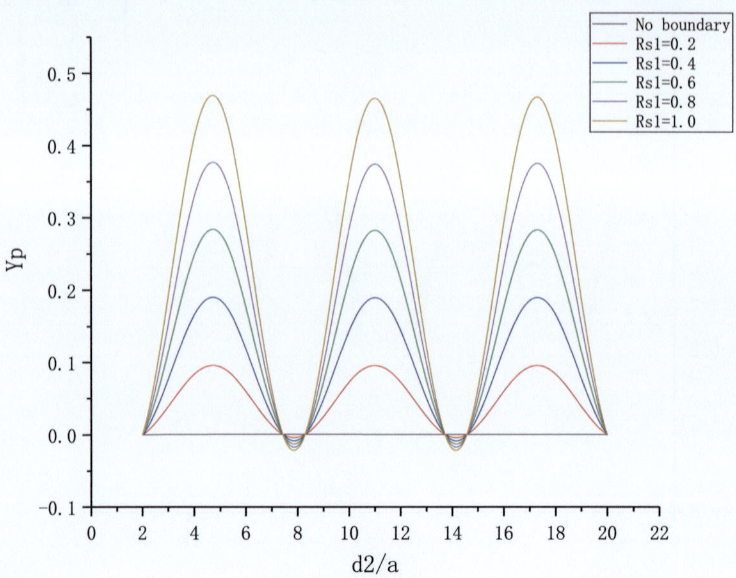

(b)

Fig. 7.33 Variation of the acoustic radiation force function of a rigid sphere and an oleic acid sphere with d_2/a under plane wave incidence. **a** Rigid sphere. **b** Oleic acid sphere

$$\Phi_{ref1} = \Phi_0 e^{-i\omega t} \sum_{n=0}^{\infty} R_s e^{j2kd1}(-1)^n g_n (2n+1) i^n j_n(kr) P_n(\cos\theta) \tag{7.119}$$

$$\Phi_{ref2} = \Phi_0 e^{-i\omega t} \sum_{n=0}^{\infty} R_s e^{j2kd2}(-1)^n g_n (2n+1) i^n j_n(kr) P_n(\cos\theta) \tag{7.120}$$

As for the reflections of the particle scattered waves from the two boundaries, they can be represented as

$$\Phi_{scat,ref1} = \Phi_0 e^{-i\omega t} \sum_{n=0}^{\infty} R_{s1}(-1)^n g_n (2n+1) i^n a_n h_n^{(1)}(k\overline{r}) P_n\left(\cos\overline{\theta}\right) \tag{7.121}$$

$$\Phi_{scat,ref2} = \Phi_0 e^{-i\omega t} \sum_{n=0}^{\infty} R_{s2}(-1)^n g_n (2n+1) i^n a_n h_n^{(1)}(k\overline{r}) P_n\left(\cos\overline{\theta}\right) \tag{7.122}$$

where g_n is the beam factor when the incident wave is expanded in spherical functions.

The total sound field in the fluid medium is the superposition of all the above sound waves, which can be represented as

$$\Phi_{total} = \Phi_0 e^{-i\omega t} \left\{ \sum_{n=0}^{\infty} A_n g_n i^n j_n(kr) P_n(\cos\theta) + \sum_{n=0}^{\infty} a_n g_n i^n h_n^{(1)}(kr) P_n(\cos\theta) \right\} \tag{7.123}$$

where,

$$\begin{aligned} A_n = {} & 1 + R_{s1} e^{i2kd1}(-1)^n + R_{s2} e^{i2kd2}(-1)^n \\ & + \frac{R_{s1}}{(2n+1) g_n i^n} \sum_{m=0}^{\infty} g_m (-1)^m (2m+1) a_m i^m Q_{mn} \\ & + \frac{R_{s2}}{(2n+1) g_n i^n} \sum_{m=0}^{\infty} g_m (-1)^m (2m+1) a_m i^m Q_{mn} \end{aligned} \tag{7.124}$$

Once the velocity potential is obtained, the distribution of the sound pressure field and velocity field can be easily obtained. According to the appropriate boundary conditions on the surface of the spherical particle, the scattering coefficient of the particle to the sound wave a_m can be obtained, which can be discussed in two cases: rigid and non-rigid, as mentioned above, without further elaboration.

As for the solution formula of the acoustic radiation force, formula (7.95) can still be used.

7.8.2 Simulation Example

We still choose a Gaussian wave with a waist width three times the wavelength for the simulation. The other parameters are the same as for the plane wave, and the acoustic radiation force function curves of the rigid sphere and oleic acid sphere are shown in Fig. 7.34.

Whether it is a rigid sphere or an oleic acid sphere, the acoustic radiation force function under the double boundary condition is significantly different from the single boundary case! For the rigid sphere, the acoustic radiation force function peak at the mid-to-low frequency is smaller, but when ka is greater than 6, a large negative acoustic radiation force function peak can appear within an appropriate frequency range, even reaching a magnitude of 50! For the oleic acid sphere, compared to the single boundary case, the peaks of the curve are more concentrated, and a large positive or negative acoustic radiation force can be obtained at certain frequencies, whether at low or high frequencies.

Next, we will briefly study the relationship between the acoustic radiation force function and the distance from the particle to the two boundaries under the double boundary. Here, only the results of the rigid column are given, as shown in Figs. 7.35 and 7.36.

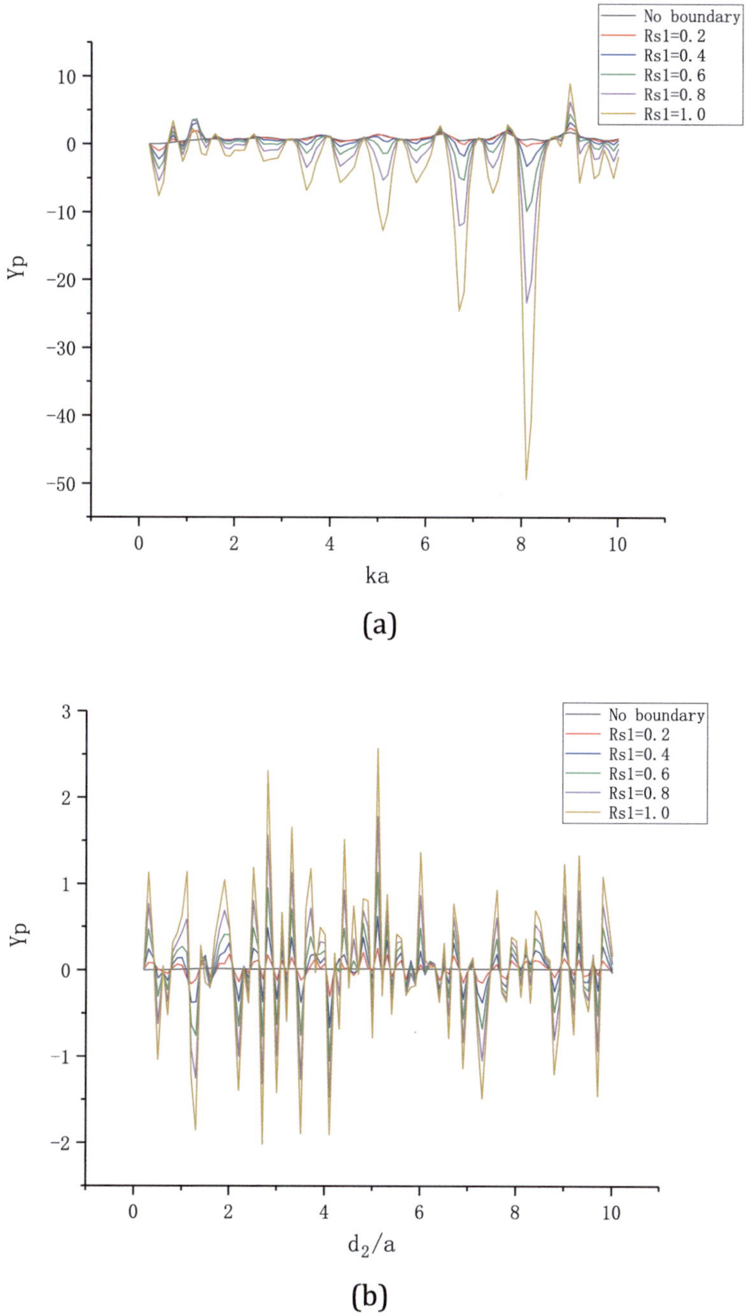

Fig. 7.34 Variation of the acoustic radiation force function for a rigid sphere and an oleic acid sphere with ka during Gaussian wave incidence. **a** Rigid sphere. **b** Oleic acid sphere

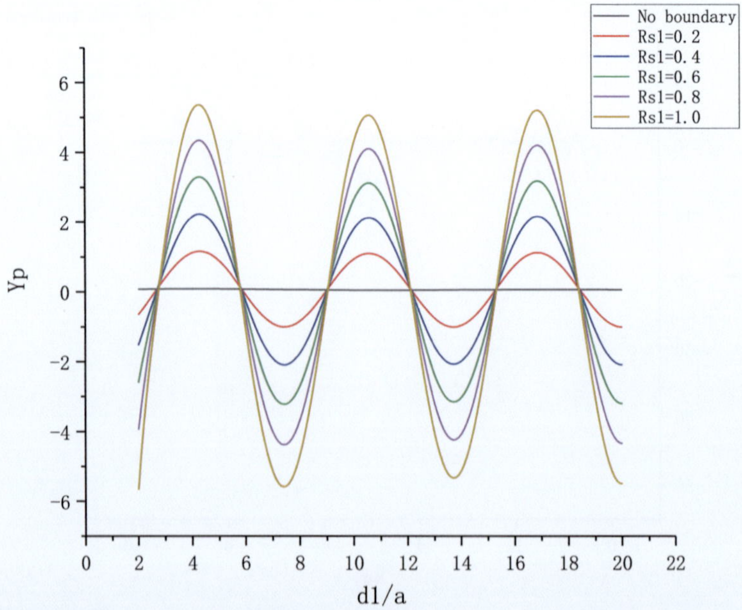

Fig. 7.35 Variation of the acoustic radiation force function for a rigid sphere with d_1/a during Gaussian wave incidence

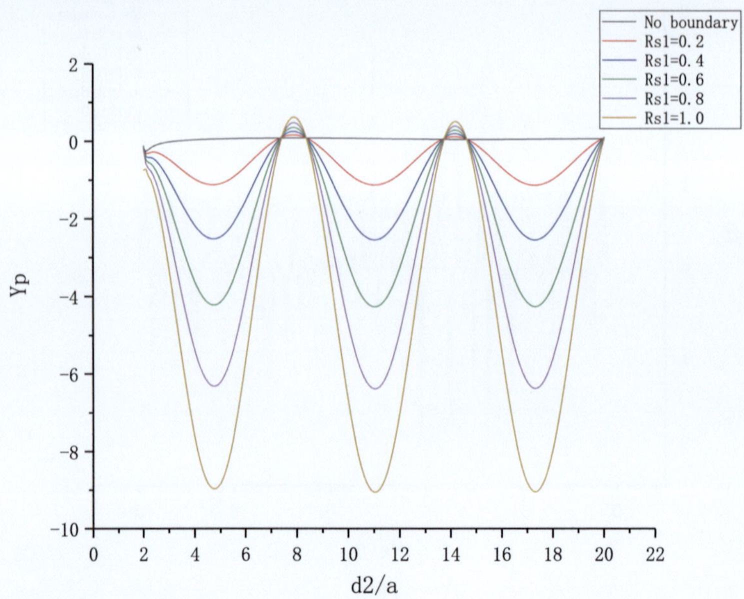

Fig. 7.36 Variation of the acoustic radiation force function for a rigid sphere with d_2/a during Gaussian wave incidence

References

1. Hailan Z (2012) Theoretical acoustics. Higher Education Press, Beijing, pp 231–296
2. Kunmiao L (2010) Mathematical physics methods. Higher Education Press, Beijing, pp 275–277
3. Zhang XF, Yun Q, Zhang GB, Sun XN (2016) Computation of the acoustic radiation force on a rigid cylinder in off-axial Gaussian beam using the translational addition theorem. Acta Acust Acust 102:334–340
4. Qiao YP, Zhang XF, Zhang GB (2017) Acoustic radiation force on a fluid cylindrical particle immersed in water near an impedance boundary. J Acoust Soc Am 141(6):4633–4641
5. Wu R (2016) Study on acoustic radiation force of spherical particles by sound waves. Institute of Acoustics, Nanjing University, Nanjing, pp 79–82
6. Qiao YP, Zhang XF, Zhang GB (2017) Acoustic radiation force on a fluid cylindrical particle immersed in water near an impedance boundary. J Acoust Soc Am 141:4633–4641
7. Qiao YP, Zhang XF, Zhang GB (2017) Axial acoustic radiation force on a rigid cylinder near an impedance boundary for on-axis Gaussian beam. Wave Motion 74:182–190
8. Qiao YP, Shi JY, Zhang XF et al (2018) Acoustic radiation force on a rigid cylinder in an off-axis Gaussian beam near an impedance boundary. Wave Motion 83:111–120
9. Mitri FG (2017) Extrinsic extinction cross-section in the multiple acoustic scattering by fluid particles. J Appl Phys 121:144904
10. Mitri FG (2017) Acoustic attraction, repulsion and radiation force cancellation on a pair of rigid particles with arbitrary cross-sections in 2D: Circular cylinders example. Ann Phys 386:1–14
11. Hasheminejad SM, Azarpeyvand M (2004) Modal vibrations of a cylindrical radiator over an impedance plane. J Sound Vib 278:461–477
12. Hasheminejad SM, Alibakhshi MA (2007) Diffraction of sound by a poroelastic cylindrical absorber near an impedance plane. Int J Mech Sci 49:1–12
13. Abramowitz M, Stegun IA (1964) Handbook of mathematical functions. National Bureau of Standards, Washington, DC, p 363
14. Mitri FM (2015) Resonance scattering and radiation force calculations for an elastic cylinder using the translational addition theorem for cylindrical wave functions. Am Inst Phys Adv 5:097205
15. Brillouin L (1964) Tensors in mechanics and elasticity. Academic Press, New York
16. Zhang XF, Yun Q, Zhang GB et al (2016) Computation of the acoustic radiation force on a rigid cylinder in off-axial Gaussian beam using the translational addition theorem. Acta Acust Acust 102:334–340
17. Maidanik G (1958) Torques due to acoustical radiation pressure. J Acoust Soc Am 30:620–623
18. Qiao YP, Zhang XF (2017) Acoustic radiation force on a rigid cylindrical particle underwater near an impedance interface. J Nanjing Univ (Nat Sci) 53(1):19–26
19. Zhiguang S, Xiaofeng Z, Guangbin Z (2013) Study on acoustic radiation force of cylindrical particles in water by Gaussian Beam. Piezoelectr Acoustoopt 35(6):792–796
20. Kunmiao L (2010) Mathematical physics methods. Higher Education Press, Beijing, pp 300–301
21. Friedman B, Russek J (1954) Addition theorems for spherical waves. Quart Appl Math 12:13–23
22. Amir KM, Mitri FG (2011) Acoustics. Acoustic radiation force on a spherical contrast agent shell near a vessel porous wall-theory. Ultrasound Med Biol 37(2):301–311
23. Dongmei C, Xiaofeng Z, Guangbin Z, Jinghua Z, Tao C (2013) Study on the acoustic radiation force of Gaussian beams on spherical particles in water. Piezoelectr Acoustoopt 35(3):329–332
24. Chen D (2014) Study on the acoustic radiation force of Gaussian beams on tiny particles in water. Key Laboratory of Ultrasonics in Shaanxi Province, Xi'an, pp 1–91
25. Majid R, Mojahed A (2016) Acoustic manipulation of oscillating spherical bodies: Emergence of axial negative acoustic radiation force. J Sound Vib 383:265–276

Chapter 8
Acoustic Radiation Force Among Multiple Particles

The phenomenon of acoustic radiation force is caused by the momentum transfer of incident waves to suspended objects [1]. It can be used to acoustically manipulate particles. Extensive research has been conducted on the acoustic radiation force applied to individual objects in previous theories and experiments [2–7]. However, in practical operations, several suspended particles may appear in an external sound field simultaneously. When multiple particles exist in the sound field, on one hand, these particles will interact with the sound field; on the other hand, there will be interactions between the particles themselves. The scattering field emerging from one particle will cause further scattering from the surrounding particles [8, 9]. In 2000, Roumeliotis et al. briefly analyzed the multiple scattering effects between two particles [10–14], then Mitri introduced it into the field of acoustic tweezers [15, 16], discussing the relationship between acoustic radiation force, particle spacing, and incident angle. In 2016, Zhang Shenwei et al. studied the acoustic radiation force on multiple particles in a plane standing wave sound field [17], J. Henrique Lopes et al. studied the acoustic radiation force on multiple particles in a liquid plane sound field [18], and in 2018, Hui Mingxin et al. also conducted research on multiple particles in a plane sound field [19].

8.1 Formula for Acoustic Radiation Force Between Two Distant Particles in a Plane Field

In a sound field, due to the energy and momentum of sound waves, objects in the sound field are subjected to the force exerted by the sound field, known as the acoustic radiation force. Moreover, due to scattering, when multiple objects exist in the sound field, the scattering waves produced by any object will exert force on other objects. This force generated by scattering can be referred to as the interaction force between

X. Liu and H. Zheng, *Acoustic Radiation Force*, Advances in Acoustics,
https://doi.org/10.1007/978-981-95-3152-3_8

objects in the sound field. This section mainly studies the interaction force between multiple particles in the medium.

When plane sound waves are incident on multiple spherical particles, for any particle n, the velocity potentials of the incident wave and the scattered wave can be represented as follows [20]:

$$\phi = i\omega t \sum_{lm} a^n_{l,m} j_l(k_0 r_n) Y_{l,m}(\theta_n, \varphi_n)_{i,n} \tag{8.1}$$

$$\phi_{s,n} = e^{-i\omega t} \sum_{lm} b^n_{l,m} h^{(1)}_l(k_0 r_n) Y_{l,m}(\theta_n, \varphi_n) \tag{8.2}$$

where $\sum_{lm} = \sum_{l=0}^{\infty} \sum_{m=-l}^{l}$, $Y_{l,m}(\theta, \varphi)$ are normalized spherical harmonic functions that can be represented by Legendre functions $Y_{l,m}(\theta, \varphi) = \zeta_l P^m_l(\cos\theta) e^{im\varphi}$, where $\zeta_l = (-1)^m \sqrt{\frac{[(2l+1)(l-m)!]}{[4\pi(l+m)!]}}$, $P^m_l(\cos\theta)$ are l order associated Legendre functions. When two particles are in the same yz plane, $\varphi = 0$, a^n_{lm} is the total incident velocity potential experienced by particle n, which consists of two parts. The first part is the velocity potential generated by the incident wave itself, and the second part is the velocity potential generated by the scattered wave from all other particles except n in space. Therefore, $a^n_{l,m}$ can be represented as:

$$a^n_{lm} = a^{n(0)}_{lm} + \sum_{q \neq n} a^{n(q)}_{lm} \tag{8.3}$$

Taking two particles, spheres n and q, in the medium as an example, as shown in Fig. 8.1, point O is the observation point location, the distances from the two particles to the observation point are r_n, r_q, and the direction vectors are $\hat{r}_n$, $\hat{r}_q$, r_{nq}, $\hat{r}_{nq}$ are the distance and direction vector from the center of q to the center of n, respectively. For convenience, we can first assume that the two particles are on the same xy plane, that is, the two particle spheres are located on the same wavefront in the plane sound field.

We first consider the velocity potential $a^{n(0)}_{lm}$ generated by the plane wave field, which is only related to the external sound field and particle n itself. This part of the velocity potential can be represented as [21]

$$a^{n(0)}_{lm} = \phi_0 i^l \sqrt{4\pi(2l+1)} e^{-ik_0 z_n} \delta_{m,0} \tag{8.4}$$

where ϕ_0 is the amplitude of the incident plane wave velocity potential, z_n is the z-coordinate of particle n, $\delta_{m,m'}$ is the Kronecker delta, which satisfies:

$$\delta_{m,m'} = \begin{cases} 0, & m \neq m' \\ 1, & m = m' \end{cases}$$

Therefore, it is clear that only when m = 0, $a^{n(0)}_{l,m}$ can a non-zero value be obtained.

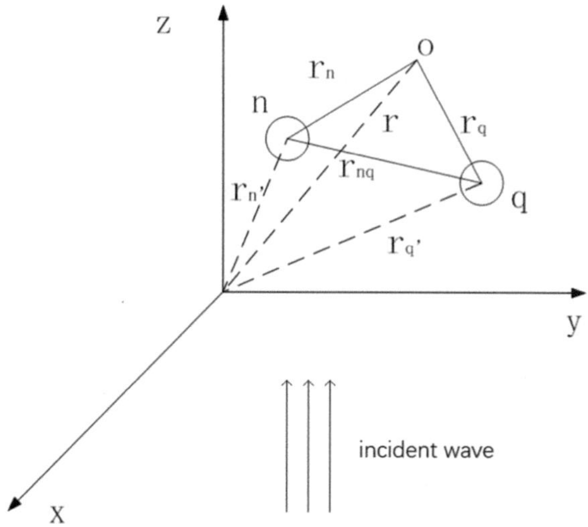

Fig. 8.1 Schematic diagram of acoustic radiation interaction between two particle spheres in a sound field

The second part, the velocity potential generated by scattering, $a_{lm}^{n(q)}$ can be obtained from the addition theorem of spherical functions [22]

$$a_{lm}^{n(q)} = \sum_{l'm'} b_{l'm'}^{q} G_{lml'm'}^{nq} \tag{8.5}$$

where

$$\begin{aligned} G_{lml'm'}^{nq} &= 4\pi \sum_{l''} i^{l+l''-l'} C_{lml''(m'-m)}^{l'm'} h_{l''}\left(k_0 r_{nq}\right) Y_{l'',m'-m}\left(\hat{r}_{nq}\right) \\ C_{lml''(m'-m)}^{l'm'} &= \int_0^{2\pi}\int_0^{\pi} Y_{l',m'}(\theta,\varphi) Y_{l,m}^{*}(\theta,\varphi) Y_{l'',m'-m}^{*}(\theta,\varphi)\sin\theta d\theta d\varphi \end{aligned} \tag{8.6}$$

There is a simple correspondence between b_{lm}^{n} and a_{lm}^{n}, that is, $b_{lm}^{n} = A_l a_{lm}^{n}$ (since all particle spheres are exactly the same, so $A_l^n = A_l^q = A_l$). By substituting this correspondence and Eq. (8.3) into Eq. (8.4), the following correspondence can be obtained:

$$a_{l,m}^{n(q)} = \sum_{l'm'} A_{l'} \left(a_{l',m'}^{q(0)} + \sum_{q' \neq q} a_{l',m'}^{q(q')} \right) G_{l,m,l',m'}^{nq} \tag{8.7}$$

Assuming that the particles in the medium are sparsely distributed, that is, the distance between each particle is sufficiently far, satisfying $k_o r_{nq} >> 1$, then the scattering waves of other particles can be regarded as a small quantity. If higher-order small quantities are not considered, Eq. (8.7) can be replaced with a simpler equation:

$$a_{l,m}^{n(q)} = \sum_{l'm'} A_{l'} a_{l',m'}^{q(0)} G_{l,m,l',m'}^{nq} \tag{8.8}$$

We define the acoustic radiation force as the integral of the average radiation stress tensor of the particle within a period $\left\langle \overleftrightarrow{S} \right\rangle$ over the entire surface S:

$$F^n = \oiint_S \left\langle \overleftrightarrow{S} \right\rangle \mathrm{ds} \tag{8.9}$$

The average stress tensor can be represented as [23]:

$$\left\langle \overleftrightarrow{S} \right\rangle = -\left(\frac{\rho_0 |v|^2}{4} - \frac{|p|^2}{4\rho_0 c_0^2} \right) \overleftrightarrow{I} + \frac{1}{2} \rho_0 \mathrm{Re}(v^* v) \tag{8.10}$$

where $\overleftrightarrow{I}$ is the unit tensor. Equation (8.10) can also be expressed in terms of velocity potential:

$$\left\langle \overleftrightarrow{S} \right\rangle = \rho_0 \left[\frac{1}{4} \nabla^2 |\phi|^2 \overleftrightarrow{I} - \frac{1}{2} \mathrm{Re}(\nabla \phi \nabla \phi^*) \right] \tag{8.11}$$

If we do not consider the boundary conditions of the medium, that is, we regard the medium as infinitely large, and the incident sound wave is incident from the positive z direction, due to symmetry, the acoustic radiation force experienced by particle n can be regarded as a collection of forces in the x and y directions, that is, $F^n =$, substituting Eqs. (8.1) and (8.2) into Eq. (8.11), F_x^n, F_y^n, F_z^n can be represented as [17, 24]:

$$F_x^n + iF_y^n = \frac{i\rho_0}{4} \sum_{lm} \begin{bmatrix} \mu_{l+1,m-1} \left(2b_{l+1,m+1}^n b_{l,m}^{n*} + b_{l+1,m-1}^n a_{l,m}^{n*} + a_{l+1,m-1}^n b_{l,m}^{n*} \right) \\ +\mu_{l+1,-m-1} \left(2b_{l,m}^n b_{l+1,m+1}^{n*} + b_{l,m}^n a_{l+1,m+1}^{n*} + a_{l,m}^n b_{l+1,m+1}^{n*} \right) \end{bmatrix} \tag{8.12}$$

$$F_z^n = \frac{\rho_0}{2} Im \left[\sum_{lm} \sqrt{\frac{(l-m+1)(l+m+1)}{(2l+1)(2l+3)}} T_l a_{l,m}^n a_{l+1,m}^{n*} \right] \tag{8.13}$$

where: $\mu_{l,m} = \sqrt{\frac{[(l-m)(l-m-1)]}{[(2l-1)(2l+1)]}}$.

We first consider F_x^n and F_y^n, substituting $b_{lm}^n = A_l a_{lm}^n$ into Eq. (8.12) we get:

$$F_x^n + iF_y^n = \frac{i\rho_0}{4} \sum_{lm} \left[T_l \mu_{l+1,m-1} a_{l,m}^{n*} a_{l+1,m-1}^n + T_l^* \mu_{l+1,-m-1} a_{l,m}^n a_{l+1,m+1}^{n*} \right] \tag{8.14}$$

where: $T_l = 2A_{l+1}A_l^* + A_{l+1} + A_l^*$ represents the total scattering situation of particle n. Since only particles n and q exist, substituting $a_{l,m}^n = a_{l,m}^{n(0)} + a_{l,m}^{n(q)}$ into Eq. (8.14) we get:

$$\begin{aligned} a_{l,m}^{n*}a_{l+1,m-1}^{n} &= a_{l,m}^{n(0)*}a_{l+1,m-1}^{n(0)} + a_{l,m}^{n(0)*}a_{l+1,m-1}^{n(q)} + a_{l,m}^{n(q)*}a_{l+1,m-1}^{n(0)} + a_{l,m}^{n(q)*}a_{l+1,m-1}^{n(q)} \\ a_{l,m}^{n}a_{l+1,m+1}^{n*} &= a_{l,m}^{n(0)}a_{l+1,m+1}^{n(0)*} + a_{l,m}^{n(0)}a_{l+1,m+1}^{n(q)*} + a_{l,m}^{n(q)}a_{l+1,m+1}^{n(0)*} + a_{l,m}^{n(q)}a_{l+1,m+1}^{n(q)*} \end{aligned} \tag{8.15}$$

From Eq. (8.4) and the properties of the Kronecker Delta function, it is known that at least one of $a_{l,m}^{n(0)}a_{l+1,m-1}^{n(0)}$ must be zero, so $a_{l,m}^{n(0)*}a_{l+1,m-1}^{n(0)} = 0$ always holds true, similarly, $a_{l,m}^{n(0)}a_{l+1,m+1}^{n(0)*} = 0$. If $m \neq 0$, $a_{l,m}^{n} = a_{l,m}^{n(q)}$ is a small quantity relative to $a_{l,0}^{n}$, if m, $m \pm 1$ are not zero, then $a_{l,m}^{n*}a_{l+1,m\pm1}^{n}$ is a high-order small quantity and can be ignored in the calculation. For $a_{l,m}^{n(q)}$, from Eq. (8.8), $a_{l,m}^{n(q)} = \sum_{l'm'} A_{l'}a_{l',m'}^{q(0)}G_{l,m,l',m'}^{nq}$, where $a_{l',m'}^{q(0)}$ also satisfies Eq. (8.4), that is, only when $m' = 0$ can a non-zero solution be obtained, so Eq. (8.8) becomes:

$$a_{l,m}^{n(q)} = \sum_{l'} A_{l'}a_{l',0}^{q(0)}G_{l,m,l',0}^{nq} \tag{8.16}$$

Similarly, $a_{l,m}^{n(q)*}a_{l+1,m-1}^{n(q)}$, $a_{l,m}^{n(q)}a_{l+1,m+1}^{n(q)*}$ is a high-order small quantity and can be ignored in the calculation, so Eq. (8.15) becomes:

$$\begin{aligned} a_{l,m}^{n*}a_{l+1,m-1}^{n} &= \sum_{l'}\left(A_{l'}a_{l,m}^{n(0)*}a_{l',0}^{q(0)}G_{l+1,m-1,l',0}^{nq} + A_{l'}^{*}a_{l',0}^{q(0)*}a_{l+1,m-1}^{n(0)}G_{l,m,l',0}^{nq*}\right) \\ a_{l,m}^{n}a_{l+1,m+1}^{n*} &= \sum_{l'}\left(A_{l'}^{*}a_{l,m}^{n(0)}a_{l',0}^{q(0)*}G_{l+1,m+1,l',0}^{nq*} + A_{l'}a_{l',0}^{q(0)}a_{l+1,m+1}^{n(0)*}G_{l,m,l',0}^{nq}\right) \end{aligned} \tag{8.17}$$

After omitting all higher-order small quantities in Eq. (8.14), we obtain:

$$F_x^n + iF_y^n = \frac{i\rho_0}{4}\sum_l \left[\begin{array}{c} T_l\left(\mu_{l+1,-1}a_{l,0}^{n*}a_{l+1,-1}^{n} + \mu_{l+1,0}a_{l,1}^{n*}a_{l+1,0}^{n}\right) \\ +T_l^*\left(\mu_{l+1,-1}a_{l,0}^{n}a_{l+1,1}^{n*} + \mu_{l+1,0}a_{l,-1}^{n}a_{l+1,0}^{n*}\right) \end{array}\right] \tag{8.18}$$

Specifically, when $l = 0$, because $-l \le m \le l$, the value of m can only be 0.

Next, consider the value of $G_{l,m,l',m'}^{nq}$. Since we have already proven $m' = 0$, so $G_{l,m,l',m'}^{nq}$ becomes

$$\begin{aligned} G_{l,m,l',0} &= 4\pi\sum_{l''} i^{l+l''-l'}C_{l,m,l'',-m}^{l',0}h_{l''}(kr_{nq})Y_{l'',-m}(\theta_{nq},\varphi_{nq}) \\ C_{l,m,l'',-m}^{l',0} &= \int_0^{2\pi}\int_0^{\pi} Y_{l',0}(\theta,\varphi)Y_{l,m}^{*}(\theta,\varphi)Y_{l'',-m}^{*}(\theta,\varphi)d\theta d\varphi \end{aligned} \tag{8.19}$$

And since particles n, q are in the same xy plane, in spherical coordinates $\theta_{nq} = \frac{\pi}{2}$ always holds, φ_{nq} is related to the position of the particle, so:

$$Y_{l'',-m}(\hat{r}_{ij}) = \zeta_{l''} P_{l''}^{-m}(0) e^{-im\varphi_{nq}} \tag{8.20}$$

Obviously, Eq. (8.19) is very complex in calculations. To facilitate calculations, we make some approximations. For two distant particles n, q, i.e., $r_{nq} > 1$, the spherical Hankel function can be approximated as $h_{l''}(kr_{nq}) = \frac{(-i)^{l''} e^{ikr_{nq}}}{ikr_{nq}} + O\left(\frac{1}{r_{nq}}\right)$, at this time:

$$G_{l,m,l',0} = \frac{e^{ikr_{nq}}}{kr_{nq}} e^{-im\varphi_{nq}} 4\pi i^{l-l'-1} \sum_{l''} C_{l,m,l'',-m}^{l',0} Y_{l'',-m}\left(\frac{\pi}{2}, 0\right) = \frac{e^{ikr_{nq}}}{kr_{nq}} e^{-im\varphi_{nq}} M_{l,m,l',0} \tag{8.21}$$

Substituting $M_{l,m,l',0}$ into Eq. (8.17) yields:

$$a_{l,m}^{n*} a_{l+1,m-1}^{n} = \frac{1}{kr_{nq}} U_{l,m},\ a_{l,m}^{n} a_{l+1,m+1}^{n*} = \frac{1}{kr_{nq}} V_{l,m} \tag{8.22}$$

where:

$$\begin{aligned} U_{l,m} &= \sum_{l'} \begin{pmatrix} A_{l'} a_{l,m}^{n(0)*} a_{l',0}^{q(0)} e^{-i(m-1)\varphi_{nq}} e^{ikr_{nq}} M_{l+1,m-1,l',0} \\ +A_{l'}^{*} a_{l',0}^{q(0)*} a_{l+1,m-1}^{n(0)} e^{im\varphi_{nq}} e^{-ikr_{nq}} M_{l,m,l',0}^{*} \end{pmatrix} \\ V_{l,m} &= \sum_{l'} \begin{pmatrix} A_{l'}^{*} a_{l,m}^{n(0)} a_{l',0}^{q(0)*} e^{i(m+1)\varphi_{nq}} e^{-ikr_{nq}} M_{l+1,m+1,l',0}^{*} \\ +A_{l'} a_{l',0}^{q(0)} a_{l+1,m+1}^{n(0)*} e^{-im\varphi_{nq}} e^{ikr_{nq}} M_{l,m,l',0} \end{pmatrix} \end{aligned} \tag{8.23}$$

Substituting (8.23) into (8.18), we get:

$$F_x^n + iF_y^n = \frac{1}{kr_{nq}} \frac{i\rho_0}{4} \sum_{l} \mu_{l+1,-1}(T_l U_{l,0} + T_l^{*} V_{l,0}) + \mu_{l+1,0}(T_l U_{l,1} + T_l^{*} V_{l,-1}) \tag{8.24}$$

Because $M_{l,m,l',0} e^{-im\varphi_{nq}} = 4\pi i^{l-l'-1} \sum_{l''} C_{l,m,l'',-m}^{l',0} Y_{l'',-m}(\theta_{nq}, \varphi_{nq})$, it can be proven through calculation that $M_{l+1,-1,l',0} e^{i\varphi_{nq}} = -M_{l+1,1,l',0} e^{-i\varphi_{nq}}$, therefore:

$$\begin{aligned} T_l U_{l,0} + T_l^{*} V_{l,0} &= \sum_{l'} \begin{bmatrix} T_l A_{l'} a_{l,0}^{n(0)*} a_{l',0}^{n(0)} e^{ikr_{nq}} M_{l+1,-1,l',0} \\ -\left(T_l A_{l'} a_{l,0}^{n(0)*} a_{l',0}^{n(0)} e^{ikr_{nq}} M_{l+1,-1,l',0}\right)^{*} \end{bmatrix} e^{i\varphi_{nq}} \\ &= \sum_{l'} 2i\,\mathrm{Im}\left(T_l A_{l'} a_{l,0}^{n(0)*} a_{l',0}^{n(0)} e^{ikr_{nq}} M_{l+1,-1,l',0}\right) e^{i\varphi_{nq}} \end{aligned} \tag{8.25}$$

$$T_l U_{l,1} + T_l^* V_{l,-1} = \sum_{l'} \begin{bmatrix} T_l A_{l'} a_{l,0}^{n(0)*} a_{l+1,0}^{n(0)} e^{-ikr_{nq}} M_{l,1,l',0}^* \\ -\left(T_l A_{l'} a_{l,0}^{n(0)*} a_{l+1,0}^{n(0)} e^{-ikr_{nq}} M_{l,1,l',0}^*\right)^* \end{bmatrix} e^{i\varphi_{nq}}$$
$$= \sum_{l'} 2i\mathrm{Im}\left(T_l A_{l'} a_{l,0}^{n(0)*} a_{l+1,0}^{n(0)} e^{-ikr_{nq}} M_{l,1,l',0}^*\right) e^{i\varphi_{nq}} \tag{8.26}$$

So:

$$F_x^n + iF_y^n = \frac{H_{nq}}{kr_{nq}} e^{i\varphi_{nq}} \tag{8.27}$$

where:

$$Z_{nq} = H_{nq} e^{i\varphi_{nq}} = \frac{i\rho_0}{4} \sum_l \left[\mu_{l+1,-1}\left(T_l U_{l,0} + T_l^* V_{l,0}\right) + \mu_{l+1,0}\left(T_l U_{l,1} + T_l^* V_{l,-1}\right)\right] \tag{8.28}$$

Specifically, because m must satisfy $-l \leq m \leq l$, when $l = 0$, the above equation only has $\mu_{l+1,-1}$ this term.

For Eq. (8.27), we can see that the real part of this equation is the component of the inter-particle interaction force in the x direction, and the imaginary part is the component in the y direction. The relationship between the two is determined by φ_{nq}, if expressed in trigonometric functions:

$$F_x^n = \frac{H_{nq}}{kr_{nq}} \cos \varphi_{nq}$$
$$F_y^n = \frac{H_{nq}}{kr_{nq}} \sin \varphi_{nq} \tag{8.29}$$

When the two particles are in the same yz plane, $\varphi_{nq} = \frac{\pi}{2}$, at this time there is no interaction force in the x direction. In fact, due to the symmetry between the particle and the sound wave in this case, it is obvious that the interaction force in the x direction is 0.

From Eq. (8.28), we can see that the value of Z_{nq} is independent of the distance between the two particles and only related to the properties of the particle and the external medium. Therefore, for H_{nq} and φ_{nq}, they are also independent of the distance between the two particles.

Next, we consider the change of F_z^n, that is, Eq. (8.13). According to Eq. (8.17), we can directly get:

$$a_{l,m}^n a_{l+1,m}^{n*} = \sum_{l'} \left(A_{l'}^* a_{l,m}^{n(0)} a_{l',0}^{n(0)*} G_{l+1,m,l',0}^{nq*} + A_{l'} a_{l',0}^{n(0)} a_{l+1,m}^{n(0)*} G_{l,m,l',0}^{nq}\right) \tag{8.30}$$

Considering the properties of Eq. (8.14) and the Kronecker delta function, in the above equation, only when m = 0 does a non-zero solution exist. Therefore,

Eqs. (8.13) and (8.30) become

$$F_z^n = \frac{\rho_0}{2} Im\left[\sum_l \sqrt{\frac{(l+1)(l+1)}{(2l+1)(2l+3)}} T_l a_{l,0}^n a_{l+1,0}^{n*}\right]$$
$$a_{l,0}^n a_{l+1,0}^{n*} = \sum_{l'} \left(A_{l'}^* a_{l,0}^{n(0)} a_{l',0}^{n(0)*} G_{l+1,0,l',0}^{nq*} + A_{l'} a_{l',0}^{n(0)} a_{l+1,0}^{n(0)*} G_{l,0,l',0}^{nq}\right) \tag{8.31}$$

Substituting the simplification of Eq. (8.21) into the above equation, we can find that the component of the force acting on the particle in the z direction can be expressed as:

$$F_z^n = \frac{1}{kr_{nq}} \frac{\rho_0}{2} Im\left[\sum_l e^{ikr_{nq}} \sqrt{\frac{(l+1)(l+1)}{(2l+1)(2l+3)}} T_l \sum_{l'} \begin{pmatrix} A_{l'}^* a_{l,0}^{n(0)} a_{l',0}^{n(0)*} M_{l+1,0,l',0}^{nq*} \\ +A_{l'} a_{l',0}^{n(0)} a_{l+1,0}^{n(0)*} M_{l,0,l',0}^{nq} \end{pmatrix}\right] \tag{8.32}$$

Equations (8.27) and (8.32) are the simplified algorithms for the interaction force between two distant particles located on the same wavefront. Next, we will study the changes in the interaction force between particles using the example of water-immersed polystyrene particles.

8.2 Acoustic Radiation Force Between Water-Immersed Polystyrene Particles

We study the interparticle forces using water-immersed polystyrene particles as an example. The physical properties of water-immersed polystyrene are as follows: the density of the external medium, i.e., water, is $\rho_0 = 1000\text{kg/m}^3$, the speed of sound in water is $c_0 = 1490\text{m/s}$, the density of the polystyrene material is $\rho_1 = 1050\text{kg/m}^3$, the longitudinal speed of sound in the material is $v_L = 2400\text{m/s}$, and the transverse speed of sound is $v_T = 1150\text{m/s}$.

By calculating Eq. (8.1), we obtained the formulas (8.27) and (8.32) for the long-distance interparticle forces. These formulas show that the magnitude of the interparticle forces is related to the external sound field and the properties of the particles themselves. We assume that the external sound field is uniform and constant, and the velocity potential ϕ_0 imposed on the particles is constant. To better study the trend of the force with the change in interparticle distance and to eliminate the influence of ϕ_0, we can define a basic force $F_0 = E_0 S_0$ generated by the sound field, where $E_0 = \rho_0 k^2 \phi_0^2/2$ is the energy density of the plane wave, $S_0 = \pi a^2/4$ is the cross-sectional area of the spherical particle in the sound field, and a is the particle radius. Therefore, the normalized interparticle force between two particles can be expressed as

$$F_{xy} = \frac{1}{F_0} \frac{H_{nq}}{2\pi (d/\lambda)}$$
$$F_z = \frac{F_z^n}{F_0} \tag{8.33}$$

where d is the distance between the two particles, and λ is the wavelength of the sound wave in the external sound field. According to Eq. (8.33), we can obtain the change in the interparticle forces with the change in interparticle distance. Here, we calculate the first two orders, i.e., l = 0,1, as shown in Fig. 8.2, which are schematic diagrams of the change in normalized interparticle forces between four different sizes of particles.

From the figure, we can see that, under a certain external sound field, as the interparticle distance increases, the interparticle forces between them generally decrease and show a sinusoidal pattern, fluctuating around $F = 0$, with the magnitude and direction of the force undergoing periodic changes. Obviously, because of the property of F_N, we can rely on acoustic radiation interaction to capture or move particles

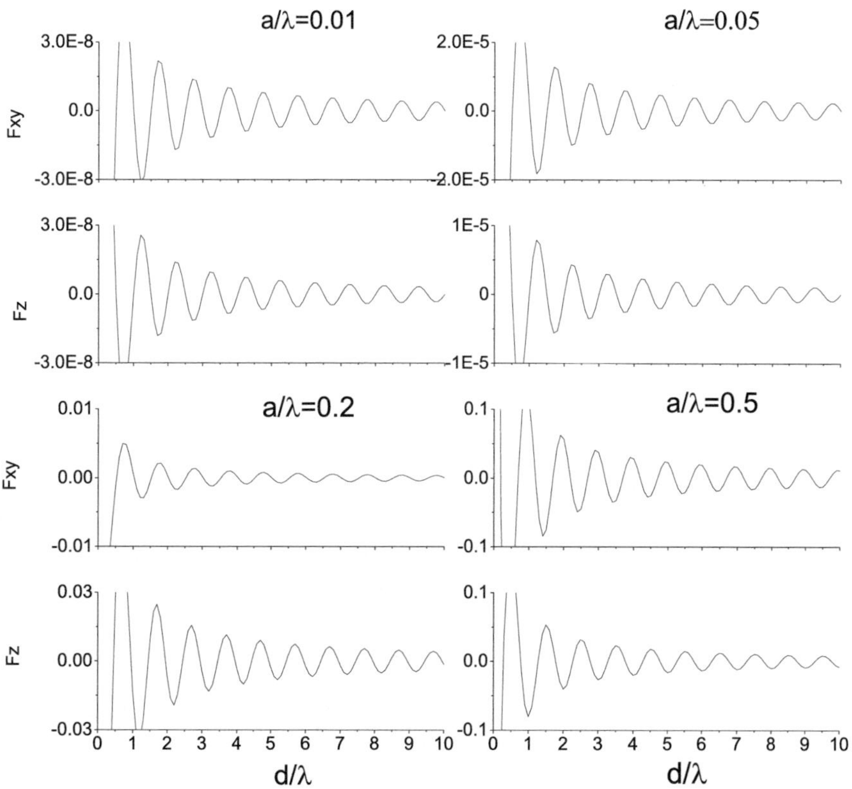

Fig. 8.2 Schematic diagram of interparticle forces between four different sizes of water-immersed polystyrene particles

in the medium. From Fig. 8.2, it can be seen that for the component of the interparticle forces that the particles experience in the xy plane, F_{xy}, under a certain external sound field, as the particle radius increases, the magnitude of F_{xy} significantly increases, but its change trend with the increase in interparticle distance d is exactly the same, similar to multiple sinusoidal curves with different amplitudes but the same phase.

Due to the approximation calculation in the previous process, i.e., the approximate substitution of the Hankel function $h_l(x)$ when x > 1, the results obtained here only fit well with the actual situation when the distance between particles is relatively far, and there is a large error when the particles are very close.

8.3 Interparticle Forces Among Multiple Particles Located on the Same Wavefront in a Medium

If there are not just two particles in the medium, but three or more independent particles, then for any one particle n, it is simultaneously subjected to the forces from all other particles. Since at long distances, the scattered sound wave is very small compared to the original sound wave in the sound field, we do not need to consider the secondary scattering caused by the scattered sound wave, then Eq. (8.33) becomes:

$$\begin{aligned} F_{xy} &= \sum_{q \neq n} \frac{1}{F_0} \frac{H_{nq}}{2\pi\left(\frac{d}{\lambda}\right)} \\ F_z &= \sum_{q \neq n} \frac{F_z^n}{F_0} \end{aligned} \tag{8.34}$$

From the above two equations, it can be seen that when the interparticle distance is relatively far (relative to the sound wave wavelength λ), the total force that any one particle receives from other particles can be regarded as a simple superposition of the forces that the remaining particles individually exert on it. That is to say, the multi-particle system in the medium can be regarded as a superposition of multiple two-particle systems. For example, three spherical particles arranged in an equilateral triangle in the medium, as shown in Fig. 8.3a. Assuming that the three spheres are located in the same xy plane, that is, they are in the same wavefront in the sound field, and the positive direction of the z-axis is perpendicular to the paper surface. The interparticle forces between the three spheres can be calculated separately using Eq. (8.34). However, it should be noted that although the distances d between the three spheres are the same, the two-particle systems formed by any two particles have different φ in the spherical coordinate system. Therefore, for each particle, although the magnitude of the interaction force in the xy plane is the same, there are differences in the specific x and y components.

Since the z-component of the interaction force between the spheres is related only to the distance between the spheres and not to their position in the xy plane, the

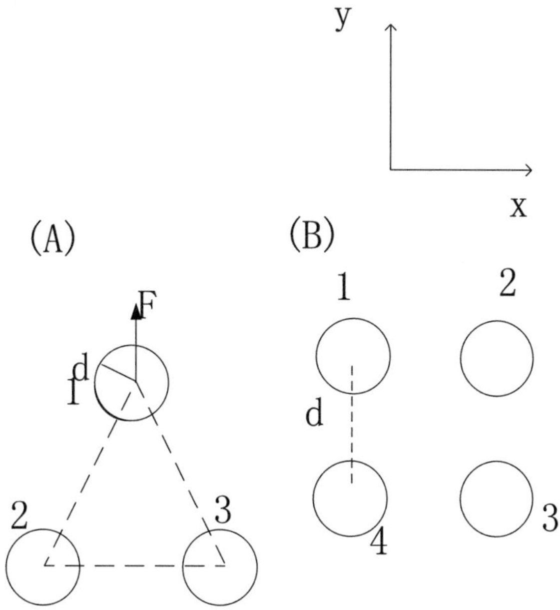

Fig. 8.3 Interaction forces between multiple small spheres on the same wavefront

z-direction interaction force experienced by the three spheres is:

$$F'_z = 2F_z(d/\lambda) \tag{8.35}$$

For the force components in the xy plane experienced by the spheres, F'_{xy}, we need to consider each sphere separately. Taking sphere 1 as an example, in the spherical coordinate system, we know that $\varphi_{12} = \frac{5\pi}{3}, \varphi_{13} = \frac{4\pi}{3}$, where φ_{12} represents the angular coordinate of the line connecting sphere 1 to 2 in the spherical coordinate system φ. According to Eq. (8.29), the x and y components of the force experienced by sphere 1 are:

$$\begin{aligned} F'_{1,x} &= F_{xy}\left(\frac{d}{\lambda}\right)\cos\varphi_{12} + F_{xy}\left(\frac{d}{\lambda}\right)\cos\varphi_{13} = 0 \\ F'_{1,y} &= F_{xy}(d/\lambda)\sin\varphi_{12} + F_{xy}(d/\lambda)\sin\varphi_{13} = -\sqrt{3}F_{xy}(d/\lambda) \end{aligned} \tag{8.36}$$

Similarly, for spheres 2 and 3, we can also obtain the components of the interaction forces they experience in the x and y directions.

$$\begin{aligned} F'_{2,x} &= \frac{3}{2}F_{xy}\left(\frac{d}{\lambda}\right), F'_{2,y} = \frac{\sqrt{3}}{2}F_{xy}\left(\frac{d}{\lambda}\right) \\ F'_{3,x} &= -\frac{3}{2}F_{xy}(d/\lambda), F'_{3,y} = \frac{\sqrt{3}}{2}F_{xy}(d/\lambda) \end{aligned} \tag{8.37}$$

For the four particle spheres forming a square with side length d as shown in Fig. 8.3b, we can also obtain the total force experienced by each particle sphere based on the interaction forces between them, but here the distances between the particle spheres are not all d.

$$\begin{aligned}
F_z' &= 2F_z(d/\lambda) + F_z\left(\sqrt{2}d/\lambda\right) \\
F_{1,x}' &= -F_{3,x}' = F_{xy}\left(\frac{d}{\lambda}\right) + \frac{\sqrt{2}}{2}F_{xy}\left(\frac{\sqrt{2}d}{\lambda}\right) \\
F_{1,y}' &= -F_{3,y}' = -F_{xy}\left(\frac{d}{\lambda}\right) - \frac{\sqrt{2}}{2}F_{xy}\left(\frac{\sqrt{2}d}{\lambda}\right) \\
F_{2,x}' &= -F_{4,x}' = -F_{xy}(d/\lambda) - \frac{\sqrt{2}}{2}F_{xy}\left(\sqrt{2}d/\lambda\right) \\
F_{2,y}' &= -F_{4,y}' = -F_{xy}(d/\lambda) - \frac{\sqrt{2}}{2}F_{xy}\left(\sqrt{2}d/\lambda\right)
\end{aligned} \tag{8.38}$$

By extension, we can easily express the total force experienced by the spheres under various arrangements, simply by adding Eqs. (8.29) and (8.33) according to the actual situation.

8.4 Interaction Forces Between Particles on Different Wavefronts

In the previous three sections, we discussed the interaction forces between particles on the same wavefront, which in the Cartesian coordinate system means that all particles have the same z-coordinate. However, due to limitations of materials, equipment, etc., this is difficult to achieve in practice. Therefore, to obtain a more generally applicable calculation formula, we need to study the interaction forces between particles on different wavefronts.

Assume two particles n and q are located in a medium, their distance is d, and the difference in z-coordinates is d_z, as shown in Fig. 8.4. In this case, according to Eq. (8.4), $a_{l,m}^{n(0)}$, $a_{l,m}^{q(0)}$ can be expressed as:

$$\begin{aligned}
a_{lm}^{n(0)} &= \phi_0 i^l \sqrt{4\pi(2l+1)} e^{-ik_0 z_n} \delta_{m,0} \\
a_{lm}^{q(0)} &= \phi_0 i^l \sqrt{4\pi(2l+1)} e^{-ik_0(z_n+d_z)} \delta_{m,0}
\end{aligned} \tag{8.39}$$

The subsequent calculation process is the same as Eqs. (8.12)–(8.23), here we first consider the changes in the x and y components. Due to the existence of d_z, Eq. (8.17) changes to:

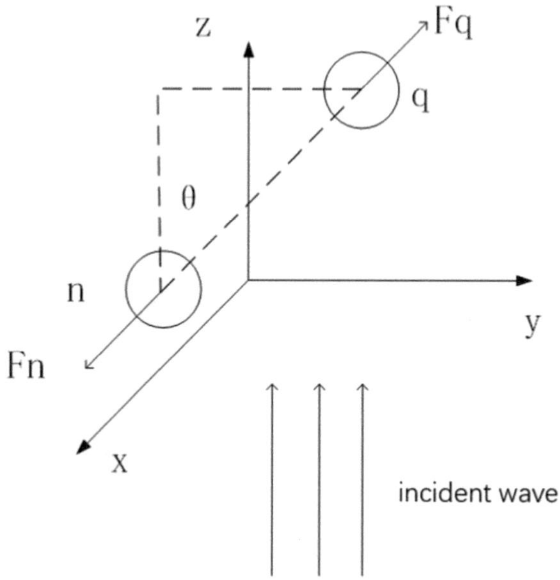

Fig. 8.4 Schematic diagram of two spherical particles located at different wavefronts

$$\begin{aligned} a_{l,m}^{n*}a_{l+1,m-1}^{n} &= \sum_{l'}\begin{pmatrix} A_{l'}a_{l,m}^{n(0)*}a_{l',0}^{n(0)}e^{-ik_0d_z}G_{l+1,m-1,l',0}^{nq} \\ +A_{l'}^{*}a_{l',0}^{n(0)*}a_{l+1,m-1}^{n(0)}e^{ik_0d_z}G_{l,m,l',0}^{nq*} \end{pmatrix} \\ a_{l,m}^{n}a_{l+1,m+1}^{n*} &= \sum_{l'}\begin{pmatrix} A_{l'}^{*}a_{l,m}^{n(0)}a_{l',0}^{n(0)*}e^{ik_0d_z}G_{l+1,m+1,l',0}^{nq*} \\ +A_{l'}a_{l',0}^{n(0)}a_{l+1,m+1}^{n(0)*}e^{-ik_0d_z}G_{l,m,l',0}^{nq} \end{pmatrix} \end{aligned} \tag{8.40}$$

Here, since $a_{l,m}^{n(0)}$ is only related to the two variables l and m and the position of the particles, and not to the physical properties of the particles themselves, $a_{l,m}^{q(0)}$ can also be written as $a_{l,m}^{n(0)}e^{-ik_0d_z}$, which does not affect the result. Transforming the Cartesian coordinate system to the spherical coordinate system, we get $d_z = d\cos\theta_{nq}$. At this time, $U_{l,m}$, $V_{l,m}$ in Eq. (8.23) is determined not only by l, m, but also related to d, θ_{nq}, so Eq. (8.23) can also be changed to

$$\begin{aligned} U'_{l,m,d,\theta_{nq}} &= \sum_{l'}\begin{pmatrix} A_{l'}a_{l,m}^{n(0)^*}a_{l',0}^{n(0)}e^{-ik_0d\cos\theta_{nq}}e^{-i(m-1)\varphi_{nq}}e^{ikr_{nq}}M_{l+1,m-1,l',0} \\ +A_{l'}^{*}a_{l',0}^{n(0)*}a_{l+1,m-1}^{n(0)}e^{ik_0d\cos\theta_{nq}}e^{im\varphi_{nq}}e^{-ikr_{nq}}M_{l,m,l',0}^{*} \end{pmatrix} \\ V'_{l,m,d,\theta_{nq}} &= \sum_{l'}\begin{pmatrix} A_{l'}^{*}a_{l,m}^{n(0)}a_{l',0}^{n(0)*}e^{ik_0d\cos\theta_{nq}}e^{i(m+1)\varphi_{nq}}e^{-ikr_{nq}}M_{l+1,m+1,l',0}^{*} \\ +A_{l'}a_{l',0}^{n(0)}a_{l+1,m+1}^{n(0)*}e^{-ik_0d\cos\theta_{nq}}e^{-im\varphi_{nq}}e^{ikr_{nq}}M_{l,m,l',0} \end{pmatrix} \end{aligned} \tag{8.41}$$

It should be noted that at this time θ_{nq} is no longer a constant $\frac{\pi}{2}$ but a variable, so $P_{l''}^{-m}(\cos\theta_{nq})$ in Eq. (8.20) is no longer $P_{l''}^{-m}(0)$. Substituting Eq. (8.41) into Eqs. (8.25) and (8.26) gives the interaction forces in the xy plane between two particles not on the same wavefront:

$$F_x^n + iF_y^n = \frac{H_{nq,\theta}}{kr_{nq}} e^{i\varphi_{nq}} \tag{8.42}$$

$$H_{nq,\theta} e^{i\varphi_{nq}} = \frac{i\rho_0}{4} \sum_l \left(\begin{array}{c} \mu_{l+1,-1}\left(T_l U'_{l,0,d,\theta_{nq}} + T_l^* V'_{l,0,d,\theta_{nq}}\right) \\ +\mu_{l+1,0}\left(T_l U'_{l,1,d,\theta_{nq}} + T_l^* V'_{l,-1,d,\theta_{nq}}\right) \end{array} \right) \tag{8.43}$$

After simplification, we can get:

$$H_{nq} = -\frac{\rho_0}{2} \sum_l \sum_{l'} \left(\begin{array}{c} \mu_{l+1,-1} Im\left(T_l A_{l'} e^{-ik_0 d \cos\theta_{nq}} a_{l,0}^{n(0)^*} a_{l',0}^{n(0)} e^{ikr_{nq}} M_{l+1,-1,l',0}\right) \\ +\mu_{l+1,0} Im\left(T_l A'_{l'} e^{-ik_0 d \cos\theta_{nq}} a_{l',0}^{n(0)^*} a_{l+1,0}^{n(0)} e^{ikr_{nq}} M_{l,1,l',0}^*\right) \end{array} \right) \tag{8.44}$$

Specifically, when $l = 0$, since m must satisfy $-l \leq m \leq l$, in the above equation, only the term $\mu_{l+1,-1}$ exists.

Next, consider the z-component of the inter-particle interaction force. As before, this component can be expressed as:

$$F_z^n = \frac{\rho_0}{2} Im\left[\sum_{lm} \sqrt{\frac{(l-m+1)(l+m+1)}{(2l+1)(2l+3)}} T_l a_{l,m}^n a_{l+1,m}^{n*} \right] \tag{8.45}$$

Simplify the above equation according to Eqs. (8.32) and (8.39):

$$F_z^n = \frac{1}{kr_{nq}} \frac{\rho_0}{2} Im\left[e^{ikr_{nq}} \sum_l \sqrt{\frac{(l+1)(l+1)}{(2l+1)(2l+3)}} T_l \sum_{l'} \left(\begin{array}{c} A_{l'}^* a_{l,0}^{n(0)} a_{l',0}^{n(0)*} e^{ik_0 d_z} M_{l+1,0,l',0}^{nq*} \\ +A_{l'} a_{l',0}^{n(0)} a_{l+1,0}^{n(0)*} e^{-ik_0 d_z} M_{l,0,l',0}^{nq} \end{array} \right) \right] \tag{8.46}$$

Similarly, in this case, the normalized components of the inter-particle interaction force in the xy-plane and the z-direction F_{xy}, F_z can be represented by Eq. (8.47)

$$\begin{aligned} F_{xy} &= \frac{1}{F_0} \frac{H_{nq}}{2\pi (d/\lambda)} \\ F_z &= \frac{F_z^n}{F_0} \end{aligned} \tag{8.47}$$

Assume as shown in Fig. 8.4, there are two spheres with a distance of d, the ratio of the sphere radius to the wavelength of the sound wave is 0.5, in the spherical coordinate system, the angle of the line connecting the two spheres is $(\theta_{nq}, \varphi_{nq})$, then according to Eq. (8.47), the normalized interaction forces F_{xy} and F_z that the spheres experience at different values of θ_{nq} can be calculated.

In Fig. 8.5, the situation of $\theta = \frac{\pi}{2}$ is the interaction force between particles on the same wavefront in Sect. 8.2, compared with the situation of $a/\lambda = 0.5$ in Fig. 8.2, it can be seen that they are indeed exactly the same. And $\theta = 0$ indicates that the two

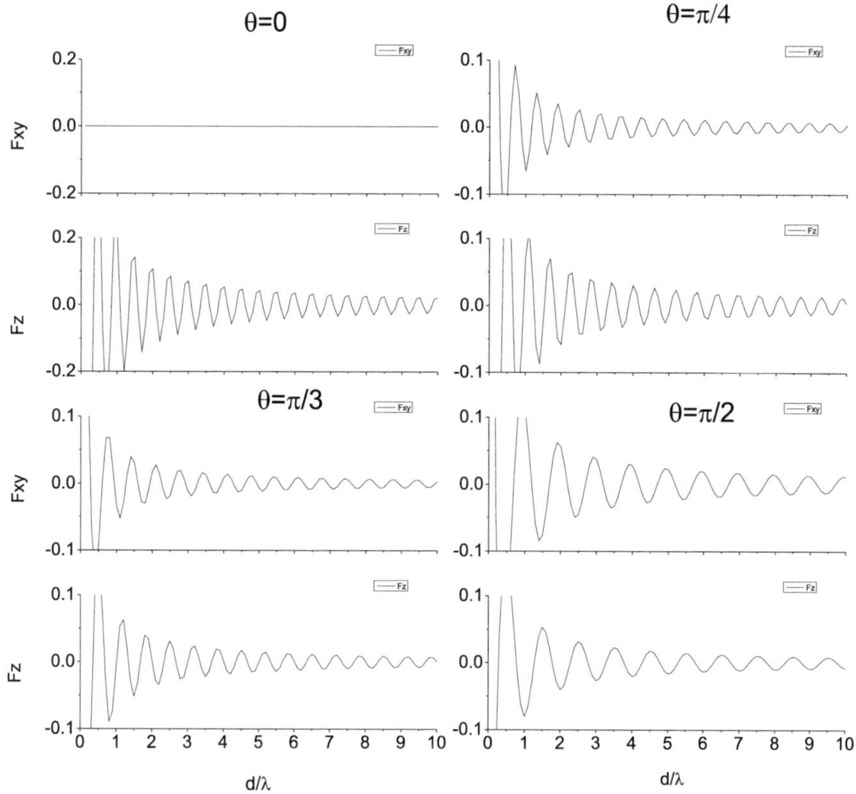

Fig. 8.5 Normalized interaction force at $\theta_{nq} = 0, \frac{\pi}{4}, \frac{\pi}{3}, \frac{\pi}{2}$

particles are on the same vertical line, and there is no interaction force component in the xy-plane between the particles.

The above calculation is the interaction force for particle n. Next, we take another particle q as the research object. For q, the calculation formula is still Eq. (8.33), but in this case, the distance between the two particles is still d, in the spherical coordinate system, the angle between the line connecting q and n is $\theta_{qn} = \pi - \theta_{nq}$, similarly, we make the corresponding $\theta_{qn} = \pi, \frac{3\pi}{4}, \frac{2\pi}{3}, \frac{\pi}{2}$, the schematic diagram of the interaction force q received is shown in Fig. 8.6

The previously calculated F_{xy} is the total force component in the xy-plane received by the particle, and the specific x, y direction components still need to consider the influence of φ in the spherical coordinates. Taking the above two particles as an example, when calculating the force received by particle n, φ is φ_{nq}, and when calculating the force received by q, φ is $\varphi_{qn} = (2\pi - \varphi_{nq})$, assuming that particles n and q are in the same y, z plane, then $\varphi_{nq} = \frac{\pi}{2}$, $\varphi_{qn} = \frac{3\pi}{2}$, substituting into Eq. (8.29), it can be known that both particles are not affected by the force in the x direction, and in the y direction, the forces they receive are in opposite directions.

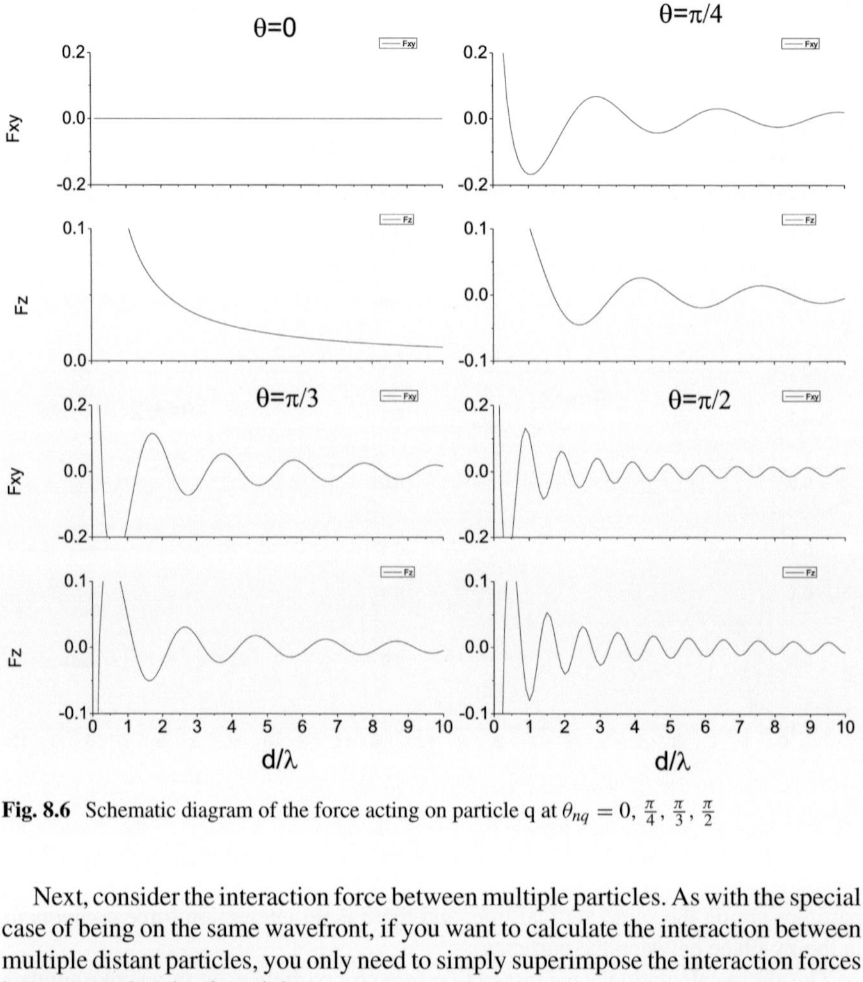

Fig. 8.6 Schematic diagram of the force acting on particle q at $\theta_{nq} = 0, \frac{\pi}{4}, \frac{\pi}{3}, \frac{\pi}{2}$

Next, consider the interaction force between multiple particles. As with the special case of being on the same wavefront, if you want to calculate the interaction between multiple distant particles, you only need to simply superimpose the interaction forces between each pair of particles.

For example, consider three particles in the medium that form an equilateral triangle with a side length of d. Assume the three particles are 1, 2, 3, where 2 and 3 are on the same wavefront, and particle 1 is above particles 2 and 3, as shown in Fig. 8.7, and the three particles are in the same yz plane, that is, $\varphi_{nq} = \frac{\pi}{2}$ or $\frac{3\pi}{2}$, due to symmetry, the x-direction interaction force received by the particles is 0.

For particle 1, the z-direction interaction force it receives can be represented as the superposition of the interaction forces produced by 2 and 3 on it. In this case $\theta_{12} = \theta_{23} = \frac{5}{6}\pi$, θ_{12} represents the angle between the line from particle 1 to 2 and the positive direction of the z-axis, the range is $[0, \pi]$, θ_{23} similarly, so the z-direction force received by particle 1 is

$$F_{1,z} = 2F_z\left(d/\lambda, \frac{5\pi}{6}\right) \tag{8.48}$$

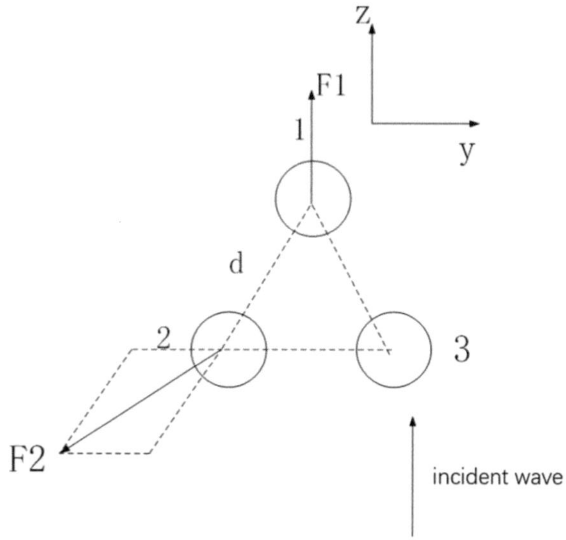

Fig. 8.7 Three-particle system on different wavefronts

The y-direction force received by particle 1 can be represented as

$$F_{1,y} = F_{xy}(d/\lambda)(\sin\varphi_{12} + \sin\varphi_{13}) \tag{8.49}$$

Given $\varphi_{12} = -\frac{3}{2}\pi$, $\varphi_{13} = \frac{1}{2}\pi$, it can be calculated that $F_{1,y} = 0$. For particle 2, $\theta_{21} = \frac{\pi}{6}$, $\theta_{23} = \frac{\pi}{2}$, $\varphi_{21} = \varphi_{23} = \frac{\pi}{2}$, so the interaction force experienced by particle 2 is

$$\begin{aligned} F_{2,z} &= F_z\left(\frac{d}{\lambda}, \frac{\pi}{6}\right) + F_z\left(\frac{d}{\lambda}, \frac{\pi}{2}\right) \\ F_{2,y} &= F_{xy}\left(d/\lambda, \frac{\pi}{6}\right) + F_{xy}\left(d/\lambda, \frac{\pi}{2}\right) \end{aligned} \tag{8.50}$$

Similarly, the interaction force experienced by particle 3 is

$$\begin{aligned} F_{3,z} &= F_z\left(\frac{d}{\lambda}, \frac{\pi}{6}\right) + F_z\left(\frac{d}{\lambda}, \frac{\pi}{2}\right) \\ F_{3,y} &= -F_{xy}\left(d/\lambda, \frac{\pi}{6}\right) - F_{xy}\left(d/\lambda, \frac{\pi}{2}\right) \end{aligned} \tag{8.51}$$

Since the formulas for calculating the interaction forces F_{xy} and F_z under various conditions have been given previously, it is easy to obtain the interaction forces experienced by each particle. As shown in Fig. 8.8, the ratio of particle radius to sound wavelength is 0.5.

As in Sect. 8.3, if the number of particles continues to increase, the expression for the interaction force will become more complex. However, as long as the distance

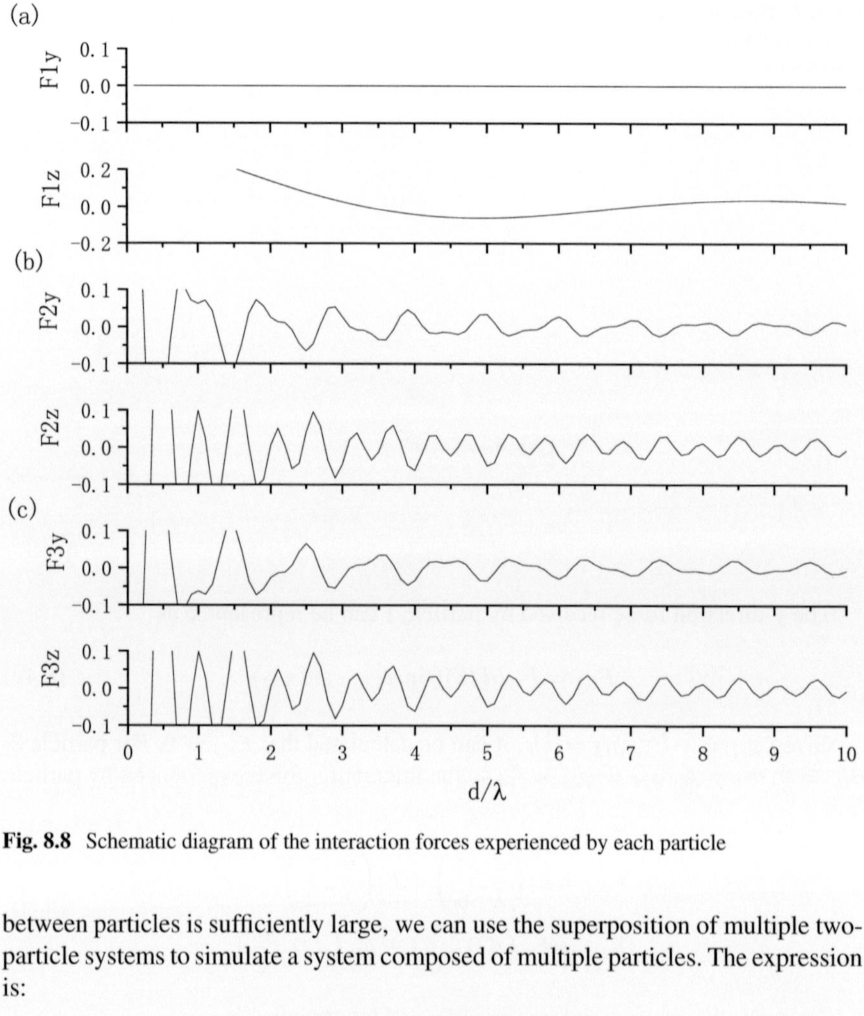

Fig. 8.8 Schematic diagram of the interaction forces experienced by each particle

between particles is sufficiently large, we can use the superposition of multiple two-particle systems to simulate a system composed of multiple particles. The expression is:

$$F_x = \sum_{q \neq n} \frac{1}{F_0} \frac{H_{nq}}{2\pi (d/\lambda)} \cos \varphi_{nq}$$

$$F_y = \sum_{q \neq n} \frac{1}{F_0} \frac{H_{nq}}{2\pi \left(\frac{d}{\lambda}\right)} \sin \varphi_{nq}$$

$$F_z = \sum_{q \neq n} \frac{F_z^n}{F_0} \tag{8.52}$$

8.5 Acoustic Radiation Interaction Force Between Sparse Particle Spheres in Gaussian Wave

As shown in Fig. 8.9, a Gaussian sound wave propagates in a fluid medium. There are multiple spherical particles in the sound field. Assuming these particle spheres are identical, the direction of the incident sound wave is z positive direction. Taking two particles p and q as examples, r_{pq} represents the distance between them, and $\hat{r}_{pq}$ represents the direction vector from p to q. The coordinate difference of the two particles z is dz. For any one of the particles p, the velocity potentials of the incident wave and scattered wave can be represented as:

$$\phi_{i,p} = \phi_0 e^{-i\omega t} \sum_{nm} a^p_{n,m} g_n j_n(k_0 r_p) Y_{n,m}(\theta_p, \varphi_p) \tag{8.53}$$

$$\phi_{s,p} = \phi_0 e^{-i\omega t} \sum_{nm} b^p_{n,m} g_n h^{(1)}_n(k_0 r_p) Y_{n,m}(\theta_p, \varphi_p) \tag{8.54}$$

where, $Y_{n,m}(\theta, \varphi)$ is the normalized spherical harmonic function which can be represented by the Legendre function $Y_{n,m}(\theta, \varphi) = \zeta_n P^m_n(\cos\theta) e^{im\varphi}$, where $\zeta_n = (-1)^m \sqrt{\frac{[(2n+1)(n-m)!]}{[4\pi(n+m)!]}}$, $P^m_n(\cos\theta)$ is the n order associated Legendre function, a^p_{nm} is the coefficient of the incident velocity potential of particle p, which consists of two parts. The first part is generated by the incident Gaussian wave itself, and the second part is generated by the scattered wave produced by all other particles in space except p at p, hence:

$$a^p_{nm} = a^{p(0)}_{nm} + \sum_{q \neq p} a^{p(q)}_{nm} \tag{8.55}$$

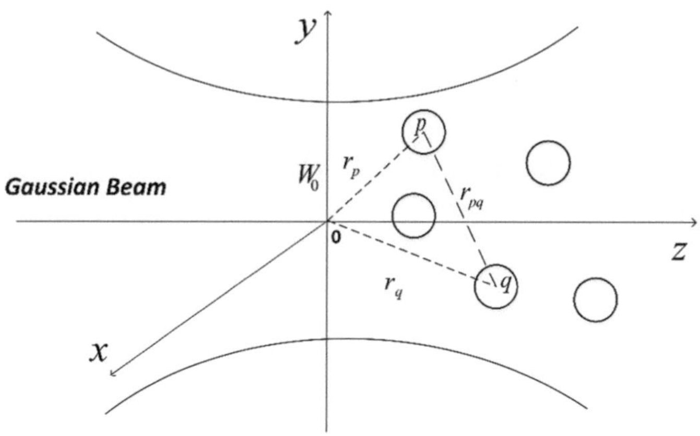

Fig. 8.9 Schematic diagram of sparse multi-particle spheres in a Gaussian wave field

For simplicity, assume that there are only two particle spheres p and q in the sound field. For the incident coefficient $a_{nm}^{p(0)}$ This part is only related to the external sound field and the particle p itself, and this part of the coefficient can be expressed as [21]:

$$a_{nm}^{p(0)} = \phi_0 i^n \sqrt{4\pi(2n+1)} e^{-ik_0 z_p} \delta_{m,0} \tag{8.56}$$

where, z_p is the z-coordinate of the particle p, $\delta_{m,m'}$ is the Kronecker delta function, satisfying $\delta_{m,m'} = 0, m \neq m'$ or $\delta_{m,m'} = 1, m = m'$, therefore, only when $m = 0$ $a_{n,m}^{p(0)}$ can obtain a non-zero value. Similarly, for particle q, there is a similar conclusion.

The second part is the velocity potential generated by scattering, as known from [13], $a_{nm}^{p(q)}$ can be obtained from the addition theorem of spherical functions [22]:

$$a_{nm}^{p(q)} = \sum_{n'm'} b_{n'm'}^{q} G_{nmn'm'}^{pq} \tag{8.57}$$

where,

$$G_{nmn'm'}^{pq} = 4\pi \sum_{n''} i^{n+n''-n'} C_{nmn''(m'-m)}^{n'm'} h_{n''}(k_0 r_{pq}) Y_{n'',m'-m}(\hat{r}_{pq})$$

$$C_{nmn''(m'-m)}^{n'm'} = \int_0^{2\pi} \int_0^{\pi} Y_{n',m'}(\theta, \varphi) Y_{n,m}^{*}(\theta, \varphi) Y_{n'',m'-m}^{*}(\theta, \varphi) \sin\theta d\theta d\varphi \tag{8.58}$$

The scattering coefficient of particle p b_{nm}^{p} is related to a_{nm}^{p}, $b_{nm}^{p} = A_n a_{nm}^{p}$, all spherical particles discussed in this section have the same structure, so the scattering factors are the same. By substituting this correspondence and $a_{nm}^{p} = a_{nm}^{p(0)} + \sum_{q \neq p} a_{nm}^{p(q)}$ into Eq. (8.56), the following relationship can be obtained:

$$a_{n,m}^{p(q)} = \sum_{n'm'} A_{n'} \left(a_{n',m'}^{q(0)} + \sum_{q' \neq q} a_{n',m'}^{q(q')} \right) G_{n,m,n',m'}^{pq} \tag{8.59}$$

For sparse particles, that is, satisfying $k_o r_{pq} >> 1$, the scattering wave of other particles can be regarded as a small quantity. Without considering higher-order small quantities, Eq. (8.59) can be simplified as:

$$a_{n,m}^{p(q)} = \sum_{n'm'} A_{n'} a_{n',m'}^{q(0)} G_{n,m,n',m'}^{pq} \tag{8.60}$$

The acoustic radiation force is the integral of the average radiation stress tensor $\langle \overleftrightarrow{S} \rangle$ of the particle over the entire surface S within a period:

$$F^p = \oiint_S \langle \overleftrightarrow{S} \rangle ds \tag{8.61}$$

Here [17],

$$\left\langle \overleftrightarrow{S} \right\rangle = -\left(\frac{\rho_1 |v|^2}{4} - \frac{|p|^2}{4\rho_1 c_1^2}\right) \overleftrightarrow{I} + \frac{1}{2}\rho_1 Re\left(v^* v\right) = \rho_1 \left[\frac{1}{4}\nabla^2 |\phi|^2 \overleftrightarrow{I} - \frac{1}{2} Re\left(\nabla\phi\nabla\phi^*\right)\right] \tag{8.62}$$

where, $\overleftrightarrow{I}$ is the unit tensor, ρ_1, c_1 are respectively the density and speed of sound of the fluid, v, p, ϕ represent the fluid velocity, pressure, and velocity potential on the particle surface respectively.

When the boundary conditions of the fluid are not considered, the fluid can be regarded as infinite. The incident Gaussian sound wave is incident from the positive direction of the z-axis, and the acoustic radiation force received by the particle p can be regarded as a collection of forces in the x, y, and z directions [25, 26], that is, $F^p = \left(F_x^p, F_y^p, F_z^p\right)$, F_x^p, F_y^p, F_z^p can be expressed as:

$$F_x^p + iF_y^p = \frac{i\rho_0}{4}\sum_{nm}\left[\begin{array}{l} \mu_{n+1,m-1} g_{n+1} g_n^* \left(2b_{n+1,m+1}^p b_{n,m}^{p*} + b_{n+1,m-1}^p a_{n,m}^{p*} + a_{n+1,m-1}^p b_{n,m}^{p*}\right) \\ +\mu_{n+1,-m-1} g_n g_{n+1}^* \left(2b_{n,m}^p b_{n+1,m+1}^{p*} + b_{n,m}^p a_{n+1,m+1}^{p*} + a_{n,m}^p b_{n+1,m+1}^{p*}\right) \end{array}\right] \tag{8.63}$$

$$F_z^p = \frac{\rho_0}{2} Im \sum_{nm}\left[\sqrt{\frac{(n-m+1)(n+m+1)}{(2n+1)(2n+3)}} T_n a_{n,m}^p a_{n+1,m}^{p^*} g_n g_{n+1}^*\right] \tag{8.64}$$

where, $\mu_{n,m} = \sqrt{\frac{[(n-m)(n-m-1)]}{[(2n-1)(2n+1)]}}$.

We first analyze F_x^p and F_y^p, substituting $b_{nm}^p = A_n a_{nm}^p$ into (8.63) can yield:

$$F_x^p + iF_y^p = \frac{i\rho_0}{4}\sum_{nm}\left[\begin{array}{l} T_n \mu_{n+1,m-1} g_{n+1} g_n^* a_{n,m}^{p*} a_{n+1,m-1}^p \\ +T_n^* \mu_{n+1,-m-1} g_n g_{n+1}^* a_{n,m}^p a_{n+1,m+1}^{p*} \end{array}\right] \tag{8.65}$$

where, $T_n = 2A_{n+1}A_n^* + A_{n+1} + A_n^*$ represents the total scattering situation of the particle p. Since there are only particles p and q, substituting $a_{np,m}^p = a_{n,m}^{p(0)} + a_{n,m}^{p(q)}$ into Eq. (8.65) yields:

$$\begin{aligned} a_{n,m}^{p*} a_{n+1,m-1}^p &= a_{n,m}^{p(0)*} a_{n+1,m-1}^{p(0)} + a_{n,m}^{p(0)*} a_{n+1,m-1}^{p(q)} + a_{n,m}^p a_{n+1,m-1}^{p(0)} + a_{n,m}^{p(q)*} a_{n+1,m-1}^{p(q)} \\ a_{n,m}^p a_{n+1,m+1}^{p*} &= a_{n,m}^{p(0)} a_{n+1,m+1}^{p(0)*} + a_{n,m}^{p(0)} a_{n+1,m+1}^{p(q)*} + a_{n,m}^{p(q)} a_{n+1,m+1}^{p(0)*} + a_{n,m}^{p(q)} a_{n+1,m+1}^{p(q)*} \end{aligned} \tag{8.66}$$

From Eq. (8.56) and the properties of the Kronecker Delta function, at least one of $a_{n,m}^{p(0)}$, $a_{n+1,m-1}^{p(0)}$ must be 0, so $a_{n,m}^{p(0)*} a_{n+1,m-1}^{p(0)} = 0$ always holds true, similarly, $a_{n,m}^{p(0)} a_{n+1,m+1}^{p(0)*} = 0$. If $m \neq 0$, $a_{n,m}^p = a_{n,m}^{p(q)}$ is a small quantity relative to $a_{n,0}^p$, if m, $m \pm 1$ are not 0, then $a_{n,m}^{p*} a_{n+1,m\pm1}^p$ is a high-order small quantity, which can be ignored in the calculation. For $a_{n,m}^{p(q)}$, from Eq. (8.60), $a_{n,m}^{p(q)} = \sum_{n'm'} A_{n'} a_{n',m'}^{q(0)} G_{n,m,n',m'}^{pq}$, where, $a_{n',m'}^{q(0)}$ also satisfies Eq. (8.54), that is, there are only non-zero solutions when $m' = 0$, so Eq. (8.60) becomes:

$$a_{n,m}^{p(q)} = \sum_{n'} A_{n'} a_{n',0}^{q(0)} G_{n,m,n',0}^{pq} \tag{8.67}$$

Similarly, $a_{n,m}^{p(q)*} a_{n+1,m-1}^{p(q)}, a_{n,m}^{p(q)} a_{n+1,m+1}^{p(q)*}$ is a high-order small quantity, which can be ignored in the calculation, then Eq. (8.65) becomes:

$$\begin{aligned} a_{n,m}^{p*} a_{n+1,m-1}^{p} &= \sum_{n'} \left(A_{n'} a_{n,m}^{p(0)*} a_{n',0}^{q(0)} G_{n+1,m-1,n',0}^{pq} + A_{n'}^{*} a_{n',0}^{q(0)*} a_{n+1,m-1}^{p(0)} G_{n,m,n',0}^{pq*} \right) \\ a_{n,m}^{p} a_{n+1,m+1}^{p*} &= \sum_{l'} \left(A_{n'}^{*} a_{n,m}^{p(0)} a_{n',0}^{q(0)*} G_{n+1,m+1,n',0}^{pq*} + A_{n'} a_{n',0}^{q(0)} a_{l+1,m+1}^{p(0)*} G_{n,m,n',0}^{pq} \right) \end{aligned} \tag{8.68}$$

After neglecting all higher-order small quantities in Eq. (8.64), we obtain:

$$F_x^p + iF_y^p = \frac{i\rho_0}{4} \sum_n \left[\begin{array}{l} T_n g_{n+1} g_n^* \left(\mu_{n+1,-1} a_{n,0}^{p*} a_{n+1,-1}^{p} + \mu_{n+1,0} a_{n,1}^{p*} a_{n+1,0}^{p} \right) \\ + T_n^* g_n g_{n+1}^* \left(\mu_{n+1,-1} a_{n,0}^{p} a_{n+1,1}^{p*} + \mu_{n+1,0} a_{n,-1}^{p} a_{n+1,0}^{p*} \right) \end{array} \right] \tag{8.69}$$

Specifically, when $n = 0$, because $-n \le m \le n$, the value of m can only be 0.

Next, consider the value of $G_{n,m,n',m'}^{pq}$. Since we have already proven $m' = 0$, $G_{n,m,n',m'}^{pq}$ becomes:

$$\begin{aligned} G_{n,m,n',0} &= 4\pi \sum_{n''} i^{n+n''-n'} C_{n,m,n'',-m}^{n',0} h_{n''}\left(kr_{pq}\right) Y_{n'',-m}\left(\theta_{pq}, \varphi_{pq}\right) \\ C_{n,m,n'',-m}^{n',0} &= \int_0^{2\pi} \int_0^{\pi} Y_{n',0}(\theta, \varphi) Y_{n,m}^{*}(\theta, \varphi) Y_{n'',-m}^{*}(\theta, \varphi) d\theta d\varphi \end{aligned} \tag{8.70}$$

To facilitate calculation, we further approximate. For sparse particles p and q, that is, satisfying $k_o r_{pq} >> 1$, the spherical Hankel function can be approximated as $h_{n''}\left(kr_{pq}\right) = \frac{(-i)^{n''} e^{ikr_{pq}}}{ikr_{pq}} + O\left(\frac{1}{r_{pq}}\right)$, at this time:

$$G_{n,m,n',0} = \frac{e^{ikr_{pq}}}{kr_{pq}} e^{-im\varphi_{pq}} 4\pi i^{n-n'-1} \sum_{n''} C_{n,m,n'',-m}^{n',0} Y_{n'',-m}\left(\frac{\pi}{2}, 0\right) = \frac{e^{ikr_{pq}}}{kr_{pq}} e^{-im\varphi_{pq}} M_{n,m,n',0} \tag{8.71}$$

After further rearrangement, we can finally obtain:

$$F_x^p + iF_y^p = \frac{H_{pq}}{kr_{pq}} e^{i\varphi_{pq}} \tag{8.72}$$

where,

$$H_{pq} = -\frac{\rho_0}{2}\sum_n\sum_{n'}\left[\begin{array}{l}\mu_{n+1,-1} Im\left(T_n A_{n'} a_{n,0}^{p(0)*} a_{n',0}^{q(0)} g_n^* g_{n+1} e^{i2\pi d\cos\theta_{pq}} e^{ikr_{pq}} M_{n+1,-1,n',0}\right)\\ +\mu_{n+1,0} Im\left(T_n A_{n'}^* a_{n',0}^{q(0)*} a_{n+1,0}^{p(0)} g_n g_{n+1}^* e^{-i2\pi d\cos\theta_{pq}} e^{-ikr_{pq}} M_{n,1,n',0}^*\right)\end{array}\right] \tag{8.73}$$

$$M_{n,m,n',0} e^{-im\varphi_{pq}} = 4\pi i^{n-n'-1}\sum_{n''} C_{n,m,n'',-m}^{n',0} Y_{n'',-m}\left(\theta_{pq},\varphi_{pq}\right) \tag{8.74}$$

For Eq. (8.71), we can see that the real part of this equation is the component of the inter-particle interaction force in the x-direction, while the imaginary part is the component in the y-direction. The relationship between the two is determined by φ_{pq}.

When the centers of the two particles are in the same yz plane and there is symmetry between the particles and the sound wave, at this time $\varphi_{pq} = \frac{\pi}{2}$, i.e., the interaction force in the x-direction is zero.

Next, we consider the change of F_z^p, that is, Eq. (8.64). According to Eq. (8.68), we can directly obtain:

$$a_{n,m}^p a_{n+1,m}^{p*} \sum_{n'}\left(A_{n'}^* a_{n,m}^{p(0)} a_{n',0}^{q(0)*} e^{-i2\pi d\cos\theta_{pq}} G_{n+1,m,n',0}^{pq*} + A_{n'} a_{n',0}^{p(0)} a_{n+1,m}^{q(0)*} e^{i2\pi d\cos\theta_{pq}} G_{n,m,n',0}^{pq}\right) \tag{8.75}$$

Considering Eq. (8.56) and the properties of the Kronecker delta function, in the above equation, only when $m = 0$ exists a non-zero solution, so Eq. (8.64) becomes:

$$F_z^p = \frac{\rho_0}{2} Im\left[\sum_n \sqrt{\frac{(n+1)(n+1)}{(2n+1)(2n+3)}} T_n a_{n,0}^p a_{n+1,0}^{p*} g_n g_{n+1}^*\right] \tag{8.76}$$

where,

$$a_{n,0}^p a_{n+1,0}^{p*} = \sum_{n'}\left(A_{n'}^* a_{n,0}^{p(0)} a_{n',0}^{q(0)*} e^{-i2\pi d\cos\theta_{pq}} G_{n+1,0,n',0}^{pq*} + A_{n'} a_{l',0}^{p(0)} a_{n+1,0}^{q(0)*} e^{i2\pi d\cos\theta_{pq}} G_{n,0,n',0}^{pq}\right) \tag{8.77}$$

By substituting the simplified form of Eq. (8.70) into the above equation, we can obtain the z-direction component of the force acting on the particle as:

$$F_z^p = \frac{1}{kr_{pq}}\frac{\rho_0}{2} Im\left[\sum_n e^{ikr_{pq}} \sqrt{\frac{(n+1)(n+1)}{(2n+1)(2n+3)}} T_n g_n g_{n+1}^* \sum_{n'}\left(\begin{array}{l}A_{n'}^* a_{n,0}^{p(0)} a_{n',0}^{p(0)*} e^{-i2\pi d\cos\theta_{pq}} M_{n+1,0,n'',0}^{pq*}\\ +A_{n'} a_{n',0}^{p(0)} a_{l+1,0}^{p(0)*} e^{i2\pi d\cos\theta_{pq}} M_{n,0,n',0}^{pq}\end{array}\right)\right] \tag{8.78}$$

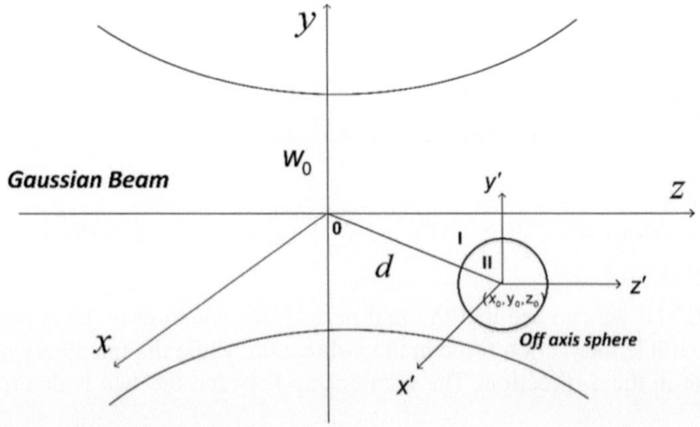

Fig. 8.10 Schematic diagram of Gaussian wave scattering off-axis spherical particles

Finally, let's explain the Gaussian beam beam factor g_n in Eqs. (8.73) and (8.76). In the first chapter, we have already solved the Gaussian beam factor at the center of the beam. However, for the multi-particle situation, the particles are generally in an off-axis state (see Fig. 8.10), and its beam factor value will change. For the off-axis beam factor situation, we can refer to the situation of off-axis cylindrical particles in the Gaussian wave field [27], using the spherical function translation addition theorem, we can obtain:

$$g_n = \begin{cases} \sum_{m=-\infty}^{\infty}\sum_{j=0}^{m} \frac{\Gamma\left(\frac{m}{2}+1\right)}{\Gamma\left(\frac{m}{2}+\frac{1}{2}\right)}\cdot\frac{\Gamma\left(\frac{m}{2}+j+\frac{1}{2}\right)}{\left(\frac{m}{2}-j\right)!j!}\cdot\left(-4Q_0s^2\right)^j Q_0\exp(-ikz_0)\cdot Q_{mn}, \\ \sum_{m=-\infty}^{\infty}\sum_{j=0}^{m} \frac{\Gamma\left(\frac{(m-1)}{2}+1\right)}{\Gamma\left(\frac{(m-1)}{2}+\frac{3}{2}\right)}\cdot\frac{\Gamma\left(\frac{(m-1)}{2}+j+\frac{3}{2}\right)}{\left(\frac{(m-1)}{2}-j\right)!j!}\cdot \\ \left(-4Q_0s^2\right)^j\left(Q_0-Q_1-jQ_1\right)\exp(-ikz_0), \end{cases} \tag{8.79}$$

Here: $Q_0 = \frac{1}{\left(1+\frac{2iz_0}{l}\right)}$, $Q_1 = \frac{2}{\left(kl\left(i-\frac{2z_0}{l}\right)\right)^2}$, $l = kW_0^2$, $s = (kW_0)^{-1}$, $Q_{nm} = \sum_{v,\mu} S_{nv}^{m\mu}(kd)$,

$S_{nv}^{m\mu}(kd) = 4\pi(-1)^m \sum_{\sigma=|n-v|}^{n+v} i^{\sigma+n-v}\cdot\vartheta(v,\mu;n,-m;q)\cdot j_\sigma(kd)Y_\sigma^{\mu-M}(\theta_d,\varphi_d)$, ϑ are Green's coefficients [28], $d = \left(x_0^2+y_0^2+z_0^2\right)^{\frac{1}{2}}$, θ_d represent the angle between the line from the origin to the center of the particle sphere and the z axis; φ_d represents the angle between its line and the x axis.

8.6 Acoustic Radiation Force Between Polyethylene Particle Spheres in Water

Firstly, in order to verify the correctness of the above work, we numerically calculated the mutual force between two liquid spheres (olive oil spheres, the size of the liquid sphere is taken as $ka = 1$) at the waist position in the Gaussian wave field using this method. The waist width of the Gaussian wave is taken as $W_0 = 5\lambda$ (Note: When the waist width of the Gaussian wave is much larger than the size of the particle sphere, the Gaussian wave near the waist can be approximately considered as a plane wave field). The simulation results are shown in Fig. 8.11A, where kd represents the normalized distance between the particle spheres, and it is compared with the previous literature [18] (Fig. 8.12). Through comparison, it can be found that our simulation results completely coincide with the results in the literature, which shows that the above derivation process is reasonable and correct.

Next, we discussed the acoustic radiation force situation of two particle spheres and three particle spheres in different positions in the wave field, and summarized the solution of the mutual force between more particle spheres.

8.6.1 Alignment of the Line Connecting the Centers of Two Particle Spheres with the y-axis

First, we consider the case where only two particles exist in the sound field and are coaxial with the y-axis, as shown in Fig. 8.13. Two identical particle spheres p and q are separated by d (the case of $k_o r_{pq} >> 1$ was considered in previous theoretical calculations, so the above calculations only apply to sparse particles, here it is required that $d > 2a, d/\lambda > 1$), assuming that the two particles are exactly at the waist position of the Gaussian wave, and use θ_{pq}, φ_{pq} to represent their angle with the coordinate axis. Figures 8.13, 8.14, 8.15 show the variation curves of the acoustic radiation force (F_y, F_z) with the distance between the two particles (d) for

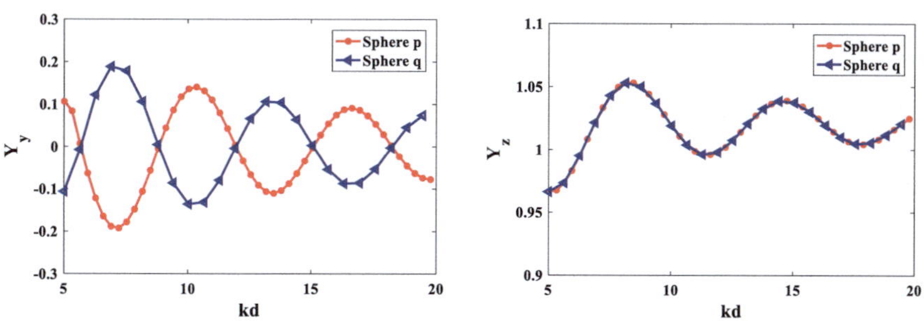

Fig. 8.11 Acoustic radiation force of liquid spheres at $W_0 = 5\lambda$

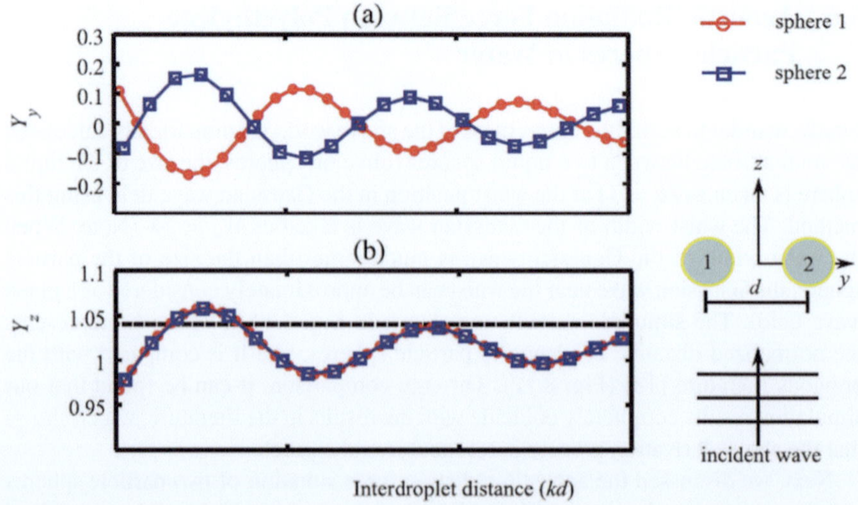

Fig. 8.12 Simulation diagram of the acoustic radiation force (Y_y, Y_z) in reference [18]

three different sizes of particle spheres under different beam widths. Where $\theta_{pq} = \frac{\pi}{2}$, $\varphi_{pq} = \frac{\pi}{2}$, according to Eq. (8.72), at this time $F_y^p = F_{xy}^p$, $F_x^p = 0$.

Figure 8.14 shows the acoustic radiation force curve of two particle spheres p and q when the diameter is $a = 0.1\lambda$. The three graphs (a), (c), and (e) in the left respectively represent the mutual force in the y direction between the two particles when the waist width is 0.5 times, 1.0 times, and 3.0 times the wavelength. From the simulation results, F_y shows a simple harmonic decay as the distance between the particles d increases, indicating that as the distance between the particles increases, the influence

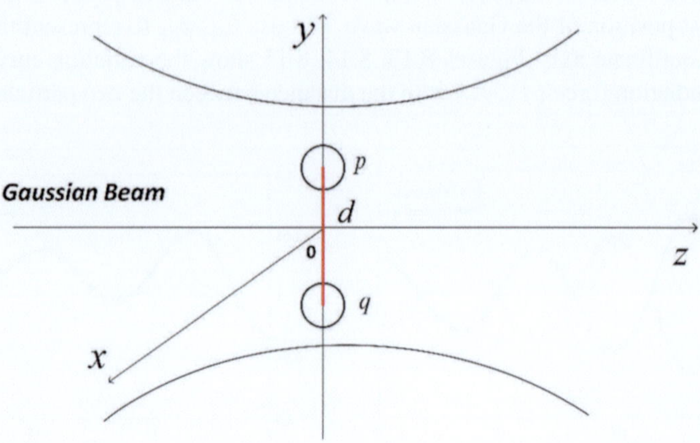

Fig. 8.13 Schematic diagram of the line connecting the centers of two particle spheres located on the y-axis and symmetric with respect to the z-axis

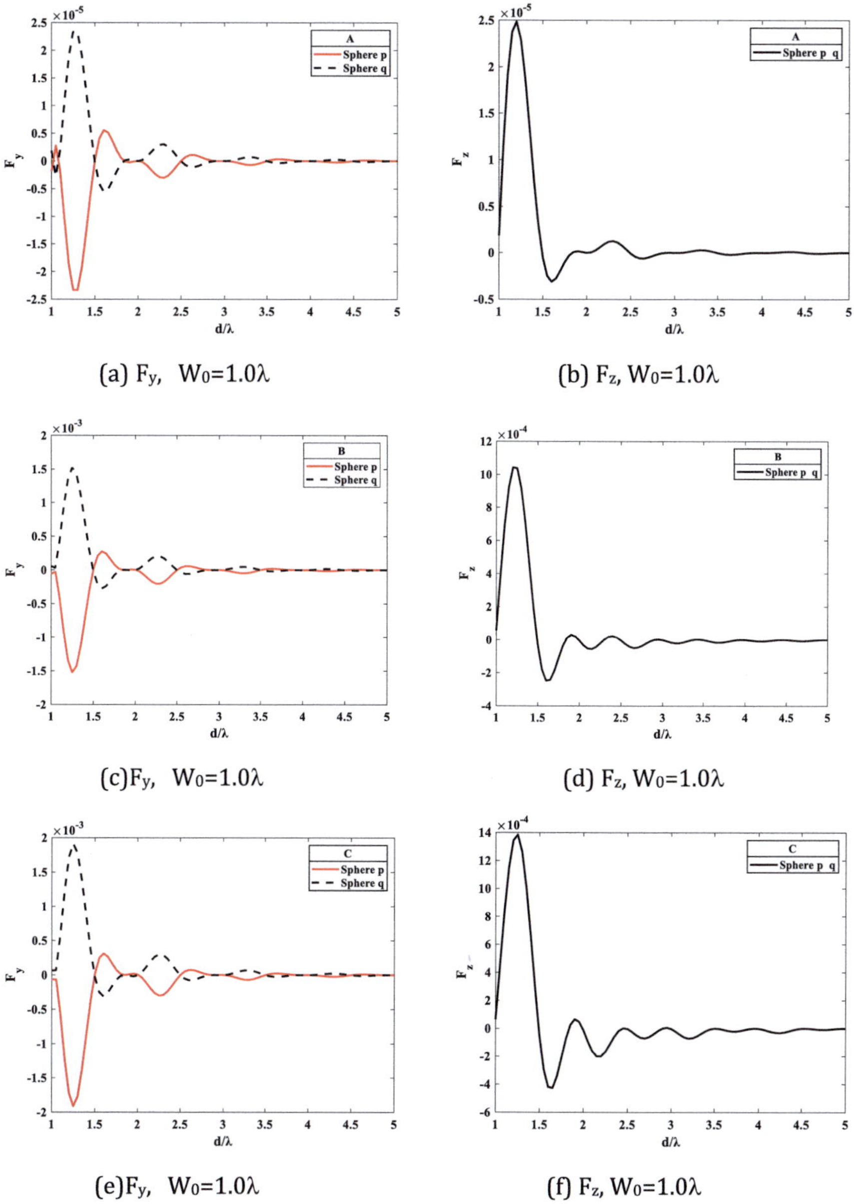

Fig. 8.14 Acoustic radiation force of polyethylene particle spheres p and q when the diameter is $a = 0.1\lambda$

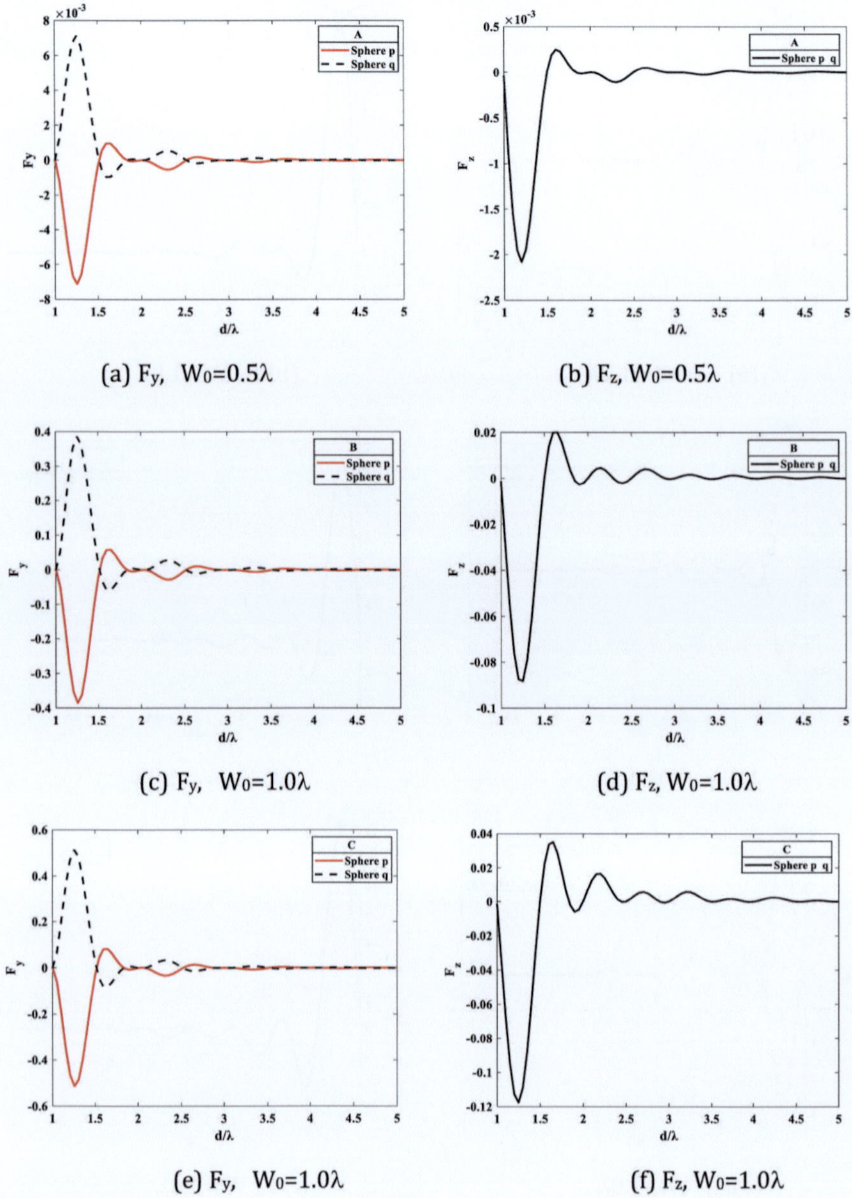

Fig. 8.15 Acoustic radiation force on polyethylene particle spheres with a diameter of $a = 0.3\lambda$ p and q

of the scattering wave between the particles will quickly decrease. Secondly, at different distances, particles p and q can generate attraction and repulsion, resulting in the phenomenon of the two particles approaching or moving away from each other, especially when the distance between the particles is small. In addition, since the particles are in a Gaussian wave field, the beam width will also have a certain impact on the force. From the three graphs (a), (c), and (f), it can be seen that as the beam width W_0 continues to increase, the force F_y also shows a significant increase. The three graphs (b), (d), and (f) on the right respectively represent the force F_z in the z direction between the two particles when the waist width is 0.5 times, 1.0 times, and 3.0 times the wavelength. From the simulation results, F_z also shows a simple harmonic decay as the distance between the particles d increases, and as the beam width W_0 increases, these characteristics are similar to those of F_y.

Next, we increased the size of the particles and studied the cases of particles with diameters of $a = 0.3\lambda$ and $a = 0.5\lambda$, as shown in Figs. 8.15 and 8.16. From the simulation results, the forces F_y and F_z have similar trends to the forces when $a = 0.1\lambda$, and as the size of the particles increases, the corresponding values also continuously increase.

8.6.2 Two Particle Spheres not Aligned with the y-axis

Next, we consider the case where the line connecting the centers of the two particle spheres is not aligned with the y-axis and has an angle, as shown in Fig. 8.17. Here, we assume that the line connecting the two particle spheres has an angle of 45° with the y-axis, and the diameter of the particle spheres is $a = 0.3\lambda$.

Figure 8.18 shows the acoustic radiation force situation when the particle spheres p and q have an angle $\theta_{pq} = 45°$ with the y-axis. The three graphs (a), (c), and (e) in the left (1) respectively represent the interaction force F_y between the two particles in the y-direction when the waist width is 0.5 times, 1.0 times, and 3.0 times the wavelength. From the simulation results, F_y oscillates and decays as the particle spacing d increases. As the particle spacing d The influence of the scattering waves between particles decreases rapidly with the increase of p. These phenomena are similar to the situation where two particles are located on the y-axis (Figs. 8.14, 8.15 and 8.16). The three graphs (b), (d) and (f) in the right side show the force between two particles in the z-direction. From the simulation results, this is different from the previous situation, the force F_z in the z-direction between particles and q is no longer equal, showing different trends. Moreover, with the increase of the beam width W_0, the force will also increase.

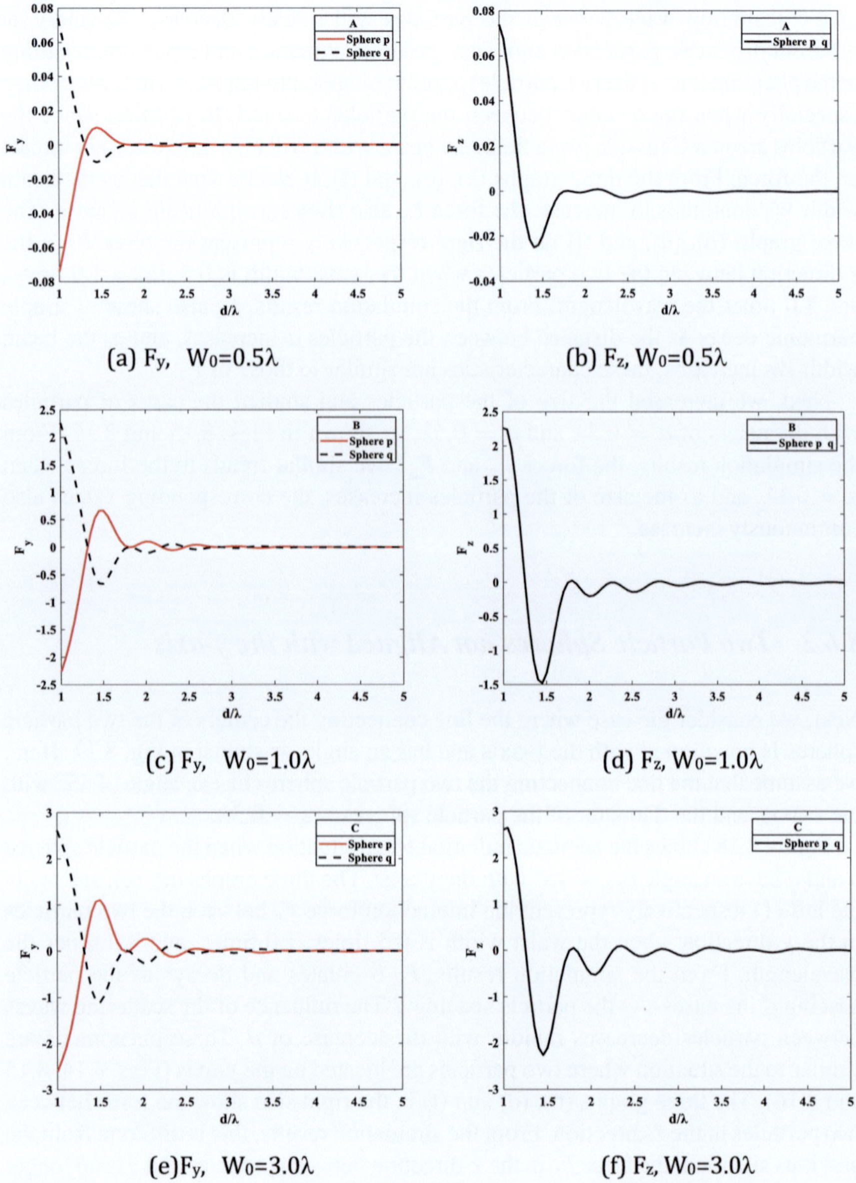

Fig. 8.16 Acoustic radiation force on polyethylene particle spheres with a diameter of $a = 0.5\lambda$ p and q

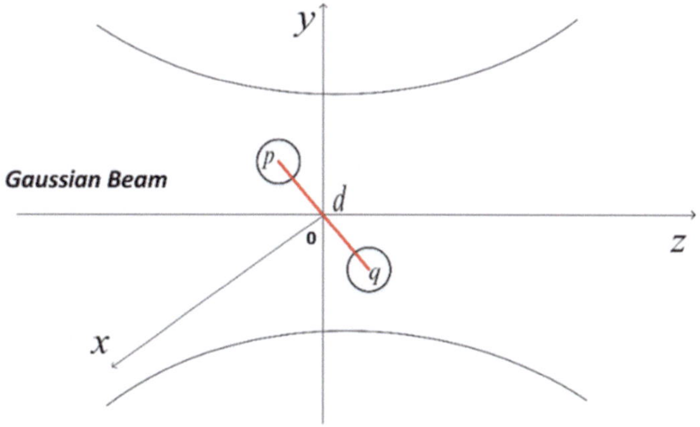

Fig. 8.17 Schematic diagram of the position when the line connecting the centers of the two particle spheres has an angle with the y-axis

8.6.3 Case of Three Particle Spheres

Finally, let's consider the situation where there are three particle spheres in the sound field, as shown in Fig. 8.19. Here, it is assumed that the line connecting the centers of particle spheres p and q is collinear with the y-axis, the center of the third sphere t is located on the z-axis, and the lines connecting the centers of the three spheres form an equilateral triangle, which is in the yoz plane. Thus, the angle $\varphi_{ij} = \frac{\pi}{2}$ between the lines connecting the two spheres and the x-axis ($i, j = p, q, t$ and $i \neq j$) can be analyzed to obtain the acoustic radiation forces p, q, and t experienced by the three particles in all directions.

For particle sphere p, we have:

$$
\begin{aligned}
F_x^p &= F_{pq,xy}\left(\frac{d}{\lambda}, \theta_{pq}\right)\cos\varphi_{pq} + F_{pt,xy}\left(\frac{d}{\lambda}, \theta_{pt}\right)\cos\varphi_{pt} = 0,\\
F_y^p &= F_{pq,xy}\left(\frac{d}{\lambda}, \theta_{pq}\right)\sin\varphi_{pq} + F_{pt,xy}\left(\frac{d}{\lambda}, \theta_{pt}\right)\sin\\
\varphi_{pt} &= F_{pq,xy}\left(\frac{d}{\lambda}, \theta_{pq}\right) + F_{pt,xy}\left(\frac{d}{\lambda}, \theta_{pt}\right),\\
F_z^p &= F_{pq,z}\left(\frac{d}{\lambda}, \theta_{pq}\right) + F_{pt,z}\left(\frac{d}{\lambda}, \theta_{pt}\right).
\end{aligned}
\tag{8.80}
$$

For particle sphere q, we have:

$$
F_x^q = F_{qp,xy}\left(\frac{d}{\lambda}, \theta_{qp}\right)\cos\varphi_{qp} + F_{qt,xy}\left(\frac{d}{\lambda}, \theta_{qt}\right)\cos\varphi_{qt} = 0,
$$

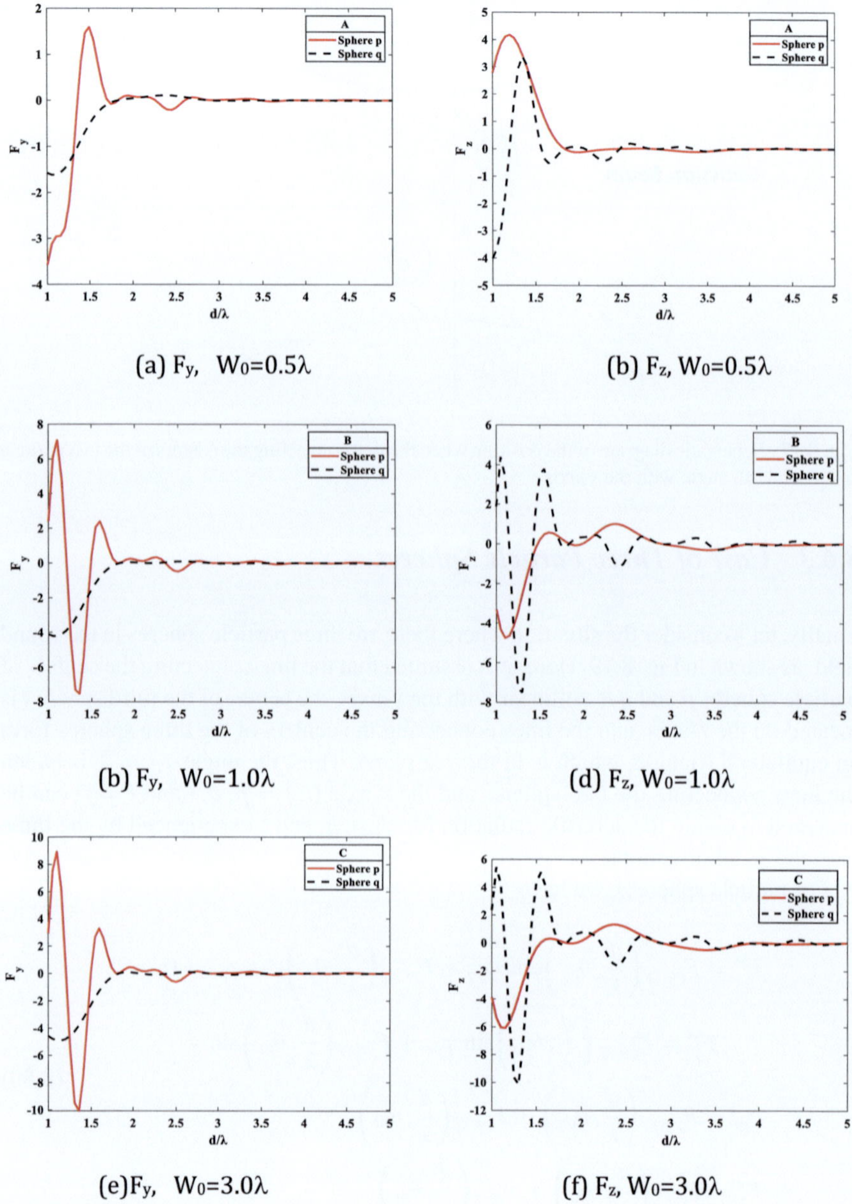

Fig. 8.18 Acoustic radiation force when the line connecting the centers of polyethylene particle spheres p and q forms an angle $\theta_{pq} = 45°$ with the y-axis

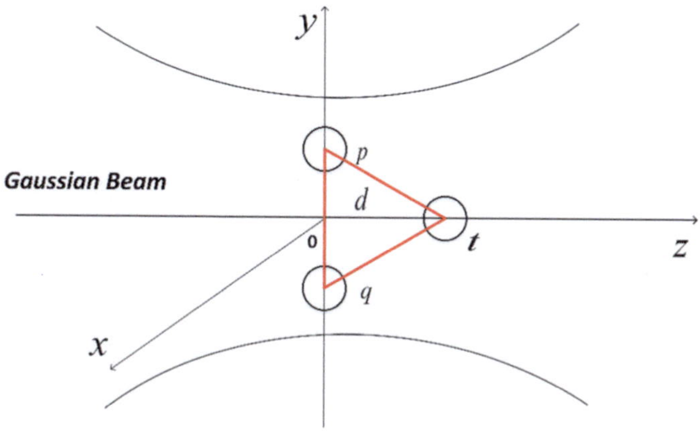

Fig. 8.19 Schematic diagram of the positions of three particle spheres

$$
\begin{aligned}
F_y^q &= F_{qp,xy}\left(\frac{d}{\lambda}, \theta_{qp}\right)\sin\varphi_{qp} + F_{qt,xy}\left(\frac{d}{\lambda}, \theta_{qt}\right)\sin\varphi_{qt} \\
&= F_{qp,xy}\left(\frac{d}{\lambda}, \theta_{qp}\right) + F_{qt,xy}\left(\frac{d}{\lambda}, \theta_{qt}\right), \\
F_z^p &= F_{qp,z}\left(\frac{d}{\lambda}, \theta_{qp}\right) + F_{qt,z}\left(\frac{d}{\lambda}, \theta_{qt}\right).
\end{aligned}
\tag{8.81}
$$

For particle sphere t, we have:

$$
\begin{aligned}
F_x^t &= F_{tp,xy}\left(\frac{d}{\lambda}, \theta_{tp}\right)\cos\varphi_{tp} + F_{tq,xy}\left(\frac{d}{\lambda}, \theta_{tq}\right)\cos\varphi_{tq} = 0, \\
F_y^t &= F_{tp,xy}\left(\frac{d}{\lambda}, \theta_{tp}\right)\sin\varphi_{tp} + F_{tq,xy}\left(\frac{d}{\lambda}, \theta_{tq}\right)\sin \\
\varphi_{tq} &= F_{tp,xy}\left(\frac{d}{\lambda}, \theta_{tp}\right) + F_{tq,xy}\left(\frac{d}{\lambda}, \theta_{tq}\right), \\
F_z^p &= F_{pq,z}\left(\frac{d}{\lambda}, \theta_{tp}\right) + F_{tq,z}\left(\frac{d}{\lambda}, \theta_{tq}\right).
\end{aligned}
\tag{8.82}
$$

After numerical calculations (where the diameter of the particle sphere is taken as 0.3 times the wavelength), the results are shown in Fig. 8.20. From the left side (a) and (c), it can be seen that the forces F_y acting on particles p and q in the y direction exhibit a simple harmonic decay as the particle spacing d increases. This indicates that as the distance between particles increases, the influence of the scattered waves between particles will continuously decrease. The force on particle t in the y direction is always zero. The two graphs (b) and (d) in the right side show the forces F_z acting on the three particles in the z axis direction. The forces on particles p and q are equal and oscillate with the increase in particle spacing d, but do not show a clear

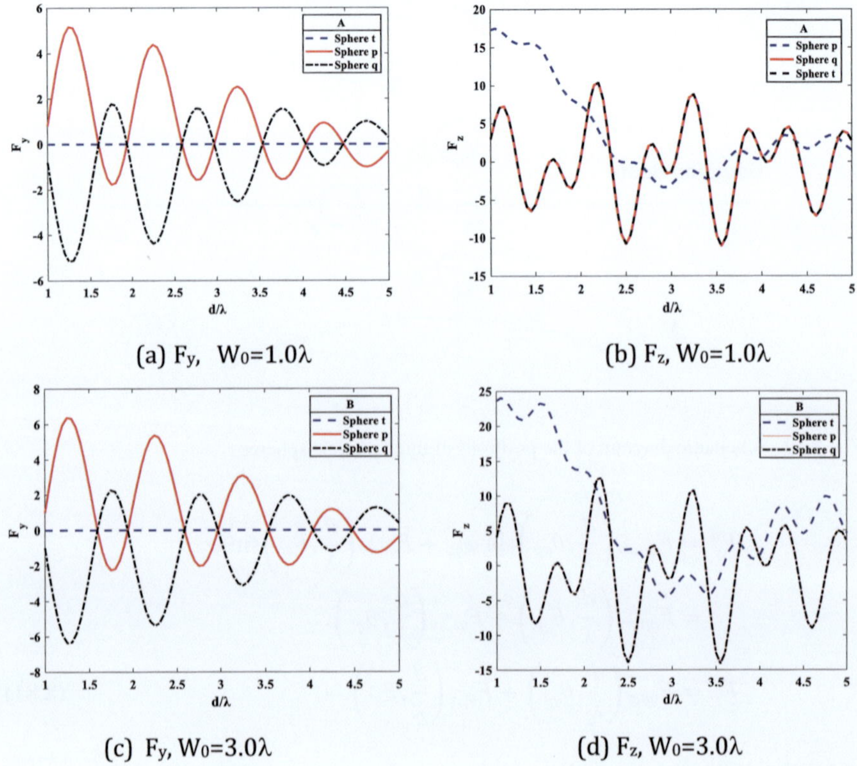

Fig. 8.20 Acoustic radiation force on three polyethylene particle spheres *p*, *q* and *t*

decreasing trend. The force on particle *t* decreases continuously with the increase in particle spacing *d*. The main reason is that as the particle spacing increases, particles *p* and *q* gradually move out of the beam center, thus the scattered waves from particles *p* and *q* received by particle *t* will gradually decrease until they eventually disappear.

For the case where there are more sparsely distributed particle spheres in the wave field, we can also obtain formulas similar to (8.80–8.82) to calculate the acoustic radiation force. This simplifies the acoustic radiation force of a multi-particle system in the sound field to the superposition of the acoustic radiation force of a two-particle system, and the acoustic radiation force received by each particle can be calculated. Once the acoustic radiation force received by the particles is obtained, we can infer the movement of the particle spheres in the sound field based on the force received by the particles, and we can control the movement of the particle spheres by changing the external sound field.

8.7 Conclusion

This section starts from the scattering of spherical particles in a Gaussian wave field, considering the relationship between the incident wave and the scattered wave in the sound field, and obtains the formula for calculating the acoustic radiation force when there are multiple sparsely distributed particles in the Gaussian wave field. This formula is applicable to particles at any position in the Gaussian wave field. The results show that when there are multiple particles in the sound field, the acoustic radiation force received by each particle is affected by other particles, and the size of this effect changes with the distance and angle between particles. When there are multiple particles in the sound field and these particles are sparsely distributed, this multi-particle system can be regarded as a superposition of a series of two-particle systems. By calculating the acoustic radiation force of each two-particle system separately, the acoustic radiation force received by any particle in the sound field can be obtained through superposition. This study on the acoustic radiation force of multiple particles will help to better manipulate tiny particles through sound waves, achieve the separation and screening of cell particles, and promote the further practice and application of acoustic tweezers.

References

1. Torr GR (1984) The acoustic radiation force. Am J Phys 52:402–408
2. Foldy LL (1945) The multiple scattering of waves. I. General theory of isotropic scattering by randomly distributed scatterers. Phys Rev (67):107–119
3. Zitron N, Karp SN (1961) Higher-order approximations in multiple scattering. I. Two-dimensional scalar case. J Math Phys (2):394–402
4. Zitron N, Karp SN (1961) Higher-order approximations in multiple scattering. II. Three-dimensional scalar case. J Math Phys (2):402–406
5. Martin PA (2006) Multiple scattering: interaction of time-harmonic waves with N Obstacles. Cambridge University Press, Cambridge, UK, 2006, pp. xii+437, ISBN: 0-521-86554-9
6. Qiao YP, Zhang XF, Zhang GB (2017) Acoustic radiation force on a fluid cylindrical particle immersed in water near an impedance boundary. J Acoust Soc Am 141:4633–4641
7. Gao S, Mao YM, Liu JH et al (2018) Acoustic radiation force induced by two Airy-Gaussian beams on a cylindrical particle. Chinese Phys B (27):014302
8. Doinikov AA (1999) Bjerknes forces between two bubbles in a viscous fluid. J Acoust Soc Am 106:3305–3312
9. Silva GT, Bruus H (2014) Acoustic interaction forces between small particles in an ideal fluid. Phys Rev E 90(6):063007
10. Embleton TFW (1962) Mutual interaction between two spheres in a plane sound fleld. J Acoust Soc Am 12:1714
11. Zheng XY, Apfel RE (1995) Acoustic interaction forces between two fluid spheres in an acoustic fleld. J Acoust Soc Am 97:2218–2226
12. Doinikov AA (1999) Bjerknes forces between two bubbles in a viscous fluid. J Acoust Soc Am 6:3305–3312
13. Fan XY, Qiu CY, Zhang SW (2016) Highly asymmetric interaction forces induced by acoustic waves in coupled plate structures. J Appl Phys 24:104301

14. Roumeliotis JA, Ziotopoulos AG, Kokkorakis GC (2001) Acoustic scattering by a circular cylinder parallel with another of small radius. J Acoust Soc Am 109:870–877
15. Mitri FG (2017) Extrinsic extinction cross-section in the multiple acoustic scattering by fluid particles. J Appl Phys 121:144904
16. Mitri FG (2017) Acoustic attraction, repulsion and radiation force cancellation on a pair of rigid particles with arbitrary cross-sections in 2D: Circular cylinders example. Ann Phys 386:1–14
17. Zhang SW, Qiu CY, Wang MD et al (2016) Acoustically mediated long-range interaction among multiple spherical particles exposed to a plane standing wave. New J Phys 18:113034
18. Lopes JH, zarpeyv A (2016) Acoustic interaction forces and torques acting on suspended spheres in an ideal fluid, IEEE Trans Ultrason Ferroelectr Freq Cont (63):186 –197
19. Mingxin H, Xiaozhou L, Jiehui L, He, Aijun, Gutian Z, Qiang Z (2018) Acoustic radiation force on multiple particles in a plane wave field. Appl Acoust 37:106–113
20. Lopes JH, Azarpeyvand M, Silva GT (2014) Acoustic interaction forces and torques acting on suspended spheres in an ideal fluid. IEEE Trans Ultrason Ferroelectr Freq Control 14:186–197
21. Colton D, Kress R (1992) Inverse acoustic and electromagnetic scattering theory. Springer-Verlag 102:1601
22. Liu ZY, Chan CT, Sheng P et al (2000) Elastic wave scattering by periodic structures of spherical objects: theory and experiment. Phys Rev B 62:2446–2457
23. Westervelt PJ (1951) The theory of steady forces caused by sound waves. J Acoust Soc Am 23:719
24. Silva GT (2011) An expression for the radiation force exerted by an acoustic beam with arbitrary wavefront. J Acoust Soc Am 130:3541
25. Silva GT (2011) An expression for the radiation force exerted by an acoustic beam with arbitrary wave front. J Acoust Soc Am 130:3541
26. Silva GT, Lopes JH, Mitri FG (2013) Off-axial acoustic radiation force of repulsor and tractor Bessel beams on a sphere. IEEE Trans Ultrason Ferroelectr Freq Control 60:1207
27. Zhang XF, Qian YB, Zhang GB et al (2016) Computation of the acoustic radiation force on a rigid cylinder in off-axial gaussian beam using the translational addition theorem. Acta Acust Acust 102:334–340
28. Silva GT, Baggio AL, Lopes JH et al (2014) Exact computations of the acoustic radiation force on a sphere using the translational addition theorem. arXiv:1210.2116

Chapter 9
Medical Applications of Acoustic Radiation Force

The multifunctionality of acoustic radiation force in manipulating particles enables them to address current challenges in the fields of biomedicine and medicine. The arrangement, separation, and sorting of cells are crucial processes in studying cell characteristics, diagnosing diseases, and treatment, especially for cancer patients. Early detection of cancer cells can greatly aid in cancer treatment. Acoustic radiation force separation of particles offers a promising alternative to traditional techniques, allowing the separation of different particles of varying sizes by adjusting the acoustic radiation force on the particles. In practical applications, it is necessary to transport biological products in a controlled manner. The method of cell transport provides a solid foundation for cell separation and sorting. Currently, a very active research area is the combination of acoustic standing waves with microfluidics. Researchers have developed many promising biological applications for separating, concentrating, and manipulating particles, especially biological cells, in a label-free environment at high throughput. The combination of acoustic radiation force and microfluidic technology is a promising tool for the development of auxiliary chip experimental equipment. By setting the sound waves in the same direction as the blood flow, different cells will experience different acoustic radiation forces during the periodic sound pressure wave nodes and antinodes, resulting in different lateral displacements, thereby sorting cancer cells from normal cells. The medical applications of acoustic radiation force are numerous, and this chapter will introduce aspects such as elasticity imaging based on acoustic radiation force, manipulation of particles, microbubbles, and neural regulation.

X. Liu and H. Zheng, *Acoustic Radiation Force*, Advances in Acoustics,
https://doi.org/10.1007/978-981-95-3152-3_9

9.1 Elasticity Imaging of Acoustic Radiation Force

9.1.1 Overview of Ultrasound Elasticity Imaging

The basic principle of ultrasound elasticity imaging is to apply an internal (including its own) or external dynamic or static, quasi-static excitation to the tissue. Under the laws of elasticity mechanics, biomechanics, and other physical laws, the tissue will produce a response. Ultrasound elasticity imaging techniques mainly include static or quasi-static compression elasticity imaging, vascular elasticity imaging technology, myocardial elasticity imaging technology, acoustic vibration imaging technology, instantaneous elasticity imaging technology, shear wave elasticity imaging technology, acoustic radiation force pulse imaging technology, supersonic shear imaging technology, and harmonic motion imaging technology [1].

Static or quasi-static elasticity imaging technology utilizes external mechanical pressure, combined with an ultrasound probe emitting ultrasound waves in the direction of compression. The time delay between the ultrasound echo signals before and after compression is analyzed to obtain the elasticity coefficient of the tissue [2].

Ultrasound instantaneous elasticity imaging technology uses a mechanical low-frequency oscillator to produce a reversible, measurable small deformation in the tissue. An ultrasound transducer then records the ultrasound echo signals of the tissue at different times. The tissue displacement is extracted through cross-correlation calculation to further obtain the elastic modulus of the tissue [1]. This technique has the advantages of being real-time, non-invasive, and not easily affected by the environment.

Acoustic Radiation Force Impulse Imaging (ARFI) utilizes focused ultrasound to generate acoustic radiation force in the tissue and tracks the tissue displacement caused by it [3]. Compared with quasi-static elasticity imaging, ARFI avoids external compression of the tissue and can quantitatively measure the elastic modulus of the tissue, requiring less from the operator.

Shear Wave Elasticity Imaging (SWE) is an emerging ultrasound elasticity imaging technology. Its principle is similar to instantaneous elasticity imaging technology and acoustic radiation force pulse imaging technology. SWE calculates the elastic modulus of the tissue by detecting the propagation speed of shear waves in the tissue. SWE has the advantages of real-time two-dimensional image guidance and intuitive image display [4].

The image analysis methods of ultrasound elasticity imaging mainly include scoring method and strain rate ratio method. As an emerging detection method, ultrasound elasticity imaging technology has huge potential in clinical applications, but a large amount of clinical research is still needed to optimize elasticity grading parameters and improve diagnostic accuracy.

9.1.2 Research Background of Ultrasound Elasticity Imaging

In 1991, Ophir and others proposed the concept of ultrasound elasticity imaging [5]. The ultrasound instantaneous elasticity imaging technology was first proposed by American scholar Paker and others [6]; Acoustic vibration imaging technology was proposed by American researchers Fatemi, Greenleaf, and others [7]; Shear wave elasticity imaging technology was proposed by American scholars Saravazyan, Emelianov, and others [8]; Supersonic shear wave technology was proposed by French scholar Fink and others [9]; The acoustic radiation force pulse imaging technology was proposed by Nightingale and others from the United States [10].

9.1.3 Application of Ultrasound Elasticity Imaging

9.1.3.1 Algorithm and System Design for Ultrasound Instantaneous Elasticity Imaging

For biological tissues, changes in the elastic modulus are often related to their pathological characteristics. There is a significant difference in elasticity between normal and diseased tissues. Therefore, measuring mechanical parameters of biological tissues such as Young's modulus using acoustic radiation force ultrasound technology is an important research direction in medical ultrasound. Based on the acoustic radiation force ultrasound instantaneous elasticity imaging technology, exploring suitable time gain compensation circuits and tissue displacement solution algorithms can enhance the accuracy and reliability of the calculation results. Design an instantaneous ultrasound system that includes a low-frequency oscillation system, an ultrasound imaging system, a low-frequency oscillator, and a single-element ultrasound transducer. Use a self-made ultrasound system to collect RF echo signals. Prepare two types of biological phantoms with different acrylamide concentrations, analyze their ultrasound RF echo signals, and compare them with the results measured by the indentation system. Experimental data show that after matched filtering, the signal-to-noise ratio of the displacement field is improved, and the propagation contour of the shear wave is enhanced; the measured shear wave speed is highly consistent with the mechanical set value [1], Fig. 9.1 shows the framework diagram of the ultrasound instantaneous elasticity imaging system.

9.1.3.2 Research on Acoustic Radiation Force Elasticity Imaging Algorithm Based on Unified Computing Architecture (CUDA)

Acoustic radiation force elasticity imaging can quantitatively measure the elastic modulus of tissues and is easy to use, but its algorithm data processing has the disadvantages of large volume and long time, and due to these disadvantages,

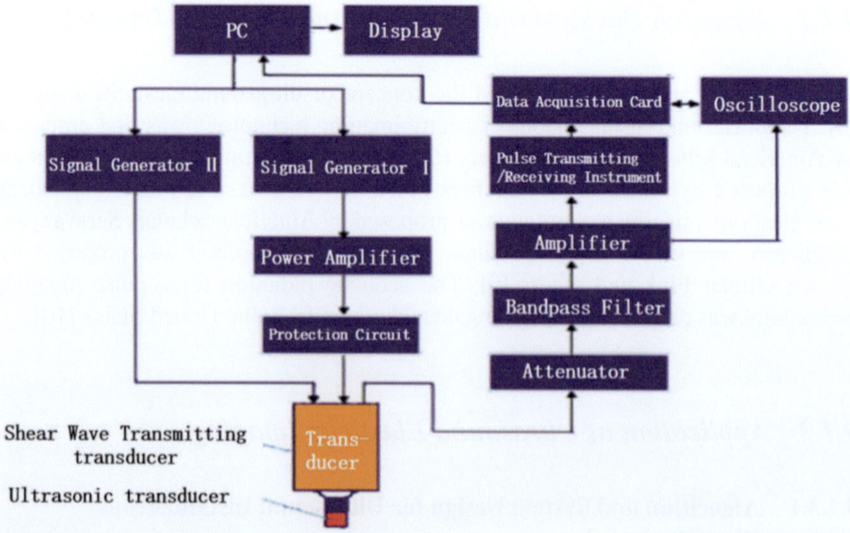

Fig. 9.1 Framework diagram of the ultrasound instantaneous elasticity imaging system [1]

acoustic radiation force elasticity imaging cannot perform accurate real-time two-dimensional imaging. GPU, as a tool widely used in medical imaging, has stronger dense computing power. The acoustic radiation force elasticity imaging algorithm includes displacement estimation, difference calculation, shear wave speed estimation, and elastic modulus estimation. Improve its algorithm, build a CUDA-based filtering algorithm, construct CUDA implementations of analytic signals and cross-correlation, CUDA-based difference algorithm, and CUDA-based Radon transform. The experiment uses a self-made elastic phantom, selects a square soft body implanted with harder material, selects a rectangular area spanning the background and high hardness area for acoustic radiation force elasticity imaging, and tests its sensitivity to materials with different elastic coefficients. Through the quality comparison of elasticity images based on CPU and GPU, the single-precision operation of GPU does not have a significant impact on image quality, and the algorithm on GPU has a faster running speed [3], Fig. 9.2 is the carrier-to-noise ratio and signal-to-noise ratio curve of GPU and CPU acoustic radiation force imaging.

9.1.3.3 Research on the Improvement of Shear Wave Speed Estimation Method Based on Radon Transform in Ultrasound Radiation Force Elasticity Imaging

Acoustic Radiation Force Impulse Imaging (ARFI) generates acoustic radiation force by focusing ultrasound beams, which acts on human soft tissues to cause local micro-displacements, and then triggers shear waves to propagate around. As a non-invasive quantitative measurement of tissue elastic modulus, ARFI has been proven to have

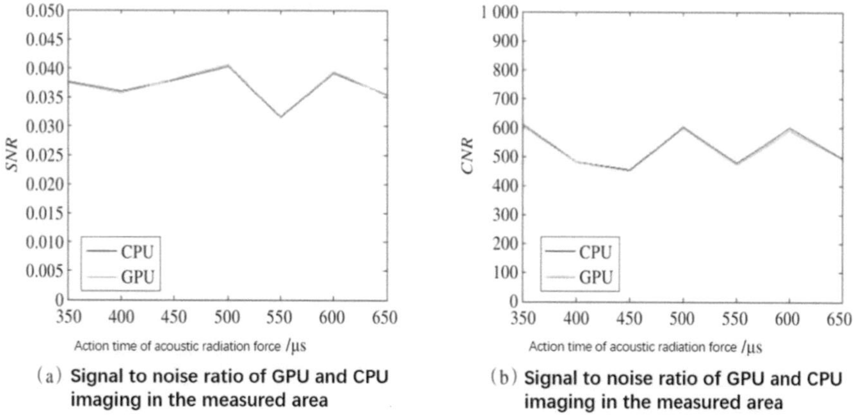

Fig. 9.2 Carrier-to-noise ratio and signal-to-noise ratio curve of GPU and CPU acoustic radiation force imaging [3]

great application potential in the diagnosis of various diseases. In the algorithm part of ARFI, there is less research on the algorithm part of shear wave speed estimation. Previously, Rouze and others proposed an improved method based on Radon transform to estimate shear wave speed, but the appearance of singular values will cause errors in the calculation results [11]. Therefore, improving this algorithm can help find a new type of shear wave estimation algorithm that balances measurement stability and calculation speed. Type (I): In the algorithm of Radon transform based on the "time (t)-lateral position (x)" coordinate displacement matrix, three methods are proposed to select a specific depth for calculation: the focal depth of generating acoustic radiation force, a few artificially selected depths, and a combination of all depths. Type (II): Based on the "time (t)-depth (z)" In the algorithm for performing Radon transformation on the displacement matrix of coordinates, the exact time when the shear wavefront passes through these lateral measurement positions is determined. Three improvements are proposed for this purpose: using Radon transformation to help determine the time, introducing the concept of threshold to determine the time in the first method, and combining the two types of algorithms to determine the time and depth. Elasticity measurement experiments and RF signal collection were carried out using the acoustic radiation force elasticity imaging system, and six improved algorithms were tested on homemade tissue-mimicking phantoms and fresh pork tissue samples. Among the six estimation methods, the three fastest in terms of computation time are Type (II) Method (2), Type (I) Method (3), and Type (I) Method (1). For all algorithms, the positions with the smallest coefficient of variation measured on the elastic phantom are at a depth of 5 cm, while on pork they are at a shallower position, and the variation values measured on pork are greater than those on the elastic phantom. The results suggest that when the depth is shallow, Method (2) or (3) in Type (I) should be used, and when it is deeper, Method (2) in Type (II) is recommended [12].

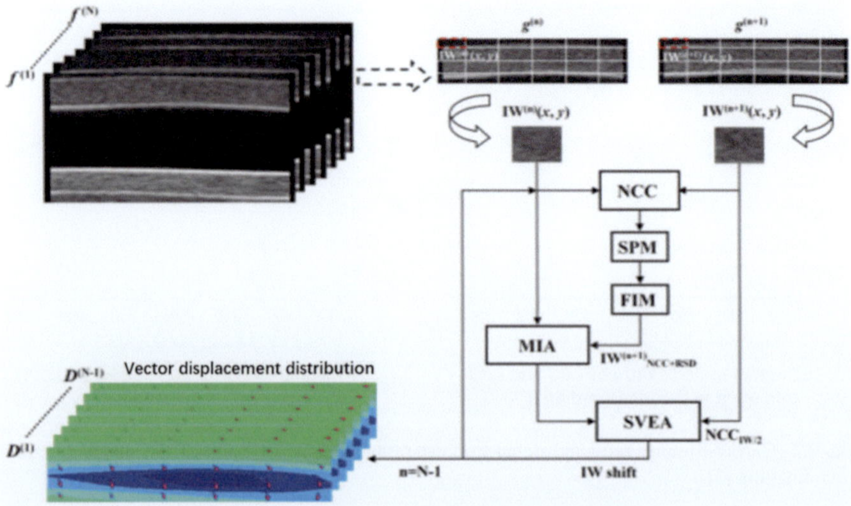

Fig. 9.3 Schematic diagram of ultrasound texture matching method [13]

9.1.3.4 Ultrasound Experimental Study on Measuring Plaque Tissue Hardness Based on Ultrasound Texture Matching Method

Plaque rupture, as a major trigger for the onset of cardiovascular and cerebrovascular diseases such as stroke, mainly consists of soft and hard tissues. The larger the proportion of soft tissue, the easier the plaque is to rupture. Texture matching is an ultrasound-based vascular elasticity imaging method that can be used to measure the elasticity distribution of plaques. Two different hardness phantoms were prepared using polyvinyl alcohol, and a pulsatile pump was used to simulate human heart motion to obtain dynamic images of the phantoms. The phantoms were irradiated using B-mode ultrasound and ultrasound texture matching method to obtain images. B-mode images are difficult to distinguish between soft and hard tissues in the phantom, while elasticity images can clearly show the difference between soft and hard tissues in the plaque [13]. Figure 9.3 is a schematic diagram of the ultrasound texture matching method.

9.1.3.5 Non-invasive Liver Fibrosis Detection Technique Based on Ultrasound Instantaneous Elasticity Imaging

It is known that the elastic modulus of liver tissue changes with the progression of liver fibrosis, making ultrasound elasticity imaging a non-invasive and rapid detection method. When a strain is applied to soft tissue, shear waves decay significantly in the ultrasound frequency band, but decay very little in the low frequency range. Ultrasound elasticity imaging utilizes the propagation of two different frequency

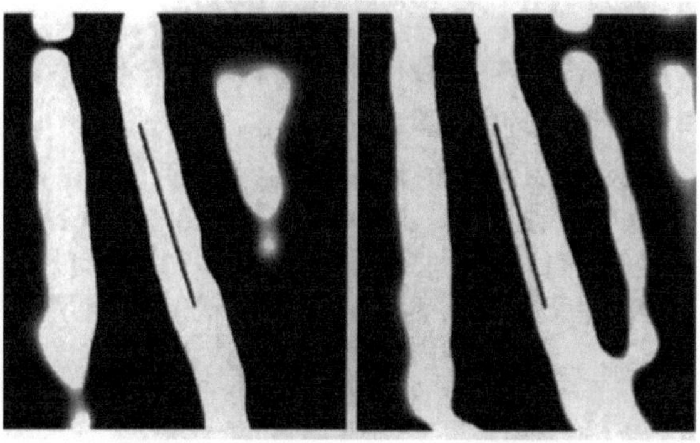

Fig. 9.4 Human experiment results, Volunteer 1 (left) and Volunteer 2 (right) [14]

and velocity sound waves in human soft tissue, and the tissue displacement changes caused by shear waves can be recorded by adjacent echo signals, from which the elastic modulus of the tissue can be calculated. Two healthy volunteers were selected, each measured 10 times. The volunteers' livers were in a normal state, confirming the feasibility of the preliminary human experiment of ultrasound instantaneous elasticity imaging [14], Fig. 9.4 shows the results of the human experiment.

9.1.3.6 Adaptive Error Compensation for Viscoelastic Measurement Based on Instantaneous Elasticity Imaging

Many current methods describing elastic tissue changes focus only on assessing tissue hardness, ignoring viscosity. Studies have shown that malignant tumors with high vascular density exhibit higher viscosity than surrounding tissues. Traditional elasticity imaging cannot distinguish between malignant and benign tumors, while ultrasound instantaneous elasticity imaging has the potential to reconstruct the viscoelasticity of the medium.

Ten metal discs of different radii were fixed on a miniature vibrating screen and placed on a polyvinyl alcohol phantom. The miniature vibrating screen vibrated at a frequency of 50 Hz, propagating shear waves in the model. The displacement caused by the shear waves was tracked and recorded. Simulation experiments showed that the radius of the disc and the selected depth of the displacement field are the two main factors affecting measurement bias. When the radius of the vibrator is between 5 and 13 mm, the deviation between the simulation results and the experimental results is less than 1%, and when the radius increases to 21 mm, the deviation is greater than 5%. By decomposing the shear wave attenuation into viscous and diffraction parts, the offset correction method described in the literature can correct both shear wave speed and shear wave attenuation [15].

9.1.3.7 The Effect of Detection Depth on Real-Time Shear Wave Elasticity Imaging of the Liver

Experiments show that ultrasound elastography is highly accurate in diagnosing liver fibrosis. However, due to the many factors influencing elastography, it can lead to instability in measurements. Studying the factors affecting ultrasound elastography is crucial for improving its detection success rate and stability. We selected 52 male and 37 female healthy examinees and recorded data such as height and weight. Using a color Doppler ultrasound diagnostic device, we started from the liver capsule, recording 8 sets of data every 1 cm, and repeated this three times. The experiment showed that the detection depth significantly affects ultrasound elastography. The success rate is lower when close to the liver capsule and at greater depths, and higher at 1, 2, and 3 cm from the liver capsule [4].

9.1.3.8 Research on Acoustic Radiation Force Impulse Imaging Technology and Its Algorithm

Ultrasound elastography has the advantages of being non-invasive, real-time, widely applicable, and easy to operate, making it widely used in clinical detection. The propagation of ultrasound waves in biological tissues is complex. In research, the acoustic radiation force is generally defined as the force generated by the energy absorbed and scattered by the tissue during propagation. Since the tissue displacement caused by the propagation of shear waves is very small, it is necessary to first analyze the radio frequency signal and perform cross-correlation calculations to reduce errors. Using the acoustic radiation force impulse elastography experimental system, combined with the shear wave theory algorithm, the shear wave speed in the phantom is calculated. The experimental results and the given shear wave speed in the phantom have a very small error, verifying the feasibility of the algorithm [16].

9.1.3.9 Multi-physics Finite Element Analysis of Viscoelastic Tissue Strain Caused by Acoustic Radiation Force

Currently, elastography mainly analyzes the viscous behavior of biological tissues based on their time response to mechanical excitation. It is very important to further study the impact of tissue viscosity on imaging. Finite element method is an indispensable step in two-dimensional and three-dimensional simulation of elastography. Using this method, a system that approximates an infinite number of unknowns with a finite number of unknowns can be established through interacting micro viscoelastic mechanical units. After simulating a uniform tissue, placing it in water, and applying a certain excitation, its three-dimensional deformation and echo sound field are observed. Three-dimensional simulation is performed to display the displacement deformation related to the linear elasticity and viscoelasticity of the uniform tissue, two-dimensional simulation is performed to analyze the strain and stress of the two-dimensional section related to elastography, and one-dimensional simulation is performed to observe the strain response of different materials under the same stimulus. The simulation results show that the three-dimensional, two-dimensional, and one-dimensional analyses of elastography under different tissue properties can clearly see the differences in stress and strain, indicating that the impact of tissue viscoelasticity on imaging cannot be ignored [17]. Figure 9.5 describes the strain and stress distribution of the uniform phantom tissue under the external excitation of 50 Hz.

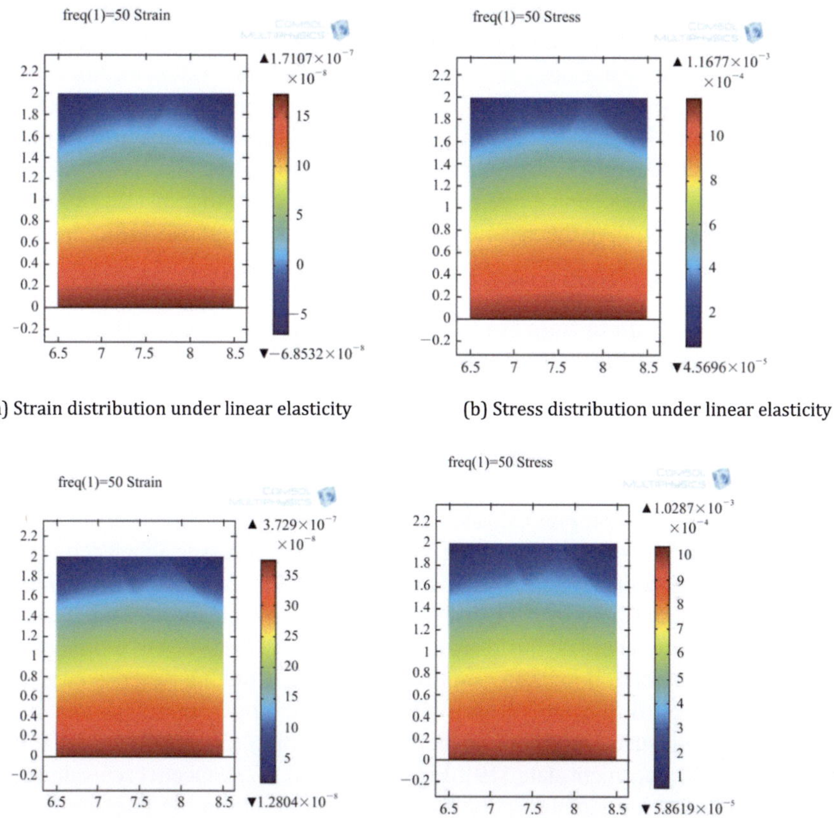

(a) Strain distribution under linear elasticity (b) Stress distribution under linear elasticity

(c) Strain distribution under viscoelasticity (d) Stress distribution under viscoelasticity

Fig. 9.5 Strain and stress distribution of uniform phantom tissue under external excitation of 50 Hz [17]

9.1.3.10 Real-Time Ultrasound Elastography Image Feature Quantification Method and Its Application Value

Current research on ultrasound elastography mainly focuses on how to construct elastographic images, but lacks research on image information and clinical diagnostic results. The five-point method and the elasticity ratio method, as the most commonly used methods for evaluating breast tumors today, limit the application of elastography technology. Therefore, a new method for evaluating breast tumors is needed for ultrasound elastography. It consists of three parts: elastographic image reconstruction, automatic segmentation of the tumor area, and extraction of elastic feature quantification. The elastic feature quantification includes the hardness rate of the tumor area, the elasticity ratio, and the elasticity score. Using a color Doppler ultrasound diagnostic device, tumor images of 251 patients were taken, including 162 benign and 89 malignant cases. The analysis of the elastic features of tumors on real-time ultrasound elastographic images showed significant differences in the three features between benign and malignant tumors, indicating that this evaluation method helps to improve the accuracy and objectivity of diagnosis [18].

9.1.3.11 Algorithm to Improve the Motion Estimation Accuracy of Quasi-Static Elastography and Its Phantom Study

In the composition of biological tissues, soft tissues account for the majority, which have anisotropic and nonlinear mechanical characteristics. In elastography, the displacement information is extracted using the radio frequency echo data before and after the tissue is compressed, and the internal elasticity distribution of the tissue can be indirectly obtained. Therefore, the estimation of tissue displacement and strain is a key issue in elastography. Using Sonix RP Experiments were conducted on a B-mode ultrasound system, using three-layer soft phantoms and phantoms containing columnar hard bodies. The results showed that the use of two-dimensional displacement estimation methods significantly reduced the interference caused by lateral motion on displacement estimation. Singular points that appeared in the experiment were successfully eliminated by segmenting and fitting the axial one-dimensional displacement data [19]. Figures 9.6 and 9.7 respectively evaluate the effects of eliminating singular points and median filtering in the phantom experiment containing columnar hard bodies, and compare the strain estimation of three differential methods.

9.1.3.12 Application of Wavelet Denoising in Instantaneous Elasticity Imaging Estimation

Accurate calculation of tissue strain based on displacement estimation is a critical issue in ultrasound elasticity imaging algorithms. Typically, biomedical ultrasound signals are low frequency, while noise signals are concentrated in the high frequency

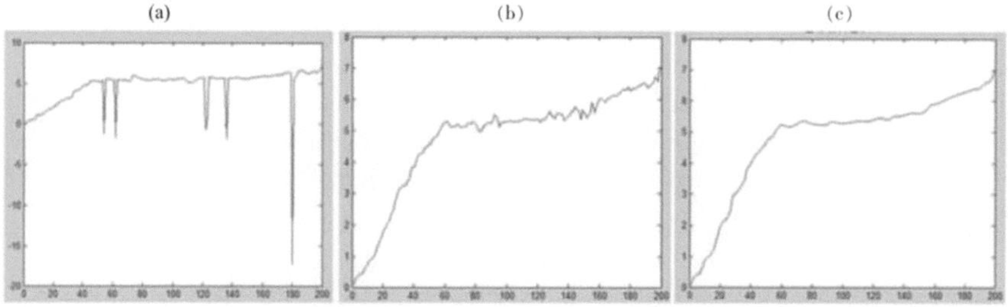

Fig. 9.6 Evaluation of the effects of eliminating singular points and median filtering in the phantom experiment containing columnar hard bodies [19]

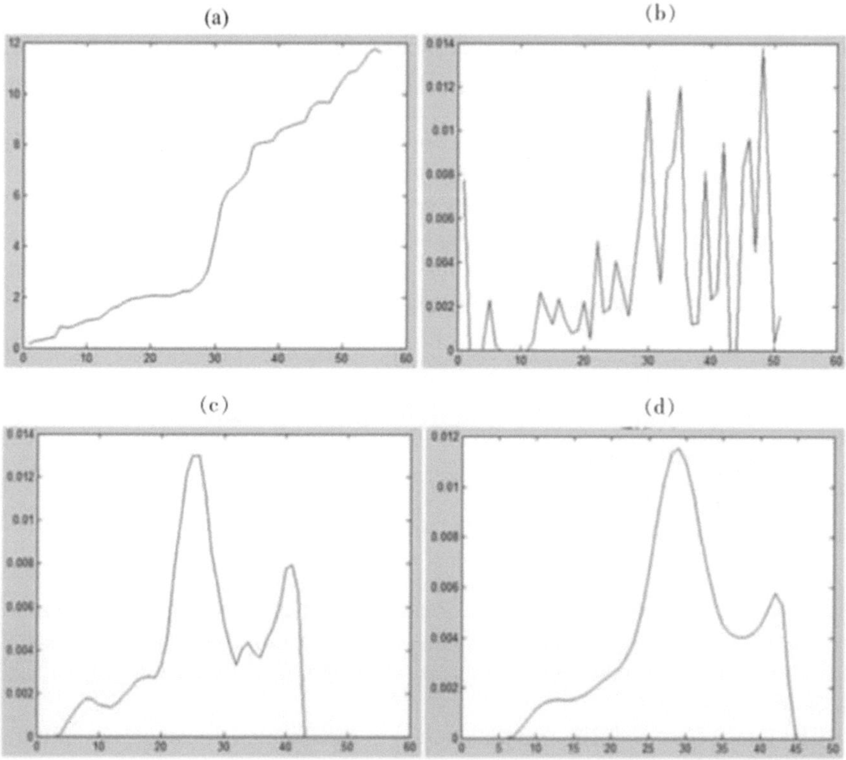

Fig. 9.7 Comparison of strain estimation by three differential methods (horizontal axis represents depth direction, vertical axis represents strain quantity) [19]

area, so wavelet denoising can be considered as a low-pass filter. Applying the Bayesian threshold method to the ultrasound instantaneous elasticity imaging system, a wavelet threshold denoising algorithm can be obtained. According to the algorithm in the literature, experiments were conducted using ultrasound data from a chemical ultrasound elasticity membrane and the liver of a healthy volunteer. The results showed that the elasticity membrane experiment achieved a good visual effect of the strain map and a higher signal-to-noise ratio, while the data in the liver experiment of the healthy volunteer was lower than the mean filter [20]. Figure 9.8 is the liver strain map.

9.1.3.13 Detection of Focused Ultrasound-Induced Ex Vivo Tissue Damage Using Quasi-Static Elasticity Imaging Technology

It has been found that B-mode imaging cannot accurately reflect the tissue damage caused by focused ultrasound. The tissue damage caused by focused ultrasound leads to changes in tissue elasticity, making elasticity imaging promising for monitoring the efficacy of focused ultrasound. Initially, phantoms containing hard bodies were used to optimize imaging effects. Fresh ex vivo uniform pig muscle tissue was then irradiated with a focused ultrasound probe to cause damage. The pig tissue was placed in a water tank and observed using the elasticity imaging system. The phantom experiment found that singular points easily appear in displacement estimation, which were eliminated by linear segment fitting to optimize the imaging effect. The ex vivo tissue experiment found that the elasticity imaging system can detect damage within the pig tissue and reflect the thermal damage situation [21].

9.2 Manipulation of Nanoparticles Based on Ultrasound Radiation Force

9.2.1 Overview of Particle Manipulation

Acoustic manipulation of particles has received widespread attention in the last 20 years. The relationship between the number of articles found in Web of Science with "acoustic radiation force" and "manipulation" as the theme keywords and the year is shown in Fig. 9.9.

The principle of acoustic manipulation is to use the mechanical effects of ultrasound waves. The reflection, refraction, absorption, etc., of particles in the sound field cause the momentum carried by the sound field to be exchanged between the sound wave and the particles, and the particles are manipulated and moved by the acoustic radiation force. The effect of the acoustic radiation force decreases rapidly with the reduction of particle size. When the particle diameter is much smaller than the wavelength, the acoustic streaming effect plays a major role. Conversely, when the particle diameter is much larger than the wavelength of the sound wave, the particle nature of the sound wave can be considered, and the strength of the sound field can be represented by the density of the sound lines.

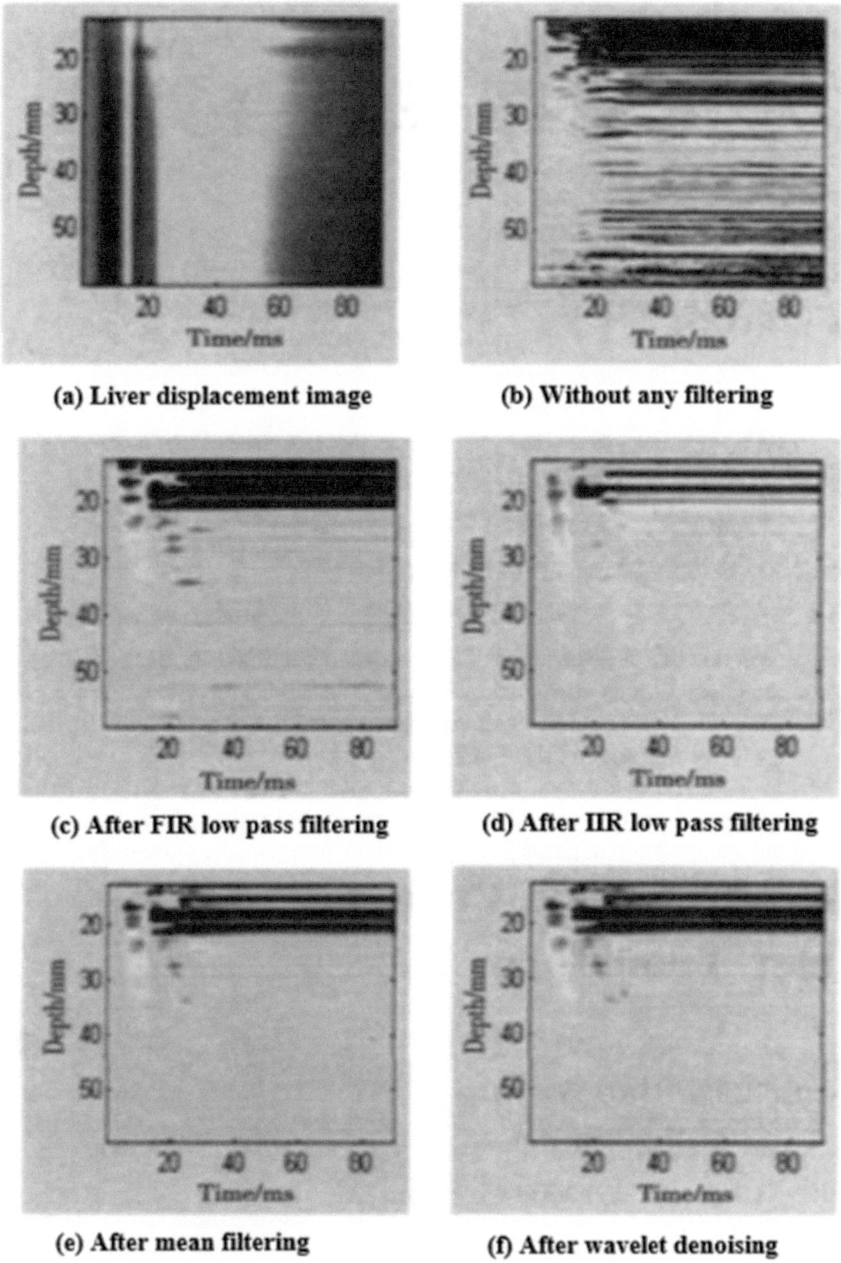

Fig. 9.8 Liver strain map [20]

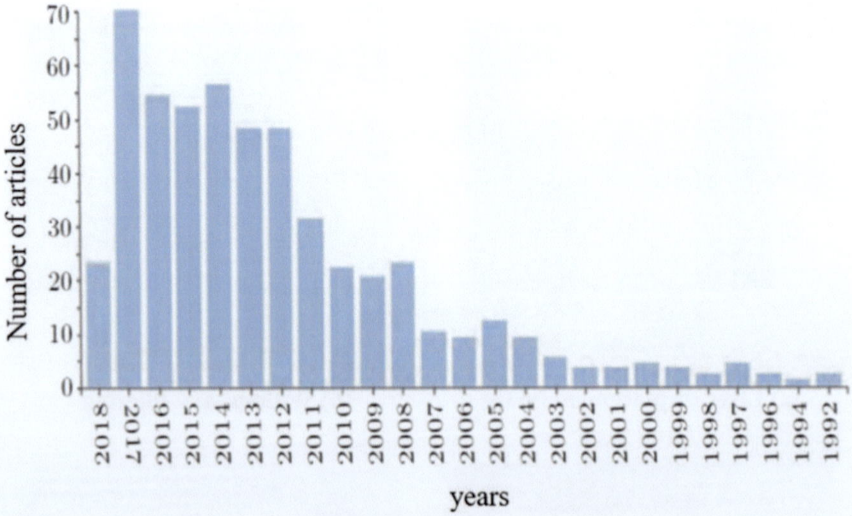

Fig. 9.9 Relationship between the number of articles with "acoustic radiation force" and "manipulation" as themes in web of science and the year [22]

Based on the two classifications of acoustic radiation force, the methods of ultrasound manipulation of particles can be divided into bulk wave manipulation and surface acoustic wave manipulation, and surface acoustic wave manipulation has the characteristics of miniaturization and energy concentration [23]. Figure 9.10 is a schematic diagram of the interaction between surface acoustic waves and liquids.

Based on the type of sound field generated by one or two transducers, methods of ultrasonic manipulation of particles can be divided into standing wave manipulation, traveling wave, and focused sound field manipulation. Because the standing wave field has the characteristic of energy extremes in space, particles can be stationed at the antinode or node positions. By controlling the phase, frequency, and other parameters of the standing wave field, the positions of the antinodes and nodes can be changed, enabling the manipulation of particles [23]. The traveling wave field generated by a single transducer has no gradient in field strength, and it is generally

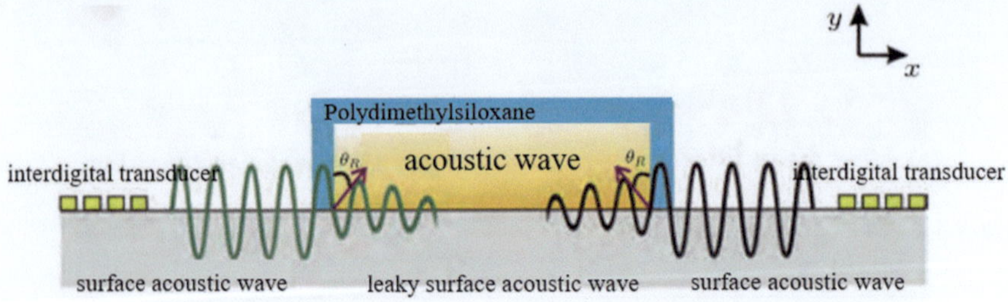

Fig. 9.10 Schematic diagram of the interaction between surface acoustic waves and liquid [23]

difficult to observe the movement of particles much smaller than the wavelength in the traveling wave field. Figure 9.11 shows the principle of moving particles with a surface acoustic standing wave, and Fig. 9.12 provides a schematic diagram of arranging and sorting particles by combining surface acoustic standing waves and traveling waves.

Because the force distribution on particles in the sound field generated by a single transducer element is fixed and difficult to manipulate flexibly, the development of technology has led to the birth of particle manipulation technology based on dynamically controlled sound fields generated by transducer arrays. If the sound parameters of each transducer array element can be controlled, in principle, a time-varying arbitrary sound field can be generated in space, enriching the types and modes of particle motion [22]. Figure 9.13 shows the arbitrary manipulation of particles by the controllable sound field generated by a large-scale planar transducer array.

Sound fields formed by resonance, interference, scattering, etc., based on structure, can also be used for acoustic manipulation. Although planar transducer arrays

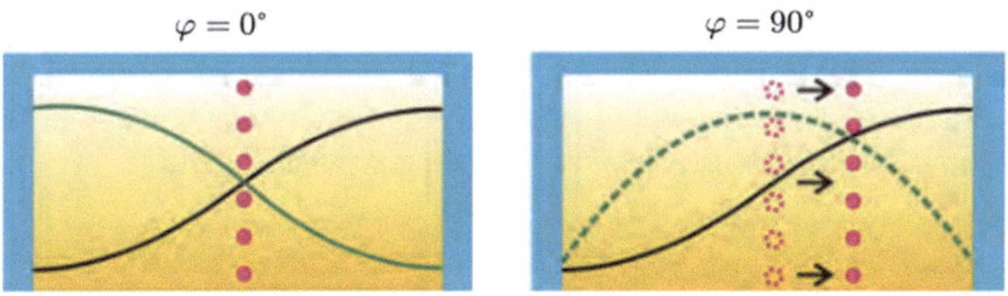

Fig. 9.11 Principle of moving particles with a surface acoustic standing wave [23]

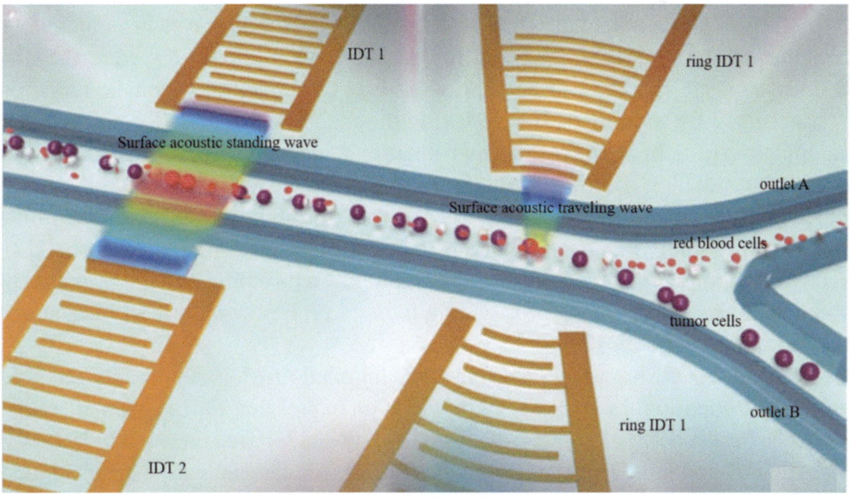

Fig. 9.12 Schematic diagram of arranging and sorting particles by combining surface acoustic standing waves and traveling waves [22]

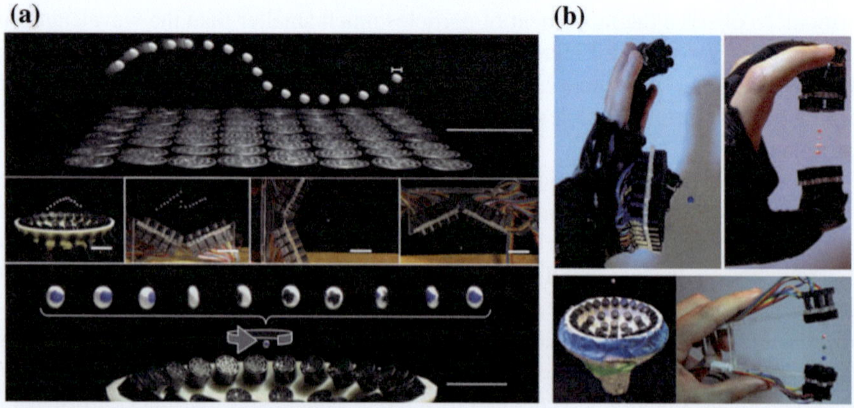

Fig. 9.13 Arbitrary manipulation of particles by the controllable sound field generated by a large-scale planar transducer array [22]

can flexibly manipulate particle motion, they are expensive to manufacture, and in specific application scenarios, the sound field pattern is basically fixed. Particle manipulation based on structural sound fields reduces cost requirements and further pushes acoustic manipulation technology into the application field.

9.2.2 History of Particle Manipulation Research

As early as 1902, Rayleigh proposed the concept of "acoustic radiation pressure" [24]. The pressure gradient generated by the motion of sound waves in a continuous medium produces acoustic radiation force, and Dayton and others were the first to study acoustic radiation force [25]. King first calculated the acoustic radiation force received by a rigid small ball, laying the theoretical foundation for acoustic manipulation [26]. In 1955, Yosioka and others extended this theory to the calculation of acoustic radiation force on wireless small balls [27], while Gor'kov and others provided an approximate expression for the force on infinitesimal particles in the sound field, simplifying the calculation method of acoustic radiation force [28]. Acoustic manipulation technology was first proposed by Junru Wu, the principle of which is to use two opposing propagating focused probes to generate a local sound field to capture particles [29].

In recent years, research on acoustic manipulation of particles has received widespread attention. In 2011, Silva provided an expression for the acoustic radiation force on a sphere under any waveform [30]; in 2013, based on Gor'kov's theory, Sapozhnikov and others supplemented the formula for calculating the acoustic radiation force of a sound beam suitable for any waveform [31]. Acoustic artificial structures have also been introduced into the regulation of acoustic radiation force.

These studies have deepened the theoretical foundation of acoustic manipulation and enriched the application prospects of acoustic manipulation.

9.2.3 Application Examples

9.2.3.1 Research on Nanoparticle Manipulation Based on Ultrasonic Radiation Force

Periodic nanomaterials possess unique optical properties due to plasmonic resonance. The refractive index of the surface of periodic nanomaterials changes with the mass variation of attached biomolecules, leading to changes in the resonance peak of the material. This dynamic change can reflect specific information about the interactions between biomolecules. Therefore, periodic nanomaterials can be used for real-time observation of interactions between proteins, peptides, and cells, and have extensive applications in the field of biosensing technology. Current methods of acoustically controlling nanomaterials mainly focus on single-scale nanowires or carbon nanotubes. The literature primarily explores the use of acoustic radiation force to control full-size nanoparticles. Experiments use surface acoustic waves to manipulate nanoparticles. Surface acoustic waves are excited by two pairs of interdigital transducers in a PDMS channel, and gold nanocages are injected into the PDMS channel. The experimental results are compared with simulation results. The literature changes the results of nanoparticle manipulation, increases the acoustic radiation force on the particles, and achieves ultrasonic radiation force control of nanoparticles [32]. Figure 9.14 shows the simulation and experimental results.

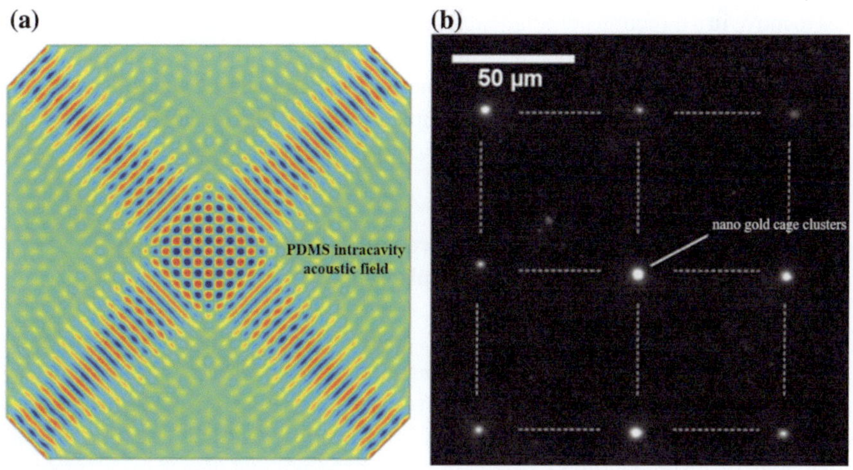

Fig. 9.14 Simulation and experimental results [32]

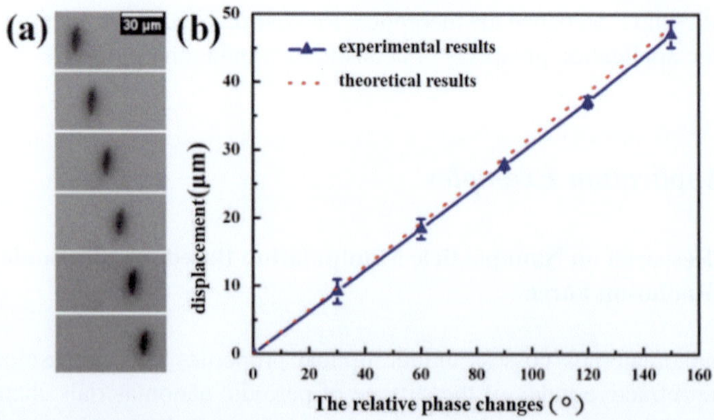

Fig. 9.15 Phase control results [33]

9.2.3.2 Two-Dimensional Manipulation Research Based on One-Dimensional Surface Acoustic Wave Chips

Currently, scholars have achieved two-dimensional manipulation of microbubbles through two-dimensional surface acoustic wave chips. Using a simpler one-dimensional surface acoustic wave chip composed of a pair of interdigital transducers, two-dimensional manipulation of particles is also achieved. The one-dimensional surface acoustic wave chip is composed of a pair of parallelly arranged tilted interdigital transducers. By changing the relative phase and frequency of the input signal of the interdigital transducer, the motion trajectory of the manipulated particles is observed. The experiment used a PDMS channel and injected microbubbles, and the operation process was recorded with a microscope. Phase control causes microbubbles to move in a direction perpendicular to the sound aperture, and frequency control causes microbubbles to move parallel to the sound aperture. The experiment is consistent with theoretical values. Therefore, a one-dimensional surface acoustic wave chip can achieve two-dimensional particle manipulation [33]. Figures 9.15 and 9.16 show the results of phase and frequency control, respectively.

9.2.3.3 Acoustic Manipulation of Micro and Nano Particles Based on Cylindrical Thin-Shell Localized Acoustic Field

Previous acoustic manipulation techniques are difficult to flexibly control, and there is an urgent need to develop new acoustic manipulation techniques. The main mechanism of the manipulation technique of micro and nano particles based on the localized strong field mode of the cylindrical thin shell in the low frequency section is used. First, numerical simulation is carried out on the experimental device to calculate the acoustic transmission spectrum and field distribution of the system, and then PS balls are used for manipulation experiments. The experimental situation is consistent with

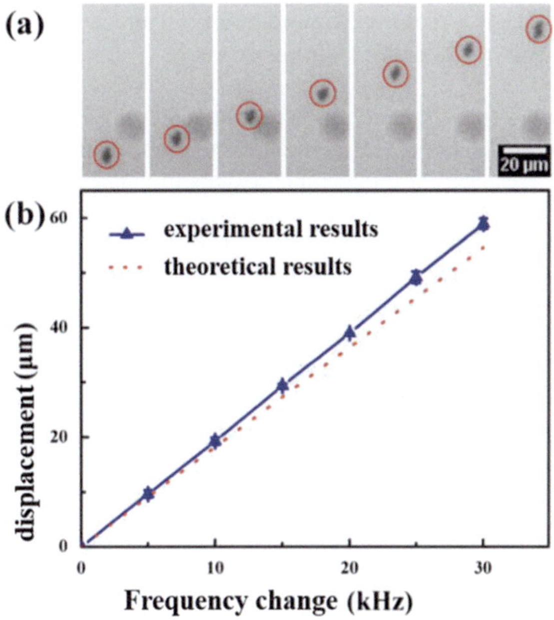

Fig. 9.16 Frequency control results [33]

numerical simulation, and the particles are captured by the cylindrical thin shell to the inner wall of the thin shell [34]. Figure 9.17 shows the result of manipulating PS microspheres.

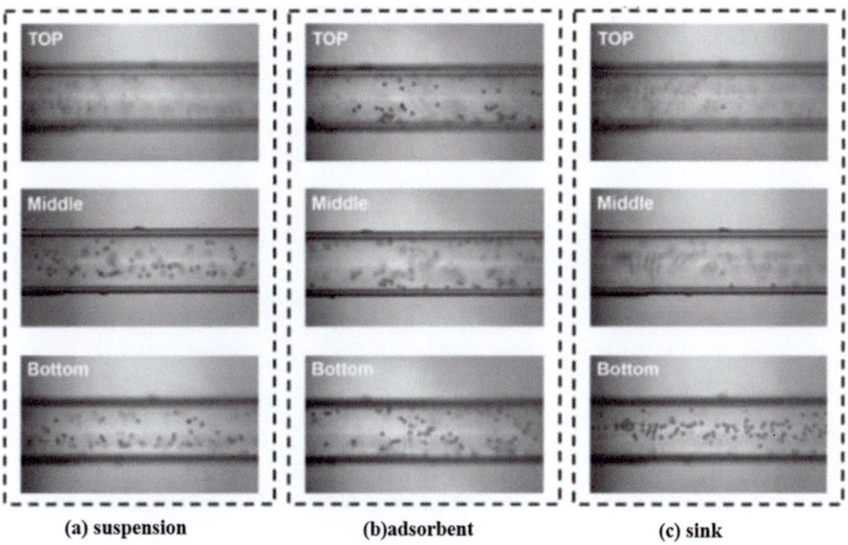

Fig. 9.17 Result of manipulating PS microspheres [34]

9.2.3.4 Acoustic Manipulation of Particles by One-Dimensional Acoustic Grating in Air

Artificial structures such as phononic crystals and acoustic metamaterials have been proven to flexibly control the propagation and distribution of sound waves. Researchers use phononic crystals to control the acoustic field in water, thereby changing the size and direction of the acoustic radiation force on particles. This paper focuses on the distribution and characteristics of the acoustic radiation force received by the surface of a one-dimensional acoustic grating. First, the projection properties and surface acoustic field distribution of the one-dimensional acoustic grating are studied, and the method of finite element combined with momentum tensor is used to study the acoustic radiation force on particles on the surface of the acoustic grating. To further obtain the force characteristics of particles on the surface of the acoustic grating, the spatial distribution of the acoustic radiation force on polystyrene foam micropillars at two resonance frequencies is studied. The simulation results show that when the resonance wavelength is equal to the grating period, the particles on its surface are attracted to the grating plate; when the wavelength is more than twice, they are attracted to the narrow slit, and the intensity is much smaller than the first one [35]. Figure 9.18 shows the force spatial distribution of PS foam micropillars.

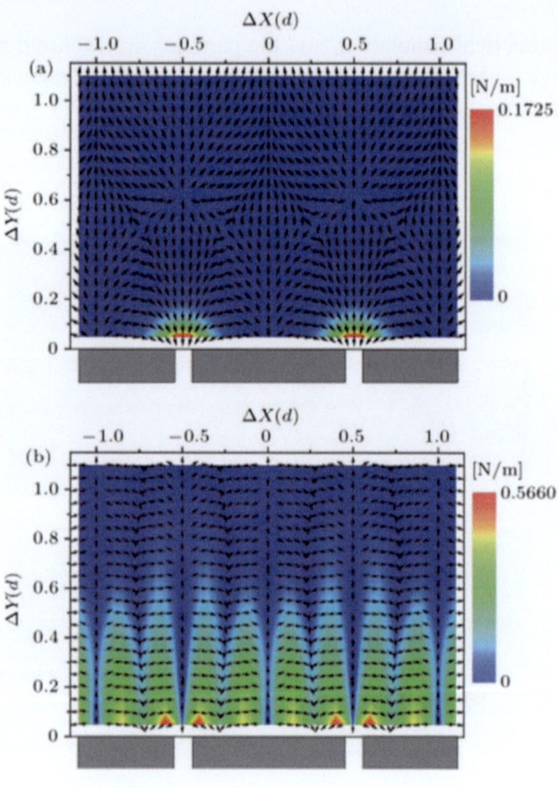

Fig. 9.18 Force spatial distribution of PS foam micropillars with a radius of 0.04d in the X direction Δ X from − 1.1d to 1.1d, Y direction Δ Y from 0.05d to 1.1d [35]. **a** Frequency is 22.7 kHz; **b** Frequency is 39.6 kHz

9.2.3.5 Study of Acoustic Radiation Force Near Sub-Wavelength Slit on Steel Column

The theoretical study of acoustic radiation force mainly involves the theory of scattered waves, with current research primarily focusing on regular scattering bodies in simple acoustic fields. Using a theory based on time-domain finite difference methods and stress tensor combinations, it is possible to study the acoustic radiation force experienced by steel columns near sub-wavelength slits. Numerical calculations of the transmission rate of the slit system show that when the plate is thin and the aperture is large, the transmission rate changes very little; when the plate is thick and the aperture is small, a noticeable resonance phenomenon appears in the low-frequency part of the transmission rate. When within the range of low-order resonance frequencies, the acoustic radiation force on the steel column is greater than that on a seamless structure. Also, within this range, as the thickness of the plate increases, the acoustic radiation force increases [36].

9.2.3.6 Three-Dimensional Acoustic Manipulation of Particles in Cylindrical Microcavities

Acoustic manipulation technology, with its non-contact, non-destructive, and highly penetrative advantages, has broad application prospects in fields such as biochemistry and environmental monitoring. In previous studies, most researchers only studied the standing wave effect in the direction of sound wave propagation, without considering transverse standing waves. Based on a cylindrical microcavity device etched from a silicon wafer, numerical simulations and experiments of three-dimensional acoustic manipulation were first conducted in the microcavity. The sound field distribution in the cylindrical cavity was simulated, and the equilibrium position of the microsphere under three-dimensional force in the cylindrical cavity was calculated. Finally, in the experiment of acoustic manipulation of microspheres in the cylindrical cavity, a stratification phenomenon was observed in the distribution of microspheres, validating the theoretical calculations. Theoretical calculations show that when gravity, buoyancy, and lateral forces are not considered, PS microspheres will form two Bessel circles at different heights on the x–y plane, which is consistent with the experiment. This research can provide experimental support for using cylindrical microcavities as carriers to manipulate micro-particles [37], Fig. 9.19 shows the experimental results of acoustic manipulation of microspheres in the cylindrical cavity.

9.2.3.7 Research on Cell Capture Based on Single Microbubble Resonance

Because microfluidics operates in a laminar flow state with a small Reynolds number, it is difficult to manipulate particles and mix liquids. The literature suggests using the second-order radiation force of ultrasound to manipulate particles and capture cells.

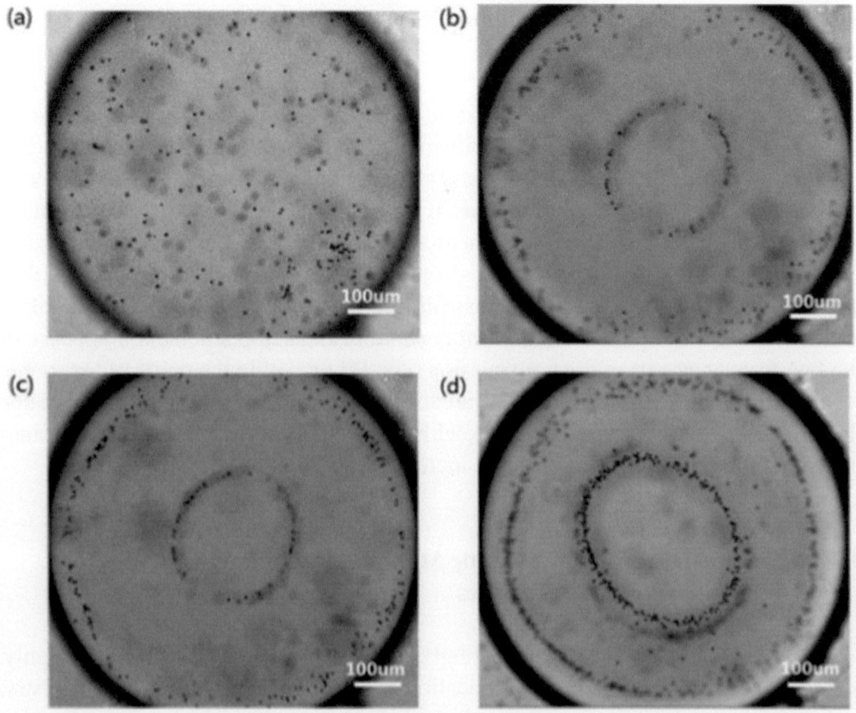

Fig. 9.19 Experimental results of acoustic manipulation of microspheres in cylindrical cavity [37]

First, a circular hole with a diameter of 40 μm is made on PDMS, and a microbubble is formed when the liquid flows under the second-order acoustic radiation force. The prepared chip and probe are placed in a circular channel, and a mixture of PS solution and water is added. Under a microscope, the movement of PS balls is observed. A suspension of mouse brain microvascular endothelial cells is injected, and the effect of vibrating microbubbles on cells is observed. The presence of secondary harmonics can be observed in the presence of microbubbles. Under the action of ultrasound, the fluid forms a symmetrical vortex-like acoustic microflow under periodic rapid expansion and contraction. Microbubbles in the channel capture cells, and the vibration of the microbubbles has almost no effect on the activity of the microbubbles [38], Fig. 9.20 shows the trajectory of the PS ball movement.

9.2.3.8 Research on Rapid Microfluidic Acoustic Mixing Method Based on Microbubble Resonance

Rapid mixing of laminar flow liquids is an important part of sample pretreatment in microfluidic control systems, but it is currently difficult to achieve rapid and uniform mixing of liquids. However, by using an external sound field to drive the liquid, the

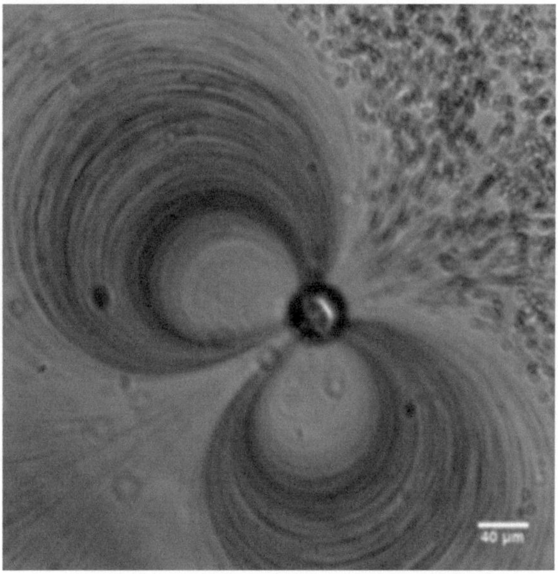

Fig. 9.20 Trajectory of PS ball movement [38]

fluid forms an acoustic flow after absorbing the sound wave energy, and the shear force of this acoustic flow can also achieve the purpose of mixing liquids [39]. This method does not change the nature of the original fluid, but it has the disadvantages of low energy absorption efficiency and significant thermal effects. First, a bottom micropore structure is designed, and the frequency range required for the experiment is determined by calculation. Experiments are conducted in an open circular PDMS channel. PS ball solution is injected into the channel, and the movement of the balls is recorded using a high-speed camera under a microscope. Two types of liquids are injected, and the degree of solution mixing is observed and calculated under the condition of microbubble vibration. In the presence of microbubbles, secondary harmonic components appear in the spectrum, and the microbubbles are in a stable cavitation state. The microflow caused by the vibration of the microbubbles causes the balls to move in a symmetrical vortex-like manner in the channel. This vortex microflow is conducive to the uniform mixing of the two solutions, and the thermal effect is not significant under this condition, Fig. 9.21 shows the acoustic microflow generated by the vibration of the microbubbles, and Fig. 9.22 shows the relative mixing index under different energies.

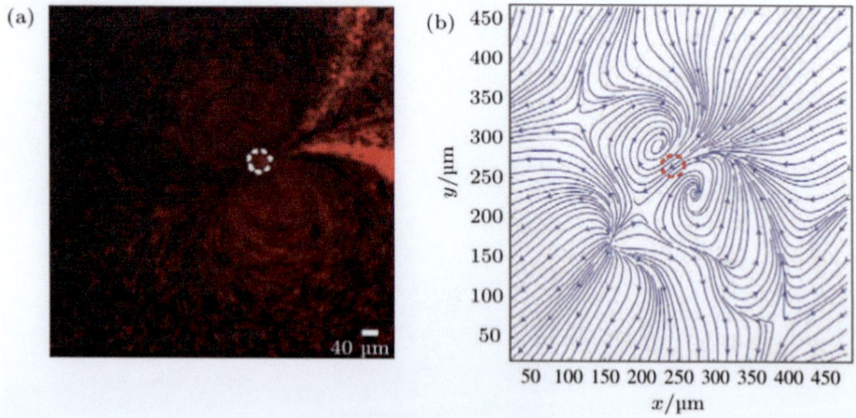

Fig. 9.21 Acoustic microflow generated by microbubble vibration [39]

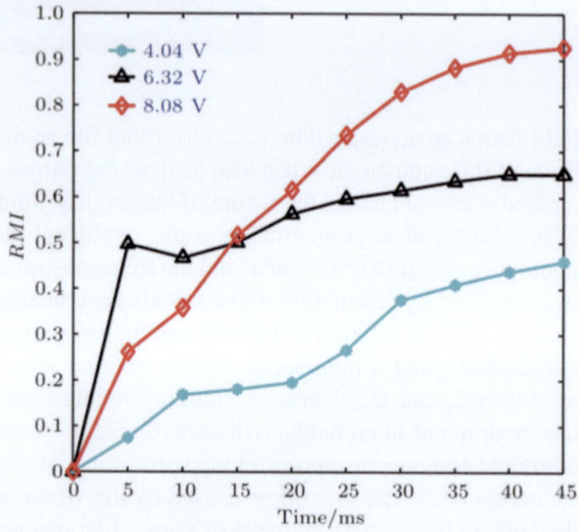

Fig. 9.22 Relative mixing index under different energies [39]

9.3 Acoustic Radiation Force Manipulation of Microbubbles: Ultrasound Molecular Imaging

9.3.1 Basic Principles of Ultrasound Molecular Imaging

Ultrasound imaging, as one of the important diagnostic techniques in clinical practice, traditionally uses linear scattering echo signals to detect tissue structure and characteristics, resulting in relatively low image contrast and signal-to-noise ratio. The

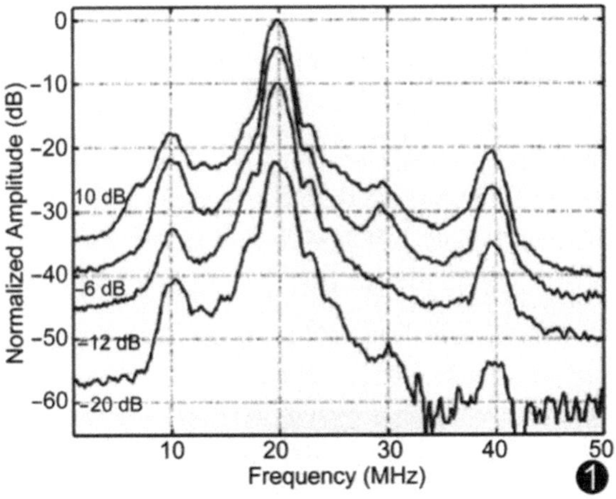

Fig. 9.23 High-frequency nonlinear response of ultrasound contrast microbubbles [41]

advent of ultrasound contrast agent microbubbles has been hailed as the third revolution in ultrasound technology. Ultrasound contrast imaging, by adding a certain dose of contrast agent, can significantly enhance the resolution, sensitivity, and specificity of ultrasound diagnosis. Ultrasound contrast agents are very small, generally less than 8 microns in diameter, and are biocompatible microbubbles with a shell structure, allowing them to move within the human vascular network [40]. Moreover, due to the compressibility of the microbubbles themselves and the mismatch of acoustic impedance between the gas in the bubble and the surrounding medium, studies have shown that microbubbles undergo nonlinear vibration and scattering in the sound field, generating subharmonics, ultraharmonics, etc., in addition to the fundamental frequency signal. Recent research has shown that this can enhance image contrast and signal-to-noise ratio [41]. Figure 9.23 shows the high-frequency nonlinear response of ultrasound contrast microbubbles.

Generally speaking, due to the small size of the microbubbles, their average diameter is much smaller than the wavelength of the ultrasound wave, so a stationary microbubble in the sound field is considered a point. Researchers using the Born approximation found that due to the strong compressibility of the bubble, the scattering cross-section of the bubble in water is much larger than that of a small iron ball of the same size [42]. At the appropriate frequency, microbubbles will resonate, and their resonance frequency increases with the radius. At lower sound pressure, microbubbles vibrate linearly; when the sound pressure amplitude reaches a certain intensity, microbubbles produce nonlinear vibrations, emitting nonlinear harmonic components [43]. Traditional ultrasound imaging systems are based on fundamental frequency imaging, but when tissue is present, the sensitivity of fundamental frequency imaging is insufficient. The nonlinear signals produced by ultrasound contrast agents under ultrasound excitation can be distinguished from the linear signals produced by tissues, so the signals from microbubbles can be imaged and detected through signal processing. Second and higher harmonic imaging relies on the second or higher harmonics produced by microbubble vibration, its advantage is high imaging resolution, but the limitation is that tissues also produce second harmonics under not high sound pressure, interfering with detection. Subharmonic imaging refers to the use of signals generally scattered by microbubbles under the action of ultrasound, and tissues do not produce subharmonic signals, so this imaging method is more sensitive, but sacrifices clarity [44]. Ultraharmonic imaging refers to signals produced by microbubbles undergoing nonlinear vibrations at 3/2, 5/2, 7/2 times the excitation signal frequency, which has a higher contrast-to-tissue ratio (CTR) and resolution compared to fundamental frequency imaging, but requires the probe to have a larger bandwidth [45]. Transient imaging refers to the use of microbubbles to produce transient responses under low-frequency ultrasound, scattering high-frequency nonlinear signals for imaging detection, this technique has a higher CTR and signal-to-noise ratio [46]. Phase inversion imaging is a technique that images by sequentially emitting two pulses with a certain time delay and opposite phases, this technique can suppress tissue signals, but sacrifices some imaging frame rate [47].

In recent years, scholars have proposed a new type of medical ultrasound imaging method, ultrasound molecular imaging. This technique involves connecting target-specific antibodies or ligands to the surface of ultrasound contrast agents to construct targeted ultrasound contrast agents, allowing the contrast agents to actively bind to target tissues, thereby observing the specific imaging of target tissues at the molecular or cellular level, reflecting changes in diseased tissues at the molecular level [48]. Because ultrasound molecular imaging technology has the ability to obtain microscopic cellular level life and early disease information, it breaks through the limitations of traditional ultrasound imaging of macroscopic anatomical structures and has become a research hotspot in recent years. Current research focuses mainly on studying microbubbles with targeted positioning as probes for ultrasound molecular imaging. Figure 9.24 shows a schematic diagram of ultrasound molecular imaging.

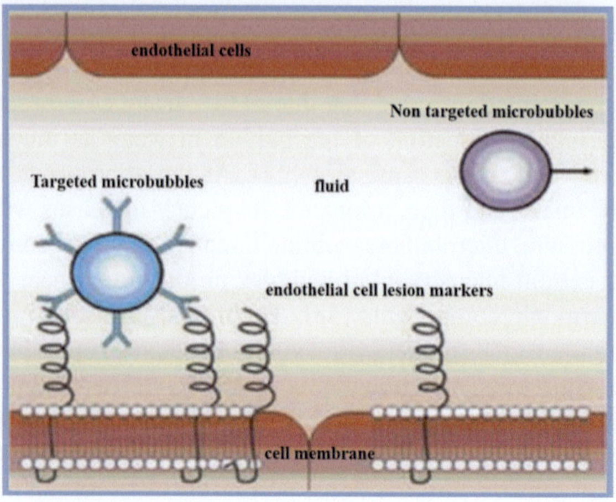

Fig. 9.24 Schematic diagram of ultrasound molecular imaging [40]

9.3.2 History of Ultrasound Molecular Imaging Research

Ultrasound contrast agents first appeared in 1968, but ultrasound contrast microbubbles are different from free microbubbles, usually the presence of a shell on the gas–liquid interface increases the stability of the microbubbles, preventing microbubble aggregation and disintegration [40]. Current theoretical research on the nonlinear acoustic phenomena of encapsulated microbubbles is mainly based on various theoretical models derived from the Rayleigh-Plesset equation. Li Fei and others compared the current theoretical models, and based on the predictive ability of the second harmonic, the simulation results of the Hoff model are closest to the experimental results [49].

Regarding the nonlinear acoustic characteristics of microbubbles, Shi et al. studied the nonlinear response of microbubbles under different sound pressure conditions [50]; Li Bin and others studied the transient nonlinear characteristics of a single nanocoated microbubble in a pulsed ultrasound field [51]; Kruse et al.'s research shows that ultrasound contrast agent microbubbles have a wide-band transient response in the low-frequency state [52]; Goertz et al. studied the nonlinear response of microbubbles under high-frequency ultrasound excitation [53]; Zhao et al. conducted preliminary research on the dynamic characteristics of adherent microbubbles [54].

9.3.3 Application Examples

9.3.3.1 Targeted BST2 Microbubbles for Ultrasound Molecular Imaging Research in Prostate Cancer

Targeted microbubble ultrasound molecular imaging, as a current hotspot in the field of molecular imaging, shows great potential in tumor molecular imaging, drug carriers, and tumor treatment with tumor-specific antibody-modified targeted microbubbles. Targeted and non-targeted microbubbles were injected into a six-well plate inoculated with mouse prostate cancer cells for in vitro imaging with an ultrasound diagnostic device. Optical microscopy showed a large amount of adhesion of targeted microbubbles on the cell surface, while non-targeted microbubbles did not show this phenomenon. The imaging of the targeted microbubble group has a strong echo, and the echo light spots are uniform, and the non-targeted group does not have this phenomenon [55].

The experiment utilized the high specificity and affinity of the "avidin–biotin" bridge connection method, successfully constructing a BST2 monoclonal antibody-coupled targeted microbubble ultrasound contrast agent. At the same time, the experiment confirmed that the contrast agent has strong adhesion to RM-1 cells and has a strong echo under ultrasound, which can be used to detect RM-1 tumor cells.

9.3.3.2 Several Key New Technologies in Ultrasound Molecular Imaging and Drug Delivery Treatment

Ultrasound molecular imaging technology has great potential in the early diagnosis and treatment of diseases such as cancer, cardiovascular and cerebrovascular diseases. The probe for ultrasound molecular imaging, also known as the "sound probe", is a micronanoparticle of targeted microbubbles or contrast agents. In order to prepare microbubbles of uniform size and increase the yield of microbubbles, the paper proposes to design a double-layer microfluidic chip with a network structure with multiple channels and arrays. Using ultrasound radiation force to increase the directional adhesion rate of targeted microbubbles. By performing ultrasound molecular imaging detection on a mouse plaque early inflammation model, the effect of ultrasound microbubble contrast agent on imaging was observed. Successfully prepared uniform microbubbles, using sound radiation force and acoustic flow effect to achieve the arrangement and capture of microbubbles. The developed targeted molecular imaging technology can detect more significant acoustic signals [56].

9.3.3.3 Research on Ultrasound Molecular Imaging and Trans-Blood–Brain Barrier Diagnosis and Treatment Technology

Since ultrasound molecular imaging can reflect the changes at the molecular level of early lesion tissues, ultrasound molecular imaging technology has been widely used in the early diagnosis and monitoring of diseases such as atherosclerosis, inflammation, and thrombosis. The blood–brain barrier is a barrier result located between blood and brain tissue, with strict selective permeability. Studies have shown that specific frequencies of ultrasound energy can safely and selectively open the blood–brain barrier, making the controlled release of small molecule and macromolecule drugs across the blood–brain barrier possible [57].

9.3.3.4 Measurement Method of Blood Shear Force in Rat Artery Model Based on Ultrasound Particle Imaging Technology

Studies have shown that shear stress (shear force) is a major factor related to the occurrence of atherosclerosis and plaque rupture [51]. Ultrasound particle imaging velocimetry has the advantages of real-time, dynamic, non-invasive, and high resolution. Using this technology, fine two-dimensional flow field velocity vectors and velocity gradient distribution can be obtained, thereby calculating the fluid shear force distribution. Ten male SD rats were selected, and the left carotid artery was detected with an ultrasound real-time molecular imaging system to measure Doppler blood flow velocity. After injecting the ultrasound contrast agent, the image was collected, and the shear force was calculated through the ultrasound particle imaging velocimetry algorithm. Compared with the results of ultrasound Doppler velocimetry,

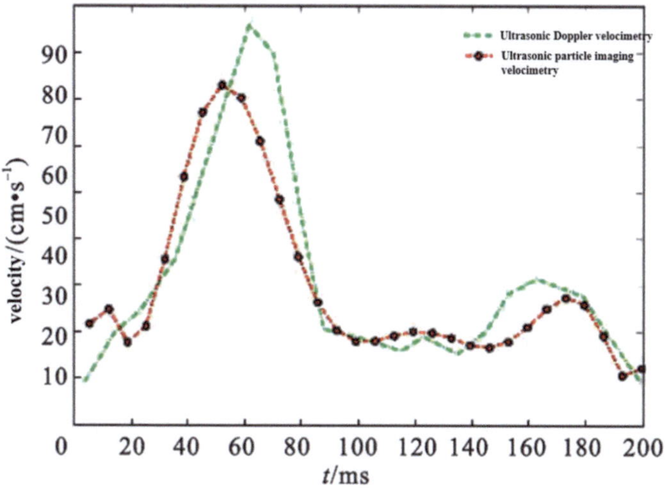

Fig. 9.25 Comparison of maximum speed measured by ultrasonic particle imaging velocimetry and ultrasonic Doppler velocimetry in the corresponding area [58]

the peak velocity curve of ultrasound particle imaging velocimetry is very close, but the peak velocity and average velocity are lower than the results of ultrasound Doppler velocimetry. Ultrasound particle imaging velocimetry provides two-dimensional full flow field velocity information, can display the velocity size and direction of blood flow at different positions, the velocimetry effect is more precise and comprehensive, and can better analyze the flow field velocity distribution situation [59]. Ultrasonic particle imaging velocimetry provides a new research method for further measuring changes in vascular shear force. Figure 9.25 compares the maximum speed measured by ultrasonic particle imaging velocimetry with the corresponding area's ultrasonic Doppler velocimetry. Figure 9.26 combines ultrasonic particle imaging velocimetry and polynomial fitting to calculate shear force.

9.3.3.5 Microfluidic Chip Manipulating Microbubble Contrast Agents with Acoustic Surface Waves

Microbubble contrast agents can be used to enhance ultrasound image contrast, drug delivery, gene therapy, and other fields. The adhesion rate between microbubbles and cells determines the effect of imaging and treatment. Acoustic surface waves, as a type of elastic wave propagating along the surface of an object, are suitable as excitation sources. Microbubbles are prepared by mechanical oscillation method, applying a 400 mw power sound wave to the PDMS composite chip when the channel depth is 50 μm. When the channel depth is 680 μm, the input power is adjusted to 270 mW. At a depth of 50 μm, the microbubbles are arranged in parallel lines (Fig. 9.27); at a depth of 680 μm, the microbubbles rotate and gather in the center of the channel

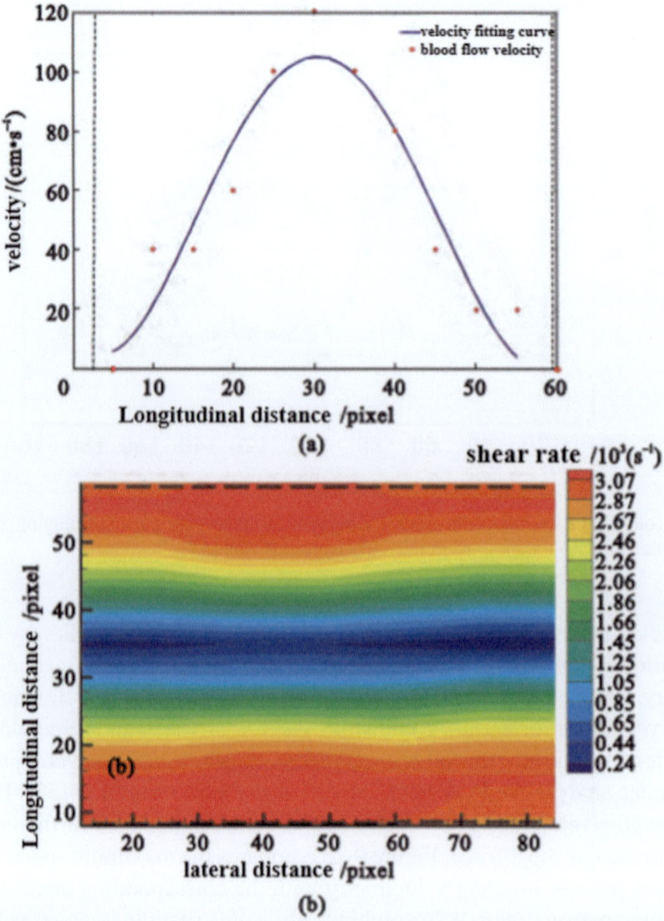

Fig. 9.26 Combining ultrasonic particle imaging velocimetry and polynomial fitting to calculate shear force [58]

(Fig. 9.28). The results show that the depth of the PDMS channel directly affects the movement behavior of the microbubbles [59]. Combining the advantages of low power consumption, miniaturization, and simple design, the PDMS composite chip provides assistance for ultrasound molecular imaging and drug delivery.

9.3.3.6 The Effect of Secondary Radiation Force on Microbubble Aggregation

When microbubbles are used as targeting agents, it is usually desired that they adhere to specific locations. However, microscopic observation found that when microbubbles move in blood vessels, they tend to move along the vessel wall, which

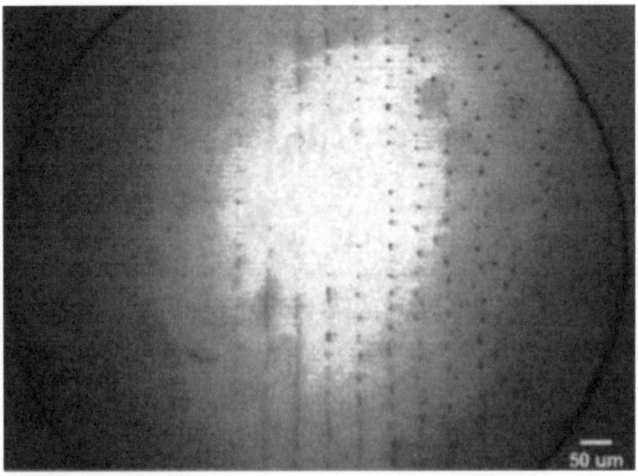

Fig. 9.27 Microbubbles arranged in parallel lines when the channel depth is 50 μm [59]

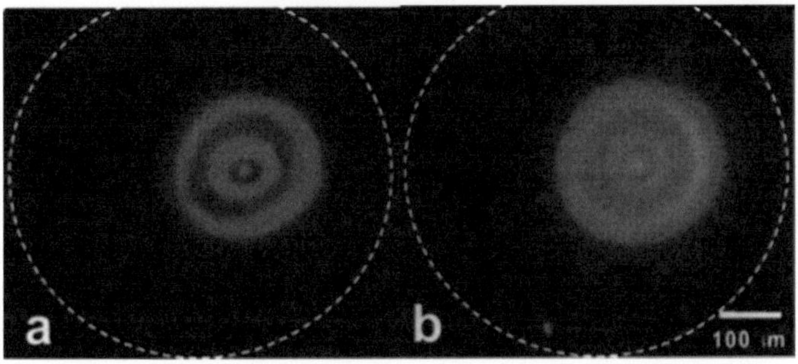

Fig. 9.28 Microbubbles gather in the center when the channel depth is 680 μm [59]

reduces the adhesion rate of the microbubbles. When ultrasound acts on microbubbles, two different types of acoustic radiation forces are generated, one of which is the secondary acoustic radiation force generated between two vibrating microbubbles. Studies have shown that under the action of primary acoustic radiation force and buoyancy, microbubbles will gather in one area, providing an opportunity for the generation of secondary acoustic radiation force. Therefore, understanding the secondary acoustic radiation force is very important for ultrasound molecular imaging.

To study the effect of secondary acoustic radiation force on microbubble aggregation, the literature studied the axial and radial motion of two interacting microbubbles, found the conditions that cause attraction between microbubbles, and explored the impact of ultrasound parameters and microbubble shell parameters on microbubble aggregation. Based on the modified Rayleigh-Plesset equation and four differential equations of radial and transverse motion, a force formula for the microbubble wall in the liquid was established. The interaction force between microbubbles was studied, and it was found that the interaction force between two microbubbles would change with the radius of the microbubbles, the distance between them, and the speed of the microbubbles. The parameters of ultrasound and microbubble shell also affect the aggregation of microbubbles. Numerical simulations show that the smaller the distance between microbubbles, the faster the aggregation; by controlling the distribution of microbubbles, when the average radius of the microbubbles deviates from the resonance radius, the aggregation of microbubbles can be suppressed; for the same purpose, low sound pressure and low elasticity ultrasound contrast agents are less likely to cause microbubble aggregation. Therefore, through theoretical calculations and numerical simulations of secondary acoustic radiation force, it was found that by selecting appropriate ultrasound parameters and microbubble shell parameters, the aggregation of microbubbles can be suppressed, which is of great significance for the further development of ultrasound molecular imaging [60].

9.3.3.7 Study on Parameter Setting of Ultrasound Radiation Force Moving Targeted Microbubbles

In the field of ultrasound imaging, the use of selective targeting contrast agents combined with specific antibodies can enhance the imaging effect of the target area and prolong the imaging time. However, the laminar flow phenomenon produced by the flow of microbubbles in blood vessels makes the adhesion rate of targeted microbubbles poor, affecting the second harmonic contrast imaging. Using ultrasound radiation force to drive targeted microbubbles is beneficial to push them to adhere to the side wall of the vessel, enhancing the imaging effect.

In the experiment, an ultrasonic device and self-made microbubble contrast agent were used to conduct experiments in static fluid, simulated human superficial lymph fluid flow conditions, and deep cavities, respectively, observing the displacement effect of microbubbles. Under static conditions, the maximum peak negative pressure that produces the maximum displacement without bursting the microbubbles is 560 kPa, and within the range less than this value, the surface affected by the microbubbles is directly proportional to the sound pressure. The maximum peak negative pressure required under simulated lymph fluid flow conditions increases, and at 1.06 MPa, three-quarters of the microbubbles can adhere to the tube wall, but 75% of them are burst. In the deep cavity state, microbubbles near the sound source will block the effect of sound radiation force on microbubbles far from the sound source. Therefore, the target target area should choose more superficial cavity [61].

9.3.3.8 Study on Enhancing the Chemotherapeutic Effect of Brain Glioma by Ultrasound Opening the Blood–Brain Barrier

When treating brain gliomas with drugs, most drugs cannot enter the lesion site due to the presence of the blood–brain barrier. Ultrasound combined with microbubbles to open the blood–brain barrier for intracranial drug delivery as a new method for treating brain gliomas has the characteristics of non-invasiveness, precision, and efficiency. An in vitro blood–brain barrier model was established, and according to the focused ultrasound parameters explored in the early stage, ultrasound and control

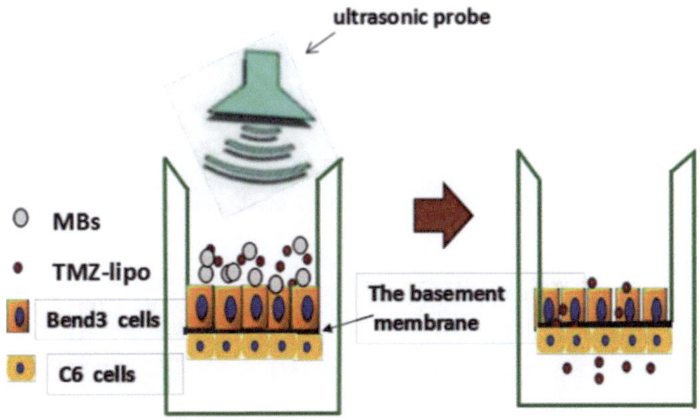

Fig. 9.29 Schematic diagram of in vitro blood–brain barrier and ultrasound action [62]

groups were set up for the experiment. After 24 h of action, the cell activity and apoptosis situation after the experiment were measured. According to the staining situation of coumarin 6 dye, ultrasound is conducive to the dye entering the lower cell layer, ultrasound also increases the concentration of the drug, and the number of apoptotic cells in the ultrasound group is more after 24 h of action. Ultrasound combined with microbubble drug treatment, increases the fluidity of the cell membrane, effectively improves the drug concentration at the lesion site, which is of great significance for drugs that are difficult to pass through the BBB [62]. Figure 9.29 is a schematic diagram of the in vitro blood–brain barrier and ultrasound action.

9.3.3.9 Ultrasound-Based Drug Delivery System

The ultrasound-based drug delivery system can use the sound penetration of ultrasound to make the drug-carrying carrier move in the blood vessels, and collect the reflection signal of the carrier to the sound wave to calculate the speed of the carrier movement. The drug carrier uses microbubbles with a diameter of 1–10 μm. The bubble is filled with air or other gases, and it is required that the solubility of this gas in the blood is lower than that of air. Compared with rigid particles, this kind of ultrasound contrast agent can increase imaging contrast and improve image quality [63].

Early studies on the effect of sound radiation force on microbubbles generally used kilohertz-level ultrasound excitation, and the volume of microbubbles could not reach the micron level. Therefore, this experiment chose megahertz-level ultrasound, and the drug delivery system mainly consists of an ultrasound probe, function amplifier, power amplifier, motor, and microscope. The experiment simulates the movement of microbubbles in blood vessels in a water tank, uses a microscope to capture the process of bubble movement, and uses MATLAB to analyze the obtained images. From the original image, it can be seen that most of the microbubbles are adsorbed

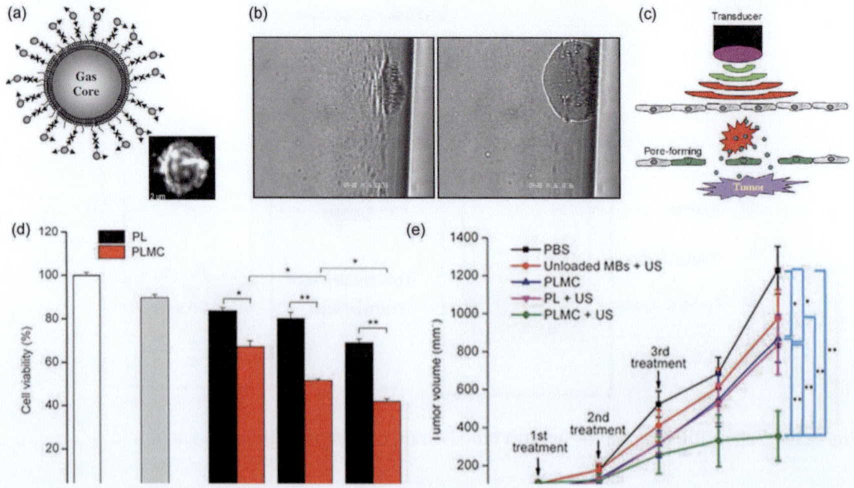

Fig. 9.30 Ultrasound microbubble drug delivery and targeted drug delivery [64]

on the tube wall and flow, and the aggregation point is the simulated lesion. However, because megahertz excitation of micron-level bubbles was used in the experiment, the real-time performance of the system is poor, the bursting of microbubbles during movement causes drug loss, and there is still a big difference between the vascular environment in living organisms and the experimental environment. In recent years, drug delivery mediated by ultrasound microbubbles has become a hot topic in drug delivery systems [64] (Fig. 9.30).

9.4 Ultrasound Neuroregulation Technology: Treating the Nervous System with Ultrasound Radiation Force

9.4.1 Basic Principles of Neuroregulation

Today, with the development of society and the trend of human aging, functional brain diseases such as depression, Parkinson's disease, and epilepsy are becoming more and more common, and the intervention and treatment of functional brain diseases are a problem in world medicine. At present, the cause of functional brain diseases still needs further research, and there is a lack of corresponding effective treatment measures. Studies have shown that the occurrence of functional brain diseases is related to the functional disorder of the specific "cortico-basal ganglia brain circuit" in the brain, and stimulating the corresponding targets can help alleviate or cure symptoms [65].

Ultrasound waves are generated by an ultrasound transducer, and the center frequency, energy intensity, pulse change rule, and the size of the irradiated target area can all be adjusted. Based on the above advantages, ultrasound neuroregulation can be widely applied to the neuroregulation field of various scales of cells, nematodes, insects, rodents, non-human primates, and humans [65, 66]. The functioning of the nervous system often requires the coordinated work of multiple brain regions and nuclei, a task that a single small single-element ultrasound probe cannot fulfill. Through array probes and phased array beamforming technology, precise calculations for transcranial ultrasound can be achieved, enabling multi-point precise positioning of the target. Ultrasound brain stimulation, as a novel, non-invasive neuroregulation technology, has enormous potential in the intervention of functional brain diseases. Over the past 30 years, a variety of pioneering neuroregulation methods have emerged, including non-invasive methods such as transcranial magnetic stimulation, transcranial electrical stimulation, and transcranial ultrasound stimulation [66].

9.4.2 History of Neuroregulation Research

In the 1920s, researchers discovered that ultrasound radiation of a frog's sciatic nerve could cause slight tremors in the gastrocnemius muscle and significantly affect the heartbeat [66]. Harvey and others pioneered the discovery that ultrasound could reversibly regulate the visual evoked potentials in cats [67]. In 1955, Fry and others found that focused ultrasound could not only treat pain and Parkinson's disease but also be used to study the structure and function of brain circuits. Subsequently, Fry and others successfully used high-intensity focused ultrasound to remove parts of the brain affected by movement disorders and chronic pain, without observable damage to surrounding tissues and blood vessels [68].

Researchers like Koroleva used ultrasound pulses on rat brains and found that pulsed ultrasound could cause stable negative potential shifts [69]; Velling and Shklyaruk used low-intensity focused ultrasound to stimulate the brains of cats and rabbits [70], finding that cortical signals would change with variations in stimulation intensity and pulse repetition frequency. These studies all demonstrate the application of large-scale ultrasound regulation in brain excitation.

Deffieux and his team were the first to conduct ultrasound regulation research on macaques, with experiments showing that low-intensity ultrasound could interfere with the processing of visual search tasks in the eye area [72]. Tyler and Yoo's team subsequently conducted experimental research on focused ultrasound regulation of the primary sensory cortex in humans [73]. Lee's group found that using focused ultrasound to stimulate the human sensory cortex could induce sensations in different limb areas [74].

In China, the Ultrasound Neuroregulation Team of the Shenzhen Institutes of Advanced Technology of the Chinese Academy of Sciences was the first to conduct research domestically, carrying out ultrasound neuroregulation research on cells, nematodes, small animals, and macaques, and making initial progress [65].

9.4.3 *Application Examples*

9.4.3.1 Experimental Study on the Therapeutic Effect of Ultrasound Radiation Force on Parkinson's Mice

As the world's second-largest degenerative disease, Parkinson's is commonly treated with oral dopamine. However, long-term use of dopamine can lead to motor abnormalities, making the exploration of non-invasive neuroregulation methods particularly important. Studies have shown that ultrasound can penetrate the skull and cause neuronal activity, and existing literature indicates that ultrasound radiation force regulation of the prefrontal cortex in rat models of depression can increase the content of brain-derived neurotrophic factor in the hippocampus, thereby improving the symptoms of rat depression. The literature will explore the therapeutic effect of ultrasound on Parkinson's model mice. First, a subacute Parkinson's mouse model induced by MPTP is prepared, an ultrasound neuroregulation device is set up, and the mice are given continuous seven-day ultrasound radiation treatment to the thalamic basal nucleus while awake. The specific ultrasound parameters are a base frequency of 3.8 MHz, pulse frequency of 1 kHz, duty cycle of 50%, duration of 1 s, pulse interval of 4 s, and total duration of 30 min. Finally, the protein changes in the substantia nigra striatum of the mice are detected using the western blotting method. The results show that the p-Akt protein in the substantia nigra of the treated mice significantly increased, indicating that ultrasound can stimulate the expression of Akt, providing neuroprotection for Parkinson's model mice [75]. Figure 9.31 is the ultrasound neuroregulation device.

9.4.3.2 The Effect of Low-Power Ultrasound Stimulation on Hippocampal Neuronal Dendritic Spines

Dendritic spines are tiny structures protruding from the branches of neuronal dendrites, serving as the primary site for synapse formation between neurons. The number, shape, and size of dendritic spines are closely related to brain function disorders such as autism, schizophrenia, and Alzheimer's disease. This study explores the effects of low-intensity ultrasound on the length of rat hippocampal neuron dendrites and the number of dendritic spines. First, primary fetal rat hippocampal neurons are cultured and transfected with green fluorescent protein plasmids. A 0.5 MHz planar transducer is used to stimulate the hippocampal neurons with a pulse repetition frequency of 500 Hz, a duty cycle of 20%, a sound pressure of 0.2 MPa, for 10 min each day, continuously for 7 days. The results show that ultrasound stimulation reduces the length of neuronal dendrites and increases the number of dendritic spines.

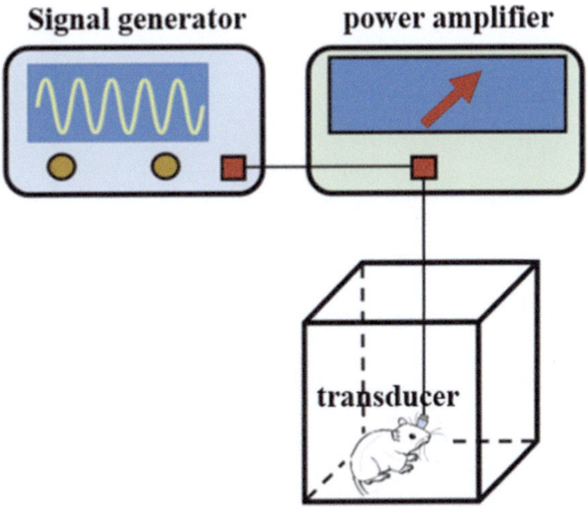

Fig. 9.31 Ultrasound neuroregulation device [75]

Further research on the effects of ultrasound stimulation with different parameters on the structure, size, and number of dendritic spines will provide a new theoretical basis for the application of ultrasound brain stimulation in the intervention and treatment of functional brain diseases [76].

9.4.3.3 Experimental Study on Low-Intensity Ultrasound Stimulation of Primary Neuronal Cilia

Primary cilia are non-motile organelles that function as signal receptors, closely related to neural development, function, and related diseases. The effects of ultrasound stimulation on primary cilia are not yet clear, and this study will investigate this. First, primary fetal rat cortical neurons are cultured and stimulated with a 0.5 MHz planar transducer. The stimulation lasts for 7 days, 10 min each day. Finally, the results are evaluated using immunofluorescence technology. Ultrasound stimulation reduces the occurrence rate of primary neuronal cilia by about 40%, demonstrating the feasibility of ultrasound stimulation as a regulator of primary neuronal cilia [77]. Figure 9.32 shows the ultrasound stimulation experimental setup.

9.4.3.4 Safety Study of Low-Intensity Focused Ultrasound Neural Regulation

Low-intensity focused ultrasound brain stimulation, as a novel and non-invasive neural regulation technology, has great potential in brain research. Studies have shown that ultrasound with different parameters can excite or inhibit neuronal activity. Twenty-eight male mice are selected and randomly divided into stimulation group A, stimulation group B, and a control group. Group A has a pulse frequency of

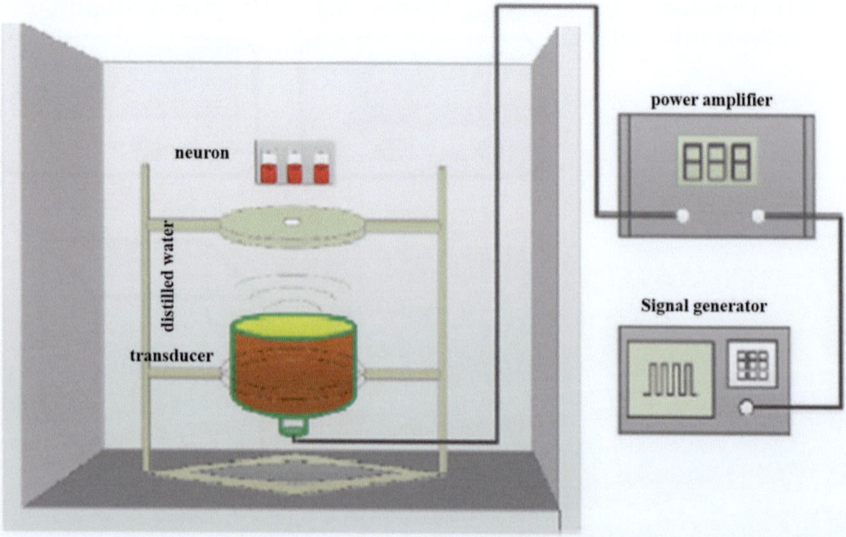

Fig. 9.32 Ultrasound stimulation experimental setup [77]

1000 Hz, a duty cycle of 50%; Group B has a pulse frequency of 10 Hz, a duty cycle of 50%. The experiment lasts for 14 days, once a day, 30 min each time. Comparing groups A, B, and the control group, no obvious bleeding or tissue edema is found. The experiment shows that the ultrasound of these two groups has a small impact on brain tissue and is within a safe range [78].

9.4.3.5 Study on Neural Regulation of Caenorhabditis Elegans Based on Surface Acoustic Waves

Caenorhabditis elegans is a classic model organism for neuroscience, containing only 302 neurons, and over 60% of its genes are homologous to human genes. Using surface acoustic waves and C. elegans, the mechanism of neural regulation can be studied. C. elegans are cultured and observed under a microscope to see if ultrasound waves stimulate them, causing behavioral responses. Under the action of ultrasound, C. elegans exhibit an avoidance behavior. As the duration of the input pulse decreases, the probability of C. elegans exhibiting avoidance behavior decreases. This shows that surface acoustic waves can effectively stimulate model organisms, and this system can be applied to the mechanism of ultrasound neural regulation [79]. Figure 9.33 shows the avoidance response.

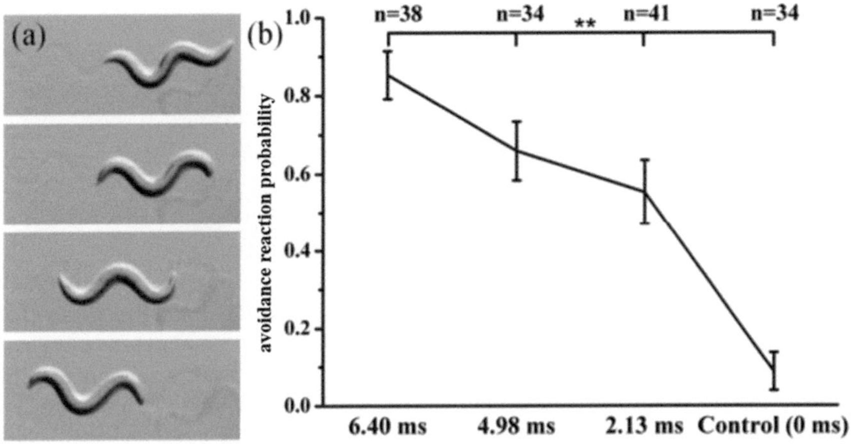

Fig. 9.33 Avoidance response [79]

9.4.3.6 Study on the Regulation Effect of Hippocampal Neurons Based on Miniature Ultrasound Chips

As a new treatment method, the mechanism of ultrasound neural regulation is still controversial. Traditional ultrasound transducers are large and subject to placement conditions. This study uses micro-nano processing technology to develop a miniature chip for ultrasound neural stimulation. The ultrasound neural stimulation chip and in vitro hippocampal brain slices are prepared. The brain slices are moved to the recording slot to record spontaneous and ultrasound-induced activity. The experiment shows that the miniature ultrasound neural stimulation chip can fit well with the in vitro brain slice patch clamp system, producing a concentrated sound field with clear boundaries. At low intensity, the chip can cause neuronal membrane depolarization, and at high intensity, it can directly cause neurons to fire action potentials. This chip is expected to provide a powerful new tool for the mechanism of ultrasound regulation [80].

9.4.3.7 Simulation Study on Ultrasound Time Reversal Transcranial Focusing Method for Non-invasive Neural Regulation Applications

In recent years, ultrasound brain intervention surgeries such as high-intensity focused ultrasound neurosurgery and low-intensity ultrasound neuroregulation techniques have become important tools for the treatment of neuropsychiatric diseases and neuroscience research [81]. To overcome the negative impact of the skull on the focusing process, Fink and others proposed the time reversal method. This method uses an ultrasound transducer to receive the ultrasound signal emitted or reflected by a strong reflector, reverses the signal on a time axis, and uses the reversed signal to stimulate the transducer to emit ultrasound [82].

In 2009, Marquet and others proposed a method of applying the time reversal method to transcranial ultrasound focusing, also known as the virtual sound source simulation method [83]. In the simulation program, a virtual sound source is placed at the expected focus position to simulate the process of its emitted sound wave propagating through the skull, thereby obtaining the sound pressure waveform at the spatial position of the transducer array surface. Further use of the time reversal method can obtain the ultrasound emission sequence. In the simulation, a simulation skull with uneven thickness and irregularity is used to make the results more accurate. In the time reversal simulation comparison experiment, different aperture probes, different sound source frequencies, and multiple focus points and continuous area comparison experiments were carried out. The experiment found that within a certain range, increasing the sound source frequency will result in worse results, while increasing the number of array elements of the probe will result in clearer focus points. In addition, the experiment also found that the irregular thickness of the skull severely interferes with the simulation results, and there is a lot of room for optimization [81].

9.4.3.8 A New Method for Visualizing the Focus of Transcranial Ultrasound Focusing in Vitro Experiments

In recent years, brain intervention techniques have become increasingly critical in the treatment of neuropsychiatric diseases, among which deep brain electrical stimulation is most widely used. However, when applied clinically, due to parameters such as the speed of sound and sound attenuation of the skull, serious phase distortion and amplitude attenuation will occur during the process of sound wave transcranial propagation. Therefore, it is very important to study the visualization method of the ultrasound focus. Using thermosensitive materials, focus visualization experiments were carried out under conditions with and without skulls. Without the skull, color borders appear on the thermosensitive film at the physical focus of the probe. With the skull, the voltage was set to 10.1 and 19.8 V respectively, and the color change was observed. From the results, it can be seen that without the skull, the film changes color immediately; with the skull, the temperature near the skull rises rapidly, and the film changes color. This phenomenon proves that the thermosensitive thin film has achieved rapid visualization of the focus of transcranial ultrasound focusing [84], Fig. 9.34 represents the initial position of the multi-point focusing experiment focus and the effect of time reversal.

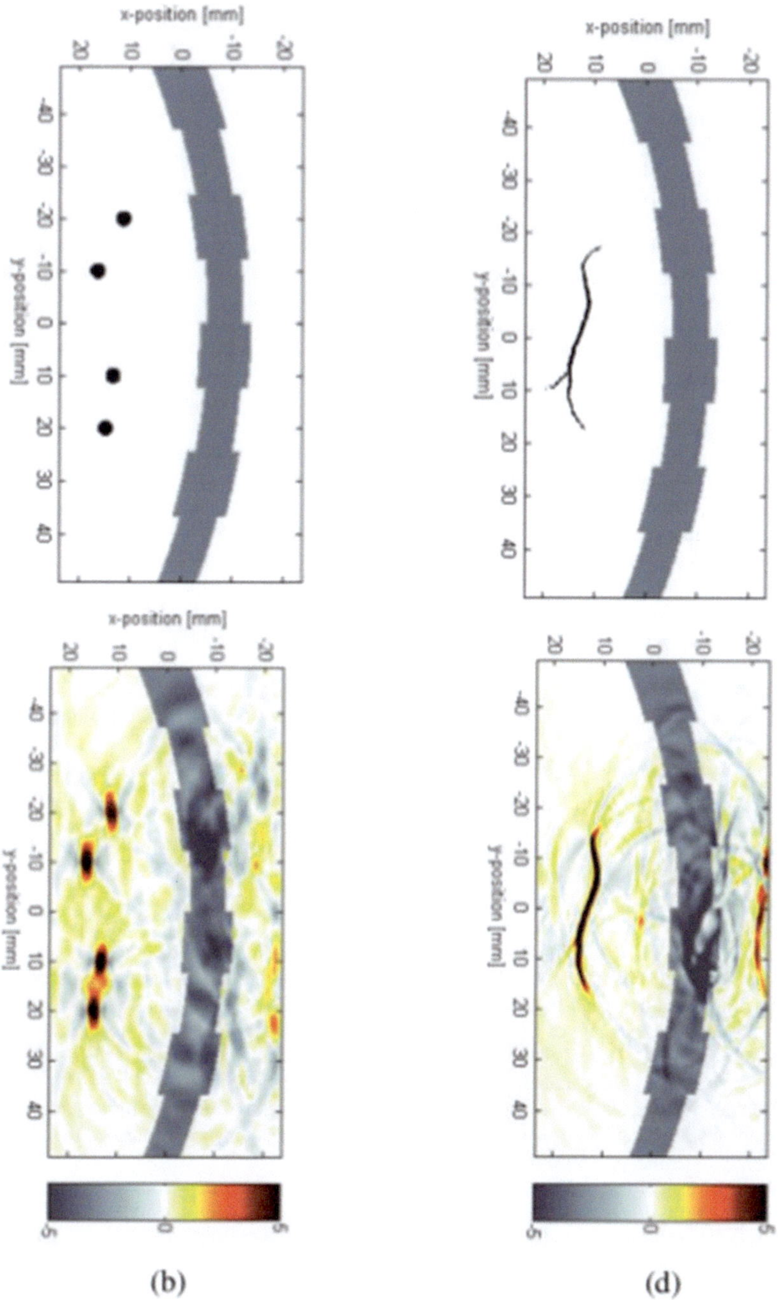

Fig. 9.34 **a**, **b** Represent the initial position of the multi-point focusing experiment focus and the effect of time reversal; **c**, **d** represent the initial position of the continuous area focusing experiment and the effect of time reversal [81]

References

1. Shen Y, Ling T, Yao H, Zheng H (2012) Ultrasound instantaneous elasticity imaging algorithm and system design. Integr Technol 1(1):93–99
2. Zheng H, Zhang W, Wang C et al (2015) Discussion on the value of ultrasound elasticity imaging in the diagnosis of breast diseases. Everyone's Health 16:75–76
3. Zeng B, Lei Y, Wang C, Zheng H (2015) Research on sound radiation force elasticity imaging algorithm based on CUDA. Comput Eng Appl 51(18):249–254
4. Jian Z, Jie Z, Rongqin Z et al (2013) The impact of detection depth on real-time shear wave elasticity imaging of the liver. In: Ultrasound medicine academic conference, Beijing, China
5. Ophir J, C spedes I, Ponnekanti H et al (1991) Elastography: a quantitative method for imaging the elasticity of biological tissues. Ultrason Imag 13:111–134
6. Lerner RM, Huang SR, Parker KJ (1990) Sonoelasticity images derived from ultrasound signals in mechanically vibrated tissues. Ultrasound Med Biol 16(3):231–239
7. Fatemi M, Greenleaf JF (1998) Ultrasound-stimulated vibro-acoustic spectrography. Science (S0036–8075) 280(3):82–85
8. Saravazyan AP, Rudenko OV, Swanson SD et al (1998) Shear wave elasticity imaging: a new ultrasonic technology of medical diagnostics. UMB (S0301–5629) 24(9):1419–1435
9. Bercoff J, Tanter M, Fink M (2004) Supersonic shear imaging: a new technique for soft tissue elasticity mapping. IEEE UFFC (S0885–3010) 51(4):396–409
10. Jeremy JD, Gianmarco FP, Mark L, Nightingale KR (2007) A parallel tracking method for acoustic radiation force impulse imaging. IEEE UFFC (S0885–3010) 54(2):301–312
11. Rouze NC, Wang MH, Palmeri ML et al (2010) Robust estimation of time-of-flight shear wave speed using a radon sum transformation. IEEE Trans Ultrason Ferroelectr Freq Control 57(12):2662–2670
12. Wang L, Jiang X, Zhang S et al (2016) Improved method for estimating shear wave speed in acoustic radiation force elasticity imaging based on Radon transform. Integr Technol 5(004):58–66
13. Kang H, Niu L, Meng L et al (2017) Ultrasound experimental study on measuring plaque tissue hardness based on ultrasound texture matching method. In: National acoustics conference of the acoustical society of China
14. Shen L, Ling T, Li Y, Zheng H (2011) Non-invasive detection of liver fibrosis based on ultrasound instantaneous elasticity imaging. In: Proceedings of the 9th youth academic conference of the acoustical society of China
15. Eskandari H, Salcudean SE, Rohling R (2008) Viscoelastic parameter estimation based on spectral analysis. IEEE Trans Ultrason Ferroelectr Freq Control 55:1611–1625
16. Shen Y, Ling T, Xiao Y et al (2011) Research on acoustic radiation force impulse elasticity imaging technology and its algorithm. In: Proceedings of the 9th youth academic conference of the acoustical society of China
17. Ming Y, Wang C, Zeng C et al (2013) Multi-physics field finite element analysis of viscoelastic tissue strain caused by acoustic radiation force. Integr Technol 5:48–52
18. Zhang X, Xiao Y, Qiu W et al (2013) Research and application of real-time ultrasound elasticity image feature quantification method. In: The 10th youth academic conference of the acoustical society of China, pp 169–170
19. Wang B, Ling T, Shen Y et al (2011) Algorithm and phantom experiment study to improve motion estimation accuracy in quasi-static elasticity imaging. Chin J Med Phys 3:2668–2673
20. Liu B, Yao H, Zheng H et al (2011) Application of wavelet denoising in instantaneous elasticity imaging estimation. Comput Simul 28(8):246–249
21. Wang B, Ling T, Shen Y et al (2011) Quasi-static elasticity imaging technology detects focused ultrasound-induced ex vivo tissue damage. Chin J Med Imag Technol 27(11):2317–2321
22. Cai F, Meng L, Li F et al (2018) Research progress on acoustically manipulated particles. Appl Acoust 5:655–663
23. Meng L, Cai F, Zhang Z et al (2011) Transportation of single cell and microbubbles by phase-shift introduced to standing leaky surface acoustic waves. Bio-microfluidics 5(4):044104

24. Lord Rayleigh FRS (1902) On the pressure of vibrations. Phil Mag 3(13–18):338–346
25. Shortencaier MJ, Dayton PA, Bloch SH et al (2004) A method for radiation-force localized drug delivery using gas-filed lipospheres. IEEE Trans Ultrason Ferroelectr Freq Control 51(7):822–831
26. King LV (1935) On the acoustic radiation pressure on sphere. Proc R Soc Lond Ser A 147(861):212–240
27. Yosioka K, Kawasima Y (1955) Acoustic radiation pressure on a compressible sphere. Acta Acust Acust 5(3):167–173
28. Gor'kov LP (1961) Forces acting on a small particle in an acoustic field within an ideal fluid. Inst Phys Probl USSR Acad Sci 140(1):88–91
29. Wu J (1991) Acoustical tweezers. J Acoust Soc Am 89:2140–2143
30. Silva GT (2018) An expression for the radiation force exerted by an acoustic beam with arbitrary wave front (L). In: Research on nanoparticle manipulation based on ultrasonic radiation force. Proceedings of the national acoustics conference
31. Sapozhnikov OA, Bailey MR (2013) Radiation force of an arbitrary acoustic beam on an elastic sphere in a fluid. J Acoust Soc Am 133(2):661–676
32. Zhou W, Chen M, Niu L et al (2018) Research on nanoparticle manipulation based on ultrasonic radiation force. In: Proceedings of the national acoustic conference
33. Zhou W, Wang K, Niu L et al (2016) Research on two-dimensional manipulation based on one-dimensional acoustic surface wave chip. In: National acoustics academic conference
34. Lin Q, Cai F, Li F et al (2019) Acoustic manipulation of micro and nano particles based on local sound field of cylindrical thin shell. In: National acoustics conference
35. Huang X, Cai F, Li W et al (2017) Acoustic manipulation of particles by one-dimensional sound grating in air. Acta Physica Sinica 4:150–155
36. Cheng X, Cai F, Zheng H (2011) Study on the acoustic radiation force of steel columns near sub-wavelength slits. In: The ninth youth academic conference of the acoustical society of China
37. Xu D, Cai F, Chen M et al (2016) Three-dimensional acoustic manipulation of particles in cylindrical microcavities. In: Proceedings of the national acoustics academic conference
38. Zhang W, Niu L, Liu X et al (2018) Research on cell capture based on single microbubble resonance. In: National acoustics conference
39. Zhao Z, Zhang W, Niu L et al (2018) Research on rapid microfluidic acoustic mixing method based on microbubble resonance. Acta Physica Sinica 67(19):156–163
40. Zheng H, Qian M (2011) Ultrasound contrast microbubbles: sensitive imaging, targeted drug delivery and therapy. China Med Dev Inform 07:1–5
41. Ling T, Zheng H (2009) Progress in research on nonlinear acoustic characteristics and imaging of ultrasound contrast microbubbles. Chin J Interven Imag Ther 4:102–106
42. de Jong N (1993) Acoustic properties of ultrasound contrast agents. Erasmus University, Rotterdam
43. Ferrara K, Pollard R, Borden M (2007) Ultrasound microbubble contrast agents: fundamentals and application to gene and drug delivery. Annu Rev Biomed Eng 9:415–447
44. Shi WT, Forsberg F, Hall AL et al (1999) Subharmonic imaging with microbubble contrast agents: initial results. Ultrason Imag 21:79–94
45. David E, Goertz EC, Andrew N et al (2005) High frequency nonlinear B-scan imaging of microbubble contrast agents. IEEE Trans Ultrason Ferroelectr Freq Control 52:65–79
46. Kruse DK, Ferrara KW (2005) A new imaging strategy using wideband transient response of ultrasound contrast agents. IEEE Trans Ultrason Ferroelectr Freq Control 52:1320–1329
47. Burns PN, Wilson SR, Hope SD (2000) Pulse inversion imaging of liver blood flow: improved method for characterizing focal masses with microbubble contrast. Invest Radiol 35:58–71
48. Wang Z (2009) Progress in research on ultrasound molecular imaging. Chin J Med Imag Technol 25(6):921–924
49. Li F, Xie L, Li D et al (2007) Comparative study on dynamic models of ultrasound contrast agent microbubbles. Chin J Med Imag Technol 23(2):295–298

50. Li B, Wan M, Wang S (2004) Study on the transient nonlinear characteristics of nano-coated contrast microbubbles in pulsed ultrasound field. Acta Acustica 29(6):525–532
51. Li B, Wan M, Wang S (2004) Transient nonlinear characteristics of nanoencapsulated contrast enhanced microbubbles in pulsed ultrasonic fields. J Acoust 29(6):525–532
52. Kruse DE, Ferrara KW (2005) A new imaging strategy using wideband transient response of ultrasound contrast agents. IEEE Trans Ultrason Ferroelectr Freq Control 52(8):1320–1329
53. Goertz DE, Cherin E, Needles A et al (2005) High frequency nonlinear B-scan imaging of microbubble contrast agents. IEEE Trans Ultrason Ferroelectr Freq Control 52(1):65–79
54. Zhao S, Ferrara KW, Dayton PA (2005) Asymmetric oscillation of adherent targeted ultrasound contrast agents. Appl Phys Lett 87(13):4103
55. Chen J, Yan F, Jin Q et al (2011) Research on ultrasound molecular imaging of BST2-targeted microbubbles for prostate cancer. In: The 9th youth academic conference of the acoustical society of China, pp 115–116
56. Zheng H, Yan F (2014) Several key new technologies in ultrasound molecular imaging and drug delivery treatment. In: The first national and second international conference on ultrasound molecular imaging, pp 36–37
57. Zheng H (2016) Research on ultrasound molecular imaging and blood-brain barrier diagnosis and treatment technology. In: National and international conference on ultrasound molecular imaging and bioeffects and treatment
58. Zhu Y, Qian M, Niu L et al (2014) Measurement method of blood shear force in rat artery model based on ultrasound particle imaging technology. J Biomed Eng 6:1355–1360
59. Meng L, Cai F, Niu L et al (2011) Microfluidic chip for manipulating microbubble contrast agents using acoustic surface waves. In: Proceedings of the 9th youth academic conference of the acoustical society of China
60. Zhang YL, Zheng HR, Tang MX et al (2011) Effect of secondary radiation force on aggregation between encapsulated microbubbles. Chin Phys B 20(11):359–364
61. Yang Y, Qiao L, Yan F et al (2015) Study on parameter setting of ultrasound radiation force displacement of targeted microbubbles. Chin J Ultrasound Med 2:170–173
62. Huang X, Yan F, Zheng H (2016) Research on enhancing the chemotherapy effect of brain glioma by ultrasound opening the blood-brain barrier. In: Proceedings of the 2016 national acoustics conference
63. Yu T, Zhang H, Zheng H (2008) Ultrasound-based drug delivery system. Chin J Med Imag Technol 24(12):2010–2012
64. Zheng H, Cai F, Yan F et al (2015) Multifunctional biomedical ultrasound: molecular imaging, drug delivery, and neuroregulation. Sci Bull 20:10
65. Li G, Qiu W, Qian M et al (2017) Ultrasound neuromodulation technology and scientific instruments. Life Sci Instrum 1:3–8
66. Naor O, Krupa S, Shoham S (2016) Ultrasonic neuromodulation. J Neural Eng 13(3):031003
67. Harvey EN (1929) The effect of high frequency sound waves on heart muscle and other irritable tissues. Am J Physio 91:284–290
68. Fry WJ, Barnard JW, Fry FJ et al (1955) Ultrasonically produced localized selective lesions in the central nervous system. Am J Phys Med Rehab 34:413–423
69. Koroleva VI, Vykhodtseva NI, Elagin VA (1986) Cortical and subcortical spreading depression in rats produced by focused ultrasound. Neurophysiology 18(1):43–48
70. Velling VA, Shklyaruk SP (1988) Modulation of the functional state of the brain with the aid of focused ultrasonic action. Neurosci Behav Physiol 18(5):369–375
71. Deffieux T, Younan Y, Wattiez N et al (2013) Low-intensity focused ultrasound modulates monkey visuomotor behavior. Curr Biol 23(23):2430–2433
72. Gavrilov L, Gersuni G, Ilyinsky O et al (1976) The effect of focused ultrasound on the skin and deep nerve structures of man and animal. Prog Brain Res 43:279–292
73. Tyler WJ (2011) Noninvasive neuromodulation with ultrasound? A continuum mechanics hypothesis. Neuroscientist 17:25–36
74. Lee W, Kim H, Jung Y et al (2015) Image-guided transcranial focused ultrasound stimulates human primary somatosensory cortex. Sci Rep 5:8743

75. Zhou H, Niu L, Meng L et al (2019) Experimental study on the therapeutic effect of ultrasonic radiation force on Parkinson's mice. In: National acoustics conference
76. Huang X, Lin Z, Niu L et al (2018) The effect of low-power ultrasound stimulation on hippocampal neuron dendritic spines. In: National acoustics conference
77. Huang X, Lin Z, Wang K et al (2017) Experimental study on low-intensity ultrasound stimulation of primary cilia of neurons. In: Acoustical society of China national acoustics conference
78. Zou J, Jia X, Bian T et al (2018) Safety study of low-intensity focused ultrasound neuromodulation. In: National acoustics conference, pp 18–19
79. Zhou W, Wang J, Huang B et al (2017) Study on neuromodulation of Caenorhabditis elegans based on surface acoustic waves. In: Acoustical society of China national acoustics conference, pp 505–506
80. Lin Z, Zhou W, Huang X et al (2017) Study on the regulation effect of hippocampal neurons based on micro-ultrasound chip. In: Acoustical society of China national acoustics conference, pp 547–548
81. Jiang X, Yang X, Xiao Y et al (2016) Simulation study of ultrasound time reversal transcranial focusing method for non-invasive neuromodulation application. National Acoustics Conference 2016:431–434
82. Fink M (1992) Time reversal of ultrasonic fields. I. Basic principles. IEEE Trans Ultrason Ferroelectr Freq Control 39:555–566
83. Marquet F, Pernot M, Aubry JF et al (2009) Non-invasive transcranial ultrasound therapy based on a 3D CT scan: protocol validation and in vitro results. Phys Med Biol 54:2597–2613
84. Yang X, Liu J, Wang C et al (2017) A new method for visualizing the focus of transcranial focused ultrasound in vitro experiments. In: Acoustical society of China national acoustics conference, pp 553–554

MIX
Papier aus verantwortungsvollen Quellen
Paper from responsible sources
FSC® C105338

If you have any concerns about our products,
you can contact us on
ProductSafety@springernature.com

In case Publisher is established outside the EU,
the EU authorized representative is:
Springer Nature Customer Service Center GmbH
Europaplatz 3, 69115 Heidelberg, Germany

Printed by Libri Plureos GmbH
in Hamburg, Germany